Polymers
Properties and Applications

13

L. T. Fan, S. K. Singh

Controlled Release

A Quantitative Treatment

With 81 Figures

Springer-Verlag
Berlin Heidelberg NewYork
London Paris Tokyo Hong Kong

Professor Liang-tseng Fan
Department of Chemical Engineering
Kansas State University
Durland Hall
Manhattan, KS 66506/USA

Dr. Satish Kumar Singh
Kabi Pharma
Gårdsvägen 6
Box 1828
S-171 26 Solna, Sweden

ISBN-13: 978-3-642-74509-6 e-ISBN-13: 978-3-642-74507-2
DOI: 10.1007/978-3-642-74507-2

Library of Congress Cataloging-in-Publication Data
Fan, L. T. (Liang-tseng), 1929
Controlled release: a quantitative treatment / L. T. Fan, S. K. Singh, p. cm. –
(Polymers, properties and applications; 13) Includes index.
I. Controlled release technology. I. Singh, S. K. II. Title. III. Series.
TP 156.C64F36 1989 660.2 - dc 19 89-4323

Softcover reprint of the hardcover 1st edition 1989

2152/3020-543210

To the Memory of
Professor Takeru Higuchi
Father of
Modern Controlled-Release Technology

Preface

The concept of controlled release has attracted increasing attention over the last two decades, with the applications of this technology proliferating in diverse fields including medicine, agriculture and biotechnology. Research and developmental efforts related to controlled release are multiplying in both industry and academia. The reason for this phenomenal growth is obvious. The use of a variety of biologically active agents, such as drugs, fertilizers and pesticides, has become an integral part of modern society. Along with the use of these reagents has evolved an awareness that their uncontrolled application almost inevitably induces harmful effects on the health of humans and their surrounding environments. To eliminate or minimize these harmful effects necessitates the controlled release of these chemicals. Moreover, the controlled release of substances, not usually considered toxic or hazardous, e.g., some catalysts and nutrients, can enhance their effectiveness.

The number and variety of controlled release systems, differing in their physical and chemical makeup, are increasing rapidly. Proliferation almost always demands correlation, generalization and unification; it requires both the development of underlying theories of their behavior and the mechanistic interpretation of their performance. This, in turn, requires a statistical and mathematical (quantitative) treatment of the scientific information and technical data pertaining to them. A quantitative treatment can also facilitate the formulation of procedures for computer-aided design of these systems through *a priori* prediction of their performance for a variety of design parameters.

Various controlled release systems are treated in this monograph; these systems are classified on the basis of the release mechanisms. A discussion is given of the release mechanisms and of the factors governing the release profile. This is followed by a review of major mathematical models that are ,applicable to the systems. These mathematical models are classified on the basis of the assumptions involved. The majority of the models are of the deterministic type; however, the heterogeneous or mesoscopic nature of some of the systems requires that a stochastic approach be employed. Thus a fairly extensive treatment of this approach to the modeling of diffusion is included at the end.

Acknowledgement

The authors' research, supported for several years by the Kansas Agricultural Experiment Station, Kansas State University, has culminated in the present monograph. The manuscript was ably typed mainly by Janet Vinduska and Peggy Hanes. The first author (LTF) also wishes to acknowledge the aid of his wife, Eva, in preparing the manuscript.

March 1989

L. T. Fan
S. K. Singh

Table of Contents

1 Introduction 1

1.1 Diffusion-Controlled Systems 3

1.1.1 Reservoir Devices 3
1.1.2 Monolithic Devices 4

1.2 Erosion or Chemical Reaction Controlled Systems 5
1.2.1 Erosion-Controlled Devices 5
1.2.2 Chemical Reaction Controlled Devices 6

1.3 Swelling-Controlled Release Systems 6
1.4 Osmotic Pumping Systems 6

References 7

2 Diffusion-Controlled Release 9

2.1 Diffusion in Polymers 11

2.1.1 Estimation of Diffusion Coefficients 14
2.1.2 Factors Influencing Diffusional Release 20

2.2 Models for Diffusion-Controlled Release Systems 44

2.2.1 Reservoir Devices 49
2.2.2 Monolithic Devices 61
2.2.3 Porous Devices 79

References 83

3 Chemical Reaction Controlled Release 89
3.1 Physically Immobilized Active Agent Systems 90
3.2 Chemically Immobilized Active Agent Systems 94
3.3 Models for Physically Immobilized Erosion-Activated Systems 102

References 107

4 Swelling-Controlled Release 110

4.1 Penetrant Transport and the Polymer Relaxation Phenomena 111

4.1.1 Transport of Penetrants in Polymers 112
4.1.2 Models for Transport of Penetrants in Glassy Polymers 120

4.2 Active Agent Release from Swellable Polymers 125
4.3 Models for Swelling-Controlled Release Systems 130
4.3.1 One-Region Models 131
4.3.2 Two-Region Models 136
References 152

5 Special Controlled-Release Systems 157
5.1 Osmotic Pumping Devices 157
5.2 Osmotically Activated Monolithic Devices 162
5.3 Externally Modulated Devices 164
References 166

6 Stochastic Model for Diffusion in Porous or Heterogeneous Polymer Matrix 167
6.1 Derivation 170
6.1.1 Statistical Model for the Porous/Heterogeneous-Polymer Matrix. 170
6.1.2 State of the Diffusing Molecule and State Transitions 172
6.1.3 Derivation of the Master Equation 175
6.1.4 Recovery of the Diffusion Equation 177
6.2 Transport Coefficients from the Diffusion Equation 184
6.2.1 Mean or Effective Diffusion Coefficient 185
6.2.2 Drift or Convective Velocity 187
6.3 Transport Coefficients in Specific Systems 188
6.3.1 Mean or Effective Diffusion Coefficient 188
6.3.2 Drift or Convective Velocity 191
6.4 Monte Carlo Simulations of Transport in Porous Media 192
6.4.1 Procedure 193
6.4.2 Results 195
6.5 Concluding Remarks. 197
References 200
Appendix 6A. Conversion of Probability to Number Concentration 202
Appendix 6B. Derivation of Equation (6.39) from Equation (6.37) 204
Appendix 6C. Algorithm and FORTRAN Code for Monte Carlo Simulation of Transport in Porous Networks 206

7 Epilogue 225
References 226

Author Index 227

Subject Index 231

1. Introduction

In the search for safe, economical, and efficient means of providing for the health and well-being of mankind, modern science has produced numerous active agents that manipulate the biological environment around and within us. Nevertheless, the use of these active agents is fraught with inefficiencies stemming from an inability to deliver these agents to their targets at the right time and in the right amounts. This results in their loss and in undesirable side effects and leads to a regimen requiring repeated treatment to produce and sustain the desired effect. With drugs, periodic dosage produces peaks and valleys in the concentration of the drug in the blood stream, possibly between harmful and ineffective levels (see Fig. 1.1). Agricultural chemicals such as pesticides, fertilizers, herbicides, and fumigants create the same problems when applied directly. To counter these problems, scientists have conventionally looked to altering the persistency and effectiveness of the reagents through modification of the reagents themselves. However, this approach tends to be difficult, time consuming, and expensive.

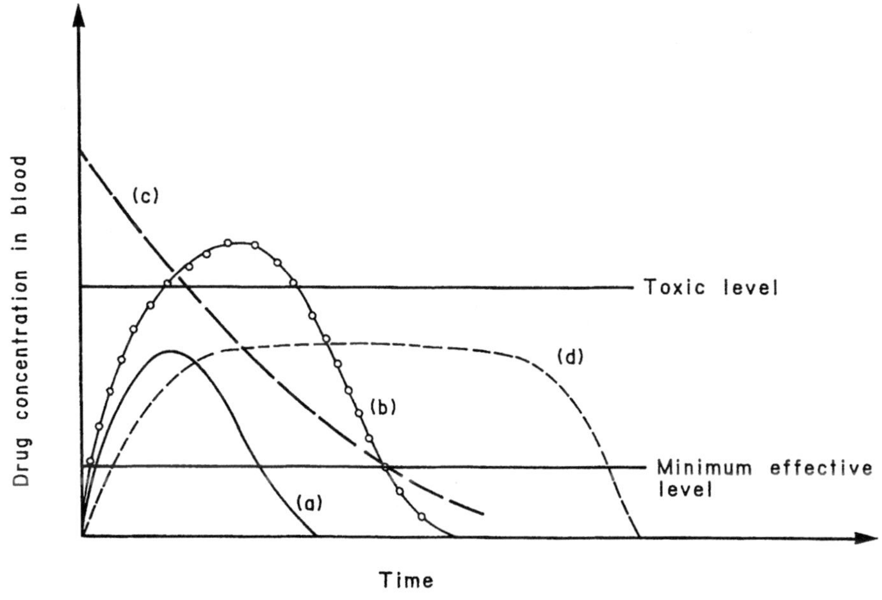

Fig. 1.1. Typical drug level in blood vs. time profiles for various modes of delivery. *a* standard oral dose, *b* oral overdose, *c* intraveneous injection, *d* controlled release dose

In recent years, increasing attention has been given to methods by which active reagents or chemicals are administered, giving rise to the field of *controlled release*. Controlled release may be defined as a technique by which active chemicals are made available to a target at a rate and duration so as to produce a desired effect. The objective of the technique is the maintenance of this desired rate or concentration level (Kydonieus, 1980). Among the advantages of this technique are more efficient utilization of the active agent, possibility of targeting, less frequent administration, and reduction in side effects. These advantages can sometimes make an otherwise unsuitable active agent attractive. This is especially true in pharmaceutics. New and powerful macromolecular drugs are being discovered or "engineered" at a rapid rate. Nevertheless, these drugs can be cumbersome to administer and often produce side effects. Controlled release can alleviate these difficulties.

In medicine, the approach of sustained-release has been utilized for a long time. This approach comprises the formulation of drug complexes, enteric formulations, suspensions, emulsions, and compressed tablets, to retard the drug's rate of availability to the body. The formulations, however, are strongly influenced by the environmental conditions and are not true controlled-release systems. Controlled release implies a predictible and reproducible release profile relatively independent of environment; this gives rise to a higher degree of control than achieved in sustained release formulations.

The basic controlled-release formulation consists of an active agent (the drug, fertilizer etc.) and a carrier (commonly a polymeric material) arranged so as to allow the active agent to be released at the target over a period of time at a controlled rate. Numerous design variations towards this objective have been proposed. Some of the common ones are (see, e.g., Cowsar, 1974; Kydonieus, 1980):

1. A capsule of polymeric material filled with the active agent in either the solid, liquid or solution form, release being controlled by diffusion through the capsule wall (reservoir devices).
2. A heterogeneous dispersion or homogeneous solution of the active agent in a polymeric matrix, release being controlled by diffusion through and/or erosion of the polymer (monolithic devices).
3. A laminate of the active agent and polymer forming a "sandwich", release being controlled by diffusion, erosion or both.
4. A swellable polymeric matrix with the active agent dispersed and/or dissolved in the polymer, release occurring as a result of swelling (or gelatinization) of the polymer in the environmental fluid and the consequent diffusion of the active agent.
5. Liquid-liquid encapsulation of the active agent in a viscous solution, release being controlled by diffusion or dilution.
6. Chemical bonding of the active agent to a polymeric backbone and subsequent release by controlled bond cleavage in the target environment.
7. Macromolecular structures of the active agent formed via ionic or covalent linkages with release controlled by hydrolysis, thermodynamic dissociation, or microbial or enzymatic degradation.
8. A capsule of active agent-impermeable polymer membrane, with release occurring

through an orifice due to osmotic pressure created by the permeation of environmental fluids.

Examples of the various kinds of devices in use can be found elsewhere (Kydonieus, 1980; Baker, 1987)

The preceding classification of controlled release systems is based on the physical design. Controlled release systems can also be classified according to their major rate controlling mechanism. This classification is more meaningful from the view point of the technologist or designer; it broadly divides the systems into those controlled by:

1. Diffusion
2. Erosion or chemical reactions
3. Swelling
4. Osmosis

The classification of a device into any of these categories is determined by the nature of interaction between the polymer and environmental fluid. Depending on the extent of this interaction, the polymer forming the device may remain unchanged or it may undergo erosion or swelling. If the polymer remains unaffected and release occurs solely through diffusion, the device falls in the category of diffusion-controlled systems. If the polymer releases the active agent by erosion of the polymer itself or a bond between the active agent and the polymer, the device may be categorized as erosion or chemical reaction-controlled. If release occurs due to swelling of the polymer caused by absorption of the environmental fluid, the device falls in the category of swelling-controlled systems. Devices activated by osmosis utilize osmotic pressure as the driving force. It must be noted that two or more of these mechanisms may be simultaneously active in some devices. Furthermore, the above classification includes only the broad and general mechanisms. Specialized devices utilizing magnets, ultrasound, and liposomes have also been created (see, e.g., Langer, 1980; Van Brunt, 1986).

1.1 Diffusion-Controlled Systems

A device is considered diffusion-controlled when the diffusion of active agent through any part of the device controls the release rate. The major types of diffusion-controlled devices are reservoir devices and monolithic devices.

1.1.1 Reservoir Devices

These devices consist of a core of the active agent surrounded by a polymeric membrane. Diffusion through the polymer is the rate limiting step. Figure 1.2 illustrates a cross-section of a spherical or cylindrical reservoir device. Among the polymers employed in the preparation of such systems are silicone rubber, ethylene-vinyl acetate copolymers, polyurethane, and polyethylene. The simplest method of preparation of such a device is to take a length of tube of the appropriate

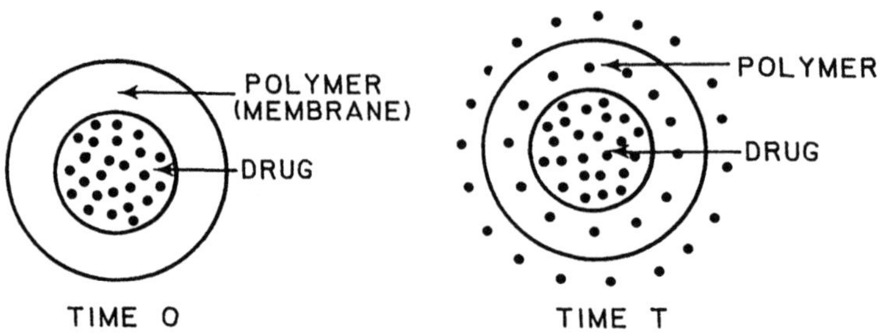

Fig. 1.2. Schematic of the cross-section of a spherical or cylindrical reservoir type diffusion-controlled release device (adapted from Langer, 1980)

polymeric material and dimensions, fill it with the active agent, and seal off the ends. More recently, growing interest in micro-, macro- and nanocapsules, has led to their utilization for drug release. Liposomes are also being explored for this purpose. These systems can also be looked upon as reservoir devices although these are never employed singly but as a population. The techniques of microencapsulation, such as coacervation, spray encapsulation and nanoencapsulation are well developed for their preparation (see, e.g., Speiser, 1976).

1.1.2 Monolithic Devices

Monolithic or matrix devices constitute the other major category of diffusion-controlled release devices. In these, the active agent is distributed in a polymer matrix (see Fig. 1.3). It can either be dispersed as a separate phase or be dissolved in the polymer. Release occurs by diffusion through the polymer; moreover, the environmental fluid may leach the active agent out of the matrix if the polymer is permeable to the fluid. If a soluble additive is mixed in the polymer matrix, fluid will enter the matrix by dissolving the additive and forming interconnected channels in the matrix. There are numerous simple ways of formulating monolithic devices. A dissolved active agent system may be obtained by soaking the polymer in a solution of the active agent and allowing it to equilibriate. Compression molding of powdered active agent and polymer will produce a dispersed

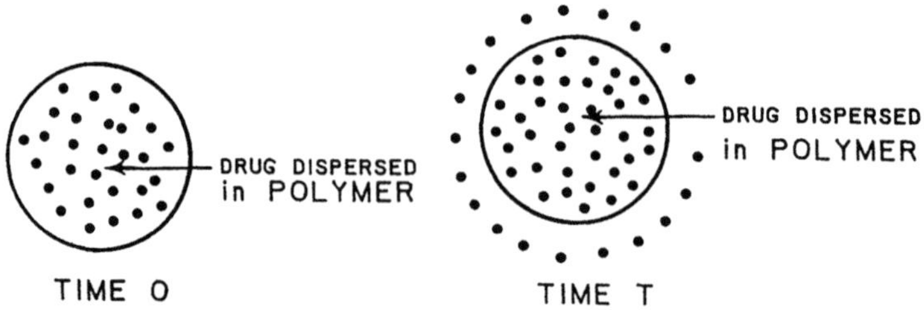

Fig. 1.3. Schematic of a typical monolithic type diffusion-controlled release device (adapted from Langer, 1980)

active agent system. Mixing the active agent with the monomer and then forming the polymer will give rise to a dissolved and/or dispersed active agent system depending on the solubility. Other techniques include solvent-casting and low-temperature casting (Langer, 1980). Polymers used as the basis for monolithic devices are mainly the same as employed for reservoir devices, such as polydimethylsiloxane (silicone rubber), ethyl cellulose, and hydroxypropyl cellulose.

1.2 Erosion or Chemical Reaction Controlled Systems

Erosion or reaction controlled release systems are delivery systems in which the polymer is not merely a passive carrier but an active participant in the release process.

1.2.1 Erosion-Controlled Devices

The active agent in these devices is physically immobilized in the polymer and is only released by erosion of the polymer as shown in the ideal erosion-controlled device in Fig. 1.4. While these devices can be made in both the reservoir and monolithic forms, the latter is more suitable for this purpose. In an ideal system, the polymeric carrier would undergo surface erosion, releasing the active agent at a rate proportional to the erosion rate. If the erosion rate is a constant and the matrix geometry is constant (e.g., slab), such systems would produce zero-order (constant) release rates. However, erosion in practical systems is seldom solely from the surface inwards. Bulk erosion leading to a loss of physical integrity of the matrix and its eventual collapse, is frequently observed.

Various polymers have been utilized in forming erodible devices, including poly-(vinyl pyrrolidone), partially esterified copolymers of methyl vinyl ether and maleic anhydride, and copolymers of lactic and glycolic acid. Recently, it has been reported that polyanhydrides, such as poly[bis (*p*-carboxy-phenoxy) propane anhydride] and poly[terepthalic acid anhydride], undergo surface erosion (Leong et al., 1985, 1986; Van Brunt, 1986). Bioerodible polymer remains a very active research area.

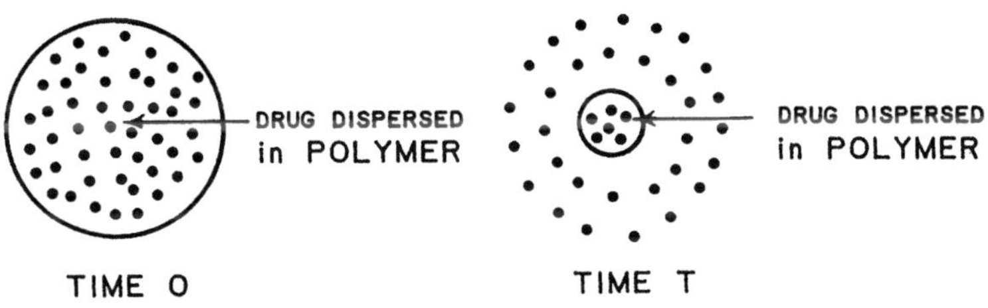

Fig. 1.4. Schematic of an ideal erosion-controlled release device (adapted from Langer, 1980)

1.2.2 Chemical Reaction Controlled Devices

In this type of system, the active agent itself is chemically bound to the polymeric backbone. Release occurs by hydrolytic or enzymatic cleavage of this bond. The polymer itself may be degradable in these systems. Such pendant chain complexes have been created with both pharmaceutical and agricultural active agents (see, e.g., Harris et al., 1976; Duncan and Kopacek, 1984).

1.3 Swelling-Controlled Release Systems

Swelling-controlled systems are formed by dissolving or dispersing the active agent in a polymeric matrix in which it is unable to diffuse to any significant extent. When this matrix is placed in an environmental fluid thermodynamically compatible with the polymer, the fluid is absorbed into the polymer, causing it to swell. The active agent in the swollen part of the matrix can then diffuse out. A schematic of an idealized swelling-controlled release system is shown in Fig. 1.5. Glassy polymers whose glass-transition temperature is lowered by the fluid will undergo a glass-to-gel transition and swell, and can be employed to form such systems. The idealized-rate of release from such systems is equal to the product of the surface area and a rate factor corresponding to the rate of advance of the boundary separating the outer swollen shell from the unswollen core. If the rate of advance of this boundary is constant, zero-order release will be obtained. A class of polymers called hydrogels [e.g., poly(hydroxy methyl methacrylate), acrylamide, and poly-(ethylene glycols)] are considered suitable for such systems (see, e.g., Roorda et al., 1986).

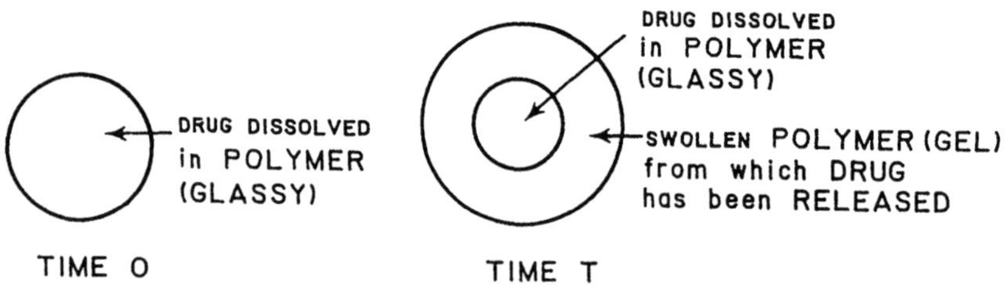

Fig. 1.5. Schematic of an idealized swelling-controlled release device (adapted from Langer, 1980)

1.4 Osmotic Pumping Systems

An osmotic pump releases its active agent at a controlled rate by utilizing osmotic pressure as the driving force. A simple osmotic pump consists of a core of compacted, solid, water-soluble active agent. The core is enclosed in a selectively water-permeable but active agent-impermeable polymer membrane with a small

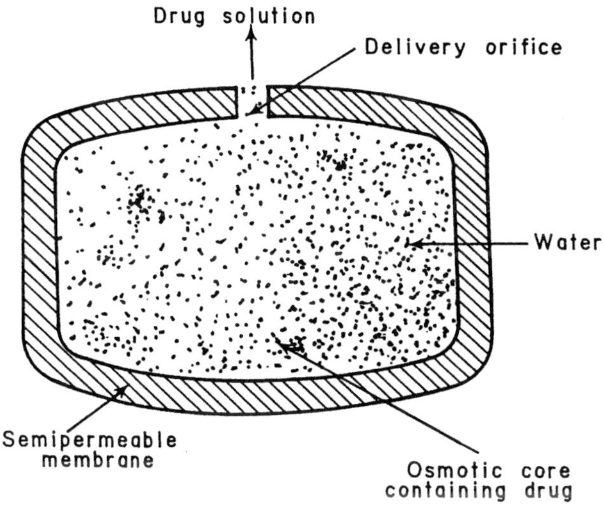

Fig. 1.6. Schematic of a simple osmotic pump (Theeuwes, 1975)

orifice. In an aqueous environment, water is transported into the core by permeation. A nonextensible polymer facilitates the build-up of hydrostatic pressure in the core, and the solution of the active agent and water is then pumped out through the orifice. Figure 1.6 illustrates a simple osmotic pump.

Numerous experimental studies are reported every year on devices of the types described above. Furthermore, new, specialized systems are being developed utilizing "exotic" concepts or techniques, such as magnetic or ultrasound modulation, viable-cell immobilization, microspheres and nanoparticles, liposomes, targeted delivery, and prodrugs. The proliferation of experimental studies requires generalization and unification through a statistical and mathematical treatment of quantitative information and technical data. In addition, a quantitative treatment can facilitate the formulation of procedures for computer-aided design of these systems through a priori prediction of their performance for a variety of design parameters. Towards this end, the present monograph attempts to broadly survey the existing mathematical models of all degrees of complexity and to categorize them. This can assist a researcher to select the optimum model for the system under study, and also assist in identifying lacunae. By necessity, the coverage presented here is less than comprehensive; thus the emphasis is on those systems that are most commonly encountered.

References

Baker, R. W.: Controlled Release of Biologically Active Agents, pp. 1—83, John Wiley and Sons, New York (1987)

Cowsar, D. R.: Introduction to Controlled Release, pp. 1–13. In: Controlled Release of Biologically Active Agents, A. C. Tanquary and R. E. Lacey (Eds.), Plenum Press, New York (1974)

Duncan, R., Kopacek, J.: Soluble Synthetic Polymers as Potential Drug Carriers, Adv. in Pol. Sci. *57*, 51–101 (1984)

Harris, F. W., Aulabaugh, A. E., Case, R. D., Dykes, M. K., Feld, W. A.: Polymers Containing Pendent Herbicide Substituents: Preliminary Hydrolysis Studies, pp. 222–230. In: Controlled Release Polymeric Formulations. D. R. Paul and F. W. Harris (Eds.), ACS Symposium Series 33, ACS Washington, D.C. (1976)

Kydonieus, A. F.: Fundamental Concepts of Controlled Release, pp. 1–19. In: Controlled Release Technologies: Methods, Theory, and Applications, Vol. I, A. F. Kydonieus (Ed.), CRC Press, Boca Raton, Florida (1980)

Langer, R.: Polymeric Delivery Systems for Controlled Drug Release. Chem. Eng. Commun. *6*, 1–48 (1980)

Leong, K. W., Amore, P. D., Marletta, M., Langer, R.: Bioerodible polyanhydrides as Drug-Carrier Matrices. II. Biocompatibility and Chemical Activity, J. Biomed. Mater. Res. *20*, 51–64 (1986)

Leong, K. W., Brott, B. C., Langer, R.: Bioerodible Polyanhydrides as Drug-Carrier Matrices. I: Characterization, Degradation, and Release Characteristics, J. Biomed. Mater. Res. *19*, 941–955 (1985)

Roorda, W. E., Bodde, H. E., De Boer, A. G., Junginger, H. E.: Synthetic Hydrogels as Drug Delivery Systems, Pharmaceutisch Weekblad, Scientific Edition *8*, 165–189 (1986)

Speiser, P.: Microencapsulation by Coacervation, Spray Encapsulation, and Nanoencapsulation, pp. 1–11. In: Microencapsulation, J. R. Nixon (Ed.), Marcel Dekker Inc., New York (1976)

Theeuwes, F.: Elementary Osmotic Pump, J. Pharm. Sci. *64*, 1987–1991 (1975)

Van Brunt, J.: Perfecting Polymers to Release Drugs, Bio/Technology *4*, 756 (1986)

2. Diffusion-Controlled Release

Molecular diffusion through polymers and synthetic membranes is an effective, simple and yet reliable means of attaining the controlled release of a variety of active agents. The principal devices utilizing this phenomenon are of the reservoir and monolithic types. In a reservoir device, the active agent is contained within the rate-controlling membrane, while in a monolithic device, the active agent is homogeneously dissolved or dispersed throughout the polymer matrix.

The principal steps in the release of an active agent from a reservoir device are illustrated in Fig. 2.1. These are diffusion within the reservoir (if not well mixed), dissolution or partition between the reservoir carrier fluid and membrane, diffusion through the membrane, partition between the membrane and elution medium, and finally, transport away from the device surface. If the reservoir contains an excess of the active agent, a steady-state concentration profile, as shown in Fig. 2.1 a, can be maintained. However, after depletion of the reservoir, the concentration profile will change, either in the pseudo-steady-state fashion shown in Fig. 2.1 b or in an unsteady fashion.

The principal steps in the release of the active agent from a monolithic device are exhibited in Fig. 2.2. The first step in the process varies depending on whether the active agent is dispersed as a separate phase or dissolved in the matrix (polymer). If it is dissolved in the matrix, the steps involved are diffusion to the surface of the matrix, partition between the matrix and elution medium, and transport away from the device surface (Fig. 2.2a). If the active agent is dispersed as a separate phase, it must first dissolve into the matrix continuum before being able to diffuse to the surface (Fig. 2.2b). This is termed the "solution-diffusion" mechanism.

The rate controlling step in both types of devices is diffusion within the polymer, although the overall release rate itself is affected by the other steps. The steps described in the preceding paragraphs assume that molecules of the active agent diffuse through the polymer continuum. However, there can be instances in which the membrane or matrix is porous, and the pores are permeated by fluid of the elution medium. In these instances, release occurs by dissolution and diffusion in this fluid. Nevertheless, the principal steps in the release mechanism remain the same.

While the reservoir and monolithic type devices are the ones primarily classified as diffusion controlled, diffusion through polymers plays a significant role in other devices or systems as well, e.g., swelling-controlled or erodible systems. Modification or design of these systems requires an understanding of the chemical and physical characteristics of diffusion in polymeric materials.

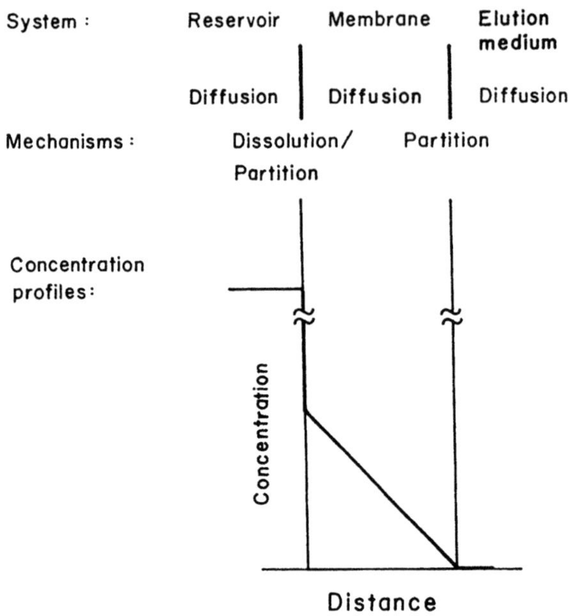

a

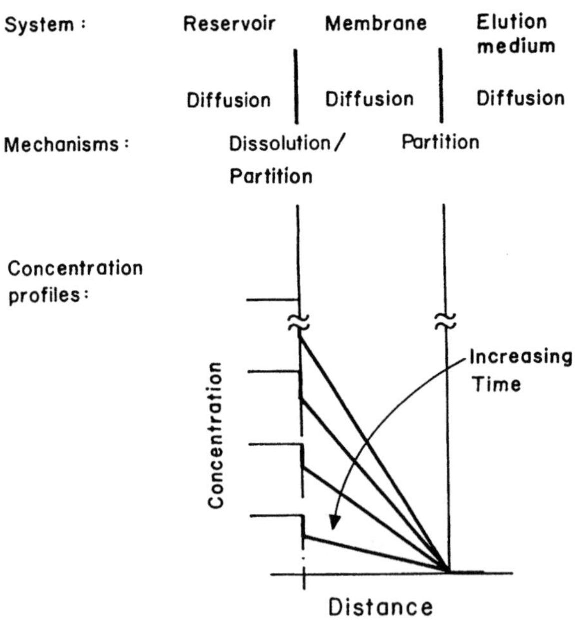

b

Fig. 2.1 a, b. Schematic representation of the functioning of a reservoir device. **a** Reservoir device with excess active agent, **b** Reservoir device without excess active agent (adapted from Baker, 1987)

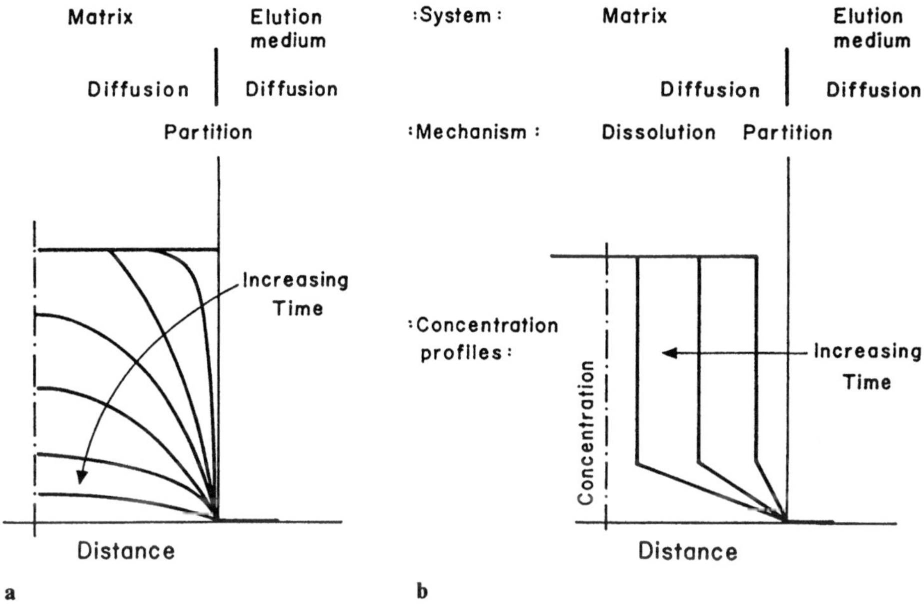

Fig. 2.2a, b. Schematic representation of the functioning of a monolithic device. **a** Monolithic device with dissolved active agent, **b** Monolithic device with dispersed active agent (adapted from Baker, 1987)

2.1 Diffusion in Polymers

Diffusion of any molecule in a polymer depends on various structural and morphological parameters. However, irrespective of the actual mechanism, diffusion can be analyzed, macroscopically, in terms of Fick's first law in which the diffusive flux, j, is proportional to the concentration gradient, ∇C,

$$j_i = -D_i \, \nabla C_i \tag{2.1}$$

where

j_i = flux of diffusing component i with respect to the mass average velocity of the system,

C_i = concentration of diffusing component i, and

D_i = diffusivity of component i in the polymer-active agent phase.

By measuring the flux relative to the mass average velocity (i.e., the rate of transfer across a section moving in such a way that no mass flows through it), the diffusivity defined above represents the intrinsic diffusion coefficient of species i (Rogers, 1965).

The diffusivity or diffusion coefficient in equation (2.1) is a constant of proportionality; it may be a function of concentration and other physical variables. The equation assumes that the driving force of diffusion is the concentration gradient. Considerations based on the thermodynamics of irreversible processes indicate that a more fundamental driving force is the gradient of chemical potential of each component (see, e.g., DeGroot, 1959). In a binary mixture, the driving force is proportional to the negative chemical potential gradient, $(-\nabla\mu)$, where μ is the chemical potential and can be written in terms of activity α_i as

$$\mu_i = RT \ln \alpha_i + \mu_i^0 . \tag{2.2}$$

The total force on all molecules of the component at any given point is proportional to $(-C_i \nabla\mu_i)$, and the resultant pure diffusional flux is (Rogers, 1965)

$$j_i = -\left(\frac{B_i C_i}{N}\right) \nabla\mu_i \tag{2.3}$$

The quantity B_i is the intrinsic mobility of a molecule of component i, and N is the Avogadro number. Substituting for $\nabla\mu_i$ from Eq. (2.2) into Eq. (2.3),

$$j_i = -\left(\frac{B_i RT}{N}\right) C_i \nabla \ln \alpha_i . \tag{2.4}$$

The activity, α_i, can be expressed, in turn, as a function of the activity coefficient, v_i, as

$$C_i \nabla \ln \alpha_i = \nabla C_i + C_i \nabla \ln v_i . \tag{2.5}$$

Thus, we have

$$j_i = \left(\frac{B_i RT}{N}\right)\left[1 + \frac{\partial \ln v_i}{\partial \ln C_i}\right] \nabla C_i . \tag{2.6}$$

Comparing Eqs. (2.1) and (2.6), we obtain

$$D_i = D_i^* \left[1 + \frac{\partial \ln v_i}{\partial \ln C_i}\right] \tag{2.7a}$$

or

$$D_i = D_i^* \left[\frac{\partial \ln \alpha_i}{\partial \ln C_i}\right] \tag{2.7b}$$

where

$$D_i^* = \left(\frac{B_i RT}{N}\right) \tag{2.7c}$$

D_i^* is the activity-based diffusion coefficient, also termed the self-diffusion coefficient; in general, it is much less concentration dependent than D_i, the intrinsic diffusion coefficient (Lightfoot, 1974). When the active agent-polymer solution is ideal, $\partial \ln v_i / \partial \ln C_i$ is zero, and the expressions in Eqs. (2.1) and (2.6) become equivalent.

The self-diffusion coefficient, D_i^*, characterizes the rate of diffusion of component i in a mixture of uniform chemical composition. In a single component system, it represents the diffusivity of isotopically or radioactively tagged molecules among unlabelled ones.

The flux in Eq. (2.1) is defined with respect to a frame of reference moving along with the mass average velocity of the system. However, it is usually extremely difficult, if not impossible, to determine this reference section experimentally. A more practical frame of reference is one fixed in the laboratory; it is particularly useful in systems exhibiting no volume change on mixing. Consider the simplest non-trivial case of a binary system. If the diffusing species are not of identical mass and size, pure diffusion of one species will be faster than that of the other. To compensate for the build-up of hydrostatic pressure on one side due to this unequal diffusion, a compensating convective flow is generated. The total rate of transfer of a component can therefore be separated into two components, one due to pure diffusion, $(-D_i \nabla C_i)$, and the other due to countering convective flow, $(C\underset{\sim}{V})$, where $\underset{\sim}{V}$ is the mean convective velocity related to the rate of movement of the reference plane defining the intrinsic diffusion coefficient. Thus, the total rate of transfer along the volume-fixed reference plane is given by

$$N_i = -D_i \nabla C_i + C_i \underset{\sim}{V} = -D_i \nabla C_i + C_i(\bar{V}_i D_i \nabla C_i + \bar{V}_j D_j \nabla C_j)\,, \quad (2.8\text{a})$$

or

$$N_i \equiv -D_{ij} \nabla C_i \quad (2.8\text{b})$$

where $\bar{V}_i$ and $\bar{V}_j$ are partial specific volumes and D_{ij} is the mutual (binary) interdiffusion coefficient. A single mutual diffusion coefficient is required for a binary system (Crank, 1975). For diffusion in solids of unchanging size or for diffusion in stationary liquids, the mutual and intrinsic diffusion coefficients are identical. Incorporation of this flux into the continuity equation yields a complete description of mass transfer in the polymer. The resultant continuity equation can be written in a generalized form as (Bird et al., 1960)

$$\frac{\partial C_i}{\partial t} + (\nabla \cdot N_i) = R_i \quad (2.9)$$

where

R_i = molar rate of "production" of component i.

The "production" term accounts for any mechanism by which the ith component is made available for diffusion in the system; dissolution of the dispersed active agent in a monolithic system is a typical example. Substituting for N_i from Eq. (2.8) into Eq. (2.9) gives

$$\frac{\partial C_i}{\partial t} + \nabla \cdot j_i + \nabla \cdot C_i \underset{\sim}{V} = R_i\,. \quad (2.10)$$

For a constant-volume system where the "production" term is absent, the familiar expression for Fick's second law is obtained as

$$\frac{\partial C_i}{\partial t} - \nabla \cdot (D_{ij} \nabla C_i) = 0 \,. \tag{2.11}$$

A detailed discussion of the relation between the mutual and intrinsic diffusion coefficients is given by Crank (1975).

So far, we have considered diffusion as being induced by the concentration (or activity) gradient. A generalized form of the Fickian diffusion equation for diffusion induced by electrostatic potential and/or pressure gradient, is available (Lightfoot, 1974); for an electrically neutral system, it is

$$N_i = -D_i \frac{C_i}{RT} [\nabla \mu_i + \bar{V}_i \nabla P + \omega_i F \nabla \Phi] + C_i \underset{\sim}{V} \tag{2.12}$$

where
$\bar{V}_i$ = partial specific volume of component i,
ω_i = ionic charge of component i,
Φ = electrostatic potential,
P = fluid pressure, and
F = Faraday constant.
With this definition of flux, substitution of Eq. (2.12) for a constant-volume system into Eq. (2.9) yields the more general continuity expression

$$\frac{\partial C_i}{\partial t} = D_{ij} \frac{C_i}{RT} [\nabla^2 \mu_i + \bar{V}_i \nabla^2 P + \omega_i F \nabla^2 \Phi] \tag{2.13}$$

which is the generalized form of Fick's second law of diffusion.

Equation (2.11), [or its more general versions, Eqs. (2.10) and (2.13)] may be solved to predict the release rate from a diffusion-controlled system. The information required for solution of this partial differential equation includes knowledge about the initial and boundary conditions and about the system parameter, D_{ij}. As stated earlier, this coefficient is dependent on various structural and morphological parameters. Experimental techniques to measure this parameter are well developed (see, e.g., Geddes and Pontius, 1959; Crank and Park, 1968; Felder and Huvard, 1980). Various models have been proposed to predict diffusion coefficients in polymers. Some of these models will be briefly examined below, along with a discussion of the influence of polymer morphological factors.

2.1.1 Estimation of Diffusion Coefficients

The structure of a solid phase may have a pronounced effect on mass transfer processes within and through this phase. Mikulecky and Caplan (1966) make a distinction between two types of situations with respect to diffusion in polymeric membranes.

i. The membrane pores are smaller than or of the same order of magnitude as the mean free path (or average jump distance) of the permeating molecules. The molecular network constituting the membrane takes part at the molecular level in frictional interactions with all of the permeating components. The membrane molecules, therefore, represent a true component of the system and must be included in the thermodynamic diffusion equations (McGregor, 1974).
ii. The membrane pores are much larger than the mean free path of the diffusing molecules. The system containing these membrane pores may be regarded as comprising two phases, of which the membrane phase represents little more than a solid "container" for the permeating fluid phase within which the transport process occurs. The membrane phase only serves to define the geometry of the permeation channels (McGregor, 1974).

Membranes of the first (small pore) type can be considered as "continuous" systems. Macroscopically homogeneous polymers may be included in this classification even though in this case the "pores" consist of nearly continuous pathways through the less ordered regions of the polymer and their dimensions may be subject to local thermal fluctuations. Langer and Peppas (1983) have classified these polymer systems as nonporous (gel) membranes where the pores are of a molecular level.

Capillary models have been employed to describe permeation through "small-pore" membranes (McGregor, 1974). However, in systems with pores having apparent radii of the order of 10 Å to 50 Å, the membrane molecule will take part in intimate molecular interactions with the permeating species. A capillary model is clearly a simplification in such a situation. For our purposes, the "small-pore" and "molecular" membranes will be analyzed as homogeneous polymer systems.

Membranes of the "large-pore" type may be more profitably analyzed by resorting to pore or capillary models. These large-pore membranes have further been divided into macroporous (0.1 μm to 1.0 μm) and microporous (100 Å to 500 Å) membranes by Langer and Peppas (1983). This distinction becomes important for diffusing species that are adsorbed from the permeating fluid to the polymer matrix inside the pores (Furusawa and Smith, 1973; Satterfield et al., 1973). In this work, the pore models and the polymeric systems to which they are applicable are analyzed with respect to heterogeneous polymer systems.

Although we have classified polymeric membranes into two main types on the basis of "pore" sizes, a real system can possess a wide distribution of pore sizes spanning both types. Transport may thus occur through both the polymer continuum (homogeneous polymer) and actual porous regions permeated by the bulk fluid (heterogeneous polymer).

We shall limit our treatment of membrane transport to passive transport of non-electrolytes only. Thus, electrokinetic phenomena such as electro-osmosis and electrophoresis, thermal gradients, facilitated transport and activated transport will not be considered.

a) Homogeneous polymer systems. The rate of diffusion of small molecules in a polymer is a complex function of a variety of factors, such as the polymer morphology, composition, and molecular weight, and of the solute concentration and size. Another important factor is temperature relative to the glass transition temperature, T_g, of the polymer. Molecular motions in polymers do not entirely cease below T_g;

the glassy polymers tend to be in a "frozen-in" non-equilibrium state. Diffusion in such systems is influenced, therefore, by the slow relaxation of this internal order. In the vicinity of T_g, the relaxation process occurs at rates comparable with those of the diffusion process, and the behavior becomes extremely complex (Park, 1968; Wang et al., 1969). At temperatures well above T_g of the system, polymer molecules in the accessible regions through which permeation occurs may participate intimately in frictional interactions with the diffusing species. The diffusion process follows the classical diffusion theory since the polymer can be analyzed as a homogeneous medium, and the basic parameter to describe mass transport is the diffusion coefficient (Duda, 1985). Of concern here are systems in their equilibrium states above their glass transition temperatures, i.e., systems with time-independent structures; anomalous behavior will be discussed elsewhere.

The polymer matrix has a strong influence on the diffusivity, since the motion of a small molecule is restricted by the three-dimensional network of polymer chains. The extent of this interaction is dependent on the concentration of the polymer-active agent system. Since a polymer employed for controlled release would be expected to be mechanically stable, the theories of interest will be the ones applicable in the concentrated polymer region. Among these theories are the approximate molecular theories (see, e.g., DiBenedetto, 1963a, 1963b; Pace and Datyner, 1979a, 1979b) or the free-volume theories (see, e.g., Fujita, 1961). Since the free-volume theories have a wider range of applicability (Duda, 1985), they will be briefly considered here.

The free-volume concept was originally employed by Cohen and Turnbull (1959) for self-diffusion in hard-sphere liquids; it considers diffusion to occur due to local density fluctuations in the liquid. These fluctuations periodically open voids of molecular dimensions into which a diffusing molecule may jump. The frequency of diffusion jumps depends on the probability of occurrence of a local void, greater in volume than a certain minimum volume. Cohen and Turnbull (1959) have proposed that the volume of a liquid is composed of two components. The first is the actual volume of the molecules and the second is the free volume, redistributed continuously by the random thermal fluctuations of the fluid. Volume changes associated with temperature variations are related to changes in this free volume, while the occupied volume is independent of the temperature (equal to the total volume at the absolute zero). Diffusion of a molecule requires that a free-volume hole of sufficient size appear adjacent to the molecule. The original theory has been modified (see, e.g., Fujita, 1961; Macedo and Litovitz, 1965) and extensively applied by Vrentas and Duda (1977, 1979) and others to polymer-solvent systems.

Vrentas and Duda (1979) have suggested that the free volume in a polymer comprises two parts, an interstitial free volume and a hole free volume. Only the hole free volume is considered to be capable of being redistributed without energy constraints and is, therefore, available for molecular migration. Based on this definition of free volume, Vrentas and Duda (1977) have proposed an expression for self-diffusion coefficient of a solvent molecule in a polymer as

$$D_1^* = D_0 \exp[-E/RT] \exp[-\gamma(\omega_1 V_1^* + \omega_2 \xi V_2^*)/V_{FH}] \tag{2.14}$$

where V_i^* is the specific critical hole free volume required for the jump of component i ($i = 1$, solvent; $i = 2$, polymer), ω_i the mass fraction of component i, γ an overlap factor since the same free volume is available to more than one molecule, V_{FH} the average hole free volume per gram of mixture, and E the energy per mole that a molecule requires to overcome the attractive forces holding it to its neighbors. At temperatures not significantly above T_g, the specific hole free volume is small. Consequently, the probability of obtaining a hole of sufficient size is significantly less than that for acquiring the necessary energy. The energy term may be absorbed into the pre-exponential factor of the self-diffusion coefficient, yielding (Vrentas and Duda, 1977)

$$D_1^* = D_{01} \exp\left[-\gamma(\omega_1 V_1^* + \omega_2 \xi V_2^*)/V_{FH}\right]. \tag{2.15}$$

A similar expression can be written for the polymer self-diffusion coefficient. Mutual diffusion coefficients may then be obtained from the self-diffusion coefficient, utilizing friction factors for the system. However, for small values of ω_1, a simplified relation

$$D_{12} = \frac{D_1^* x_2}{RT}\left(\frac{\partial \mu_1}{\partial \ln x_1}\right)_{T,P} \tag{2.16}$$

may be utilized (Vrentas and Duda, 1979). In this expression, x_1 is the mole fraction of diffusing small molecular species, and x_2 is the mole fraction of polymer in the mixture. Under the assumption of additivity of free volumes, the parameters required to predict the hole free volume can be obtained from correlations of the viscosity and density data of pure components (Duda, 1985).

Since the critical free volume required for diffusion of molecules is dependent on their size, its value relative to the average free volume of the polymer-solvent system is of consequence. For very small diffusing molecules, only a few chain segments will be involved in the creation of void spaces. This class of behavior, called the type A Fickian diffusion (Frisch, 1965), is characterized by ideally dilute sorption behavior, a concentration-independent diffusion coefficient, and an apparent activation energy independent of temperature and concentration. The type B Fickian diffusion, observed with large organic molecules, is generally characterized by concentration-dependent diffusion coefficients, temperature-dependent activation energies, and strong polymer-penetrant interactions, as evidenced by high penetrant solubility (Frisch, 1965).

b) Heterogeneous polymer systems. Semicrystalline polymers contain regions of crystalline morphology interspersed between amorphous regions. These crystalline regions have low segmental chain mobility and can be considered virtually impermeable to a diffusing molecule. Therefore, it is necessary that the molecule follow a tortuous path around these regions. In this respect, crystallites behave in a fashion similar to impermeable filler particles, serving only to obstruct diffusion.

Analogously, a highly swollen polymeric gel, above its glass transition temperature, can be regarded as a randomly cross-linked molecular network permeated by a fluid in which diffusion may occur. As the degree of swelling increases, an increasing portion of this internal fluid will possess essentially the same properties as the bulk fluid. If a diffusing molecule does not interact significantly with the polymer chains, the effect of the polymer matrix on the diffusion process may be simply one of geometrical obstruction. The average cross-sectional area available for diffusion will be less than in the bulk fluid, and in general, the paths of the diffusing molecules will also be more tortuous. Furthermore, at least in very highly swollen gels these paths would be interconnected (McGregor, 1974).

The effect of obstruction can be empirically related to the volume fraction of amorphous regions or volume fraction of the fluid in the gel ($= \Phi$), i.e.,

$$D_{\text{eff}} = f(\Phi)\,. \tag{2.17}$$

Nevertheless, the possible effect of an ordered array of obstructing particles or impermeable regions on the effective diffusivity, D_{eff}, can be calculated based on the external dimensions of the array of the particles (Barrer, 1968). If the system is very dilute in these particles (e.g., in low-crystallinity polymers or in highly swollen gels), the dependence on the particle geometry or on the nature of their distribution is slight. For $\Phi \to 1$,

$$D_{\text{eff}} = D\left(\frac{2}{3-\Phi}\right) \tag{2.18}$$

where D is the (mutual) diffusivity in the amorphous polymer or in the fluid in the gel. A generalization of this relation to include the effect of particle geometry, though only valid for dilute systems, is

$$D_{\text{eff}} = D\,\frac{\varkappa}{(\varkappa-1)-\Phi} \tag{2.19}$$

where the shape factor, $\varkappa$, ranges from 2 for random spheres to 1.1 for oblate spheroids with an axial ratio of 4:1. If the obstructions are densely packed, the detailed dimensions and geometry become increasingly important, and the void volume alone is insufficient to describe the behavior. In general, it is not possible to construct more than semiempirical relationships between the quantities D_{eff}, D, Φ, and $\varkappa$. If the constructed relationship is correct and if there are no specific interactions. D_{eff} should exhibit the same temperature dependence as D (McGregor, 1974).

Another approach is to consider these semicrystalline or highly swollen gel systems as analogous to porous media with the permeable regions serving as the pores and the impermeable regions as the walls. A vast amount of literature exists on diffusion in porous media especially in catalysis, which can be adopted here. The obstruction effect is generally accounted for by the "porosity ε", and the tortuosity of the paths, by an empirical "tortuosity τ", to yield the effective diffusivity as

$$D_{\text{eff}} = \in\frac{D}{\tau}\,. \tag{2.20}$$

The porosity can be equated to the volume fraction of amorphous regions or that of the fluid in the gel, Φ, i.e., the fractional permeable space. [In an alternative though similar approach, Michaels et al. (1964) have defined effective diffusivity in amorphous polymers as

$$D_{\text{eff}} = \frac{D\psi}{\beta} \tag{2.21}$$

where ψ ($\leqq 1$) is the "detour" ratio, and β ($\geqq 1$) the "blocking" factor measuring the reduction in mobility caused by the narrowness of the amorphous regions. The detour ratio is proportional to a power of the amorphous polymer volume fraction; β depends on the size of the diffusing species (Michaels et al. 1964, Peterlin, 1979).] When simple models are visualized for the configuration of the pore structure, τ can be related to a geometrical factor; nevertheless, in general, it is a "fudge factor" of greater or lesser sophistication (Aris, 1975).

Theoretical models of pores with constrictions have been proposed by Petersen (1958), Michaels (1959), and Currie (1960). Petersen depicts the pores as a series of hyperbolas of revolution with constrictions at the vertices. Currie considers the pores as tubes of sinusoidal form. Michaels portrays the pores as consisting of two cylindrical capillaries of different diameters and lengths, joined in series. This model was evaluated by Rowe et al. (1972) to predict potassium chloride release from tablets made by compression of varying amounts of potassium chloride with a vinyl chloride/vinyl acetate copolymer. The model has been found to predict the release data well.

Johnson and Stewart (1965) have visualized the pore structure as comprising an array of parallel cylindrical pores having the pore size distribution found experimentally. The effective flux is predicted by applying the relevant equations of mass transport to each pore size and summing the individual contributions over the complete pore size distribution. A general version of this model where the pores are not parallel has also been discussed by Aris (1975).

A model has been proposed for the bidisperse porous structure produced by pressing together particles that are themselves porous (Wakao and Smith, 1962). The diffusion flux comprises the fluxes through macropores and micropores, and the flux by series diffusion through both.

A computational model by Foster and Butt (1966) describes the tortuosities and constrictions of pores via the notion of a major pore composed of minor pores. Evans et al. (1961) have approached this problem on the basis of the so-called "dusty gas" model, in which the porous medium is visualized as an array of dust particles held stationary in space. The presence of gas-surface interactions is then incorporated into it by treating the dust particles as giant molecules in terms of the classical kinetic theory of gases.

Network models have been proposed that attempt to describe the porous medium as being composed of random structural elements (pores), chaotically connected with each other (Haring and Greenkorn, 1970; Pismen, 1974). Periodic capillary models of varying degrees of complexity have been proposed by Wakad and Nardse (1974) and Gavalas and Kim (1981).

Among the more recent approaches, Zygourakis (1987) has described release of bioactive agents from hydrophobic polymers by modeling the pore structure on a network of large spherical cavities interconnected via smaller cylindrical neck pores. Balazs et al. (1985) and Siegel and Langer (1986) have adopted Monte Carlo techniques to describe diffusion in constricted pore geometries. Percolation theories are also increasingly being employed, (see, e.g., Leuenberger et al., 1987; Saltzman and Langer, 1987).

When the preceding models for porous media are applied to highly swollen gels, an implicit assumption is that the active agent in the fluid permeating the gel does not adsorb on the polymer inside the pores. While adsorption on the walls of macropores is insignificant, this factor may become significant for micropores (Furusawa and Smith, 1973). In this case,

$$D_{\mathrm{eff}} = D \frac{\epsilon}{\tau} K_r \tag{2.22}$$

where K_r is the fractional reduction in the diffusivity within the pore when the solute and pore size are of comparable magnitude (Satterfield et al., 1973). The effect of solute partitioning between the exterior and interior of pores at equilibrium must be accounted for by an equilibrium partition coefficient.

In any of the preceding models, the diffusivity in the amorphous polymer, D, itself can be variously obtained from the available theories for homogeneous polymers, discussed earlier. For swollen gels, this can be taken to be equal to diffusivity in the bulk fluid.

The analysis presented above for semi-crystalline polymers is based on the two-phase model where the amorphous and crystalline regions do not interact except in a geometrical fashion (characterized by crystallinity, crystallite geometry, dimensions, and orientation-distribution). However, the crystalline regions also influence the amorphous regions with respect to the density and orientation of polymer segments. This is because macromolecules need not either belong completely to a crystallite or lie completely in the noncrystalline regions. The dynamic and conformative flexibility of macromolecules in the amorphous regions of a semicrystalline polymer is, therefore, smaller than in the undercooled melt (no crystalline region) at the same temperature and pressure. The two-phase model must therefore be considered as a first approximation. A more extended model postulates a third phase containing polymer segments and diffusing molecules of mobility lower than that of those in the rest of the amorphous regions. Some aspects of this model have been discussed by Fuhrmann (1979).

2.1.2 Factors Influencing Diffusional Release

The *in vitro* performance of a diffusion-controlled release device is determined by a number of factors. Some of these factors are properties of the polymer-active agent system, such as the diffusivity, partition coefficient and solubility; others are design factors such as active agent loading, device geometry and barrier thickness, that may be controlled independently.

a) Diffusivity of active agent in polymer. As outlined in the preceding subsection, except for highly swollen gels in which molecules of the solute or active agent diffuse

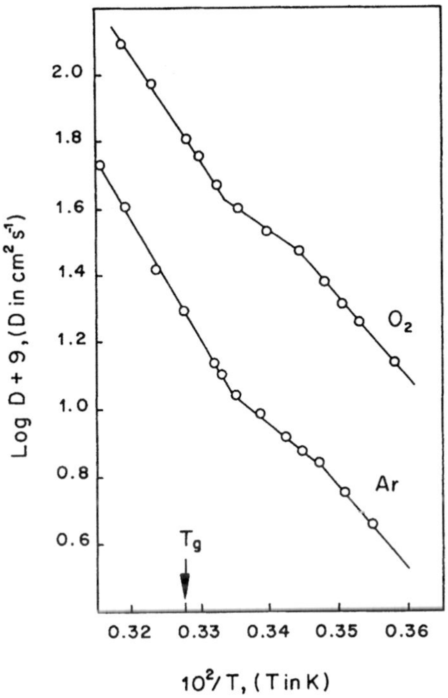

Fig. 2.3. Temperature dependence of the diffusivities of argon and oxygen in poly(vinyl acetate) (Meares, 1954)

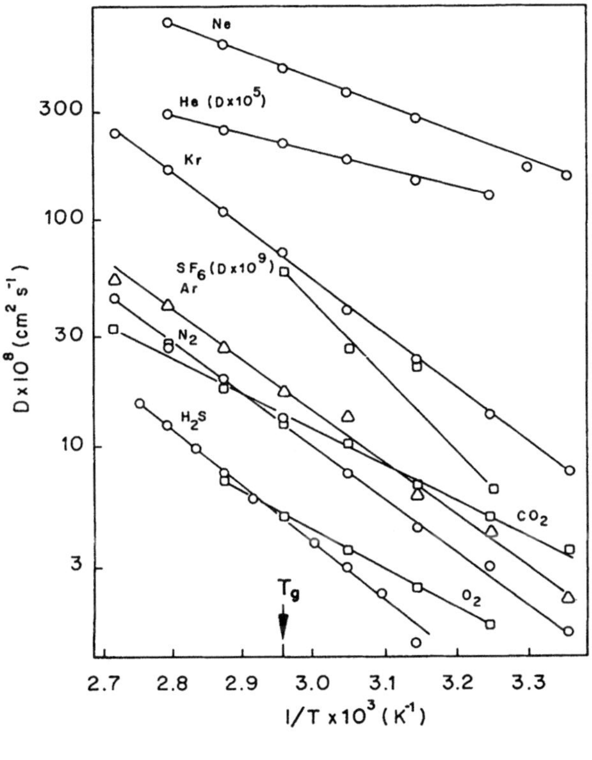

Fig. 2.4. Temperature dependence of the diffusivities of various gases in poly(ethyl methacrylate) (Stannett and Williams, 1965)

in the permeating fluid, diffusion in polymers occurs through the amorphous polymetric regions. The diffusivity is related to the mobility of the polymer chains and, thereby, to the free volume of the system. The free volume is, in turn, dependent upon a number of factors, the most prominent being the temperature of the system relative to the glass transition temperature T_g. As the system temperature is lowered and approaches T_g, the free volume available for diffusion decreases, thereby decreasing the diffusivity. Experimental studies have shown that the diffusivity can exhibit either a continuous change or a discontinuous jump at T_g. This is illustrated in Fig. 2.3 for diffusion of argon and oxygen in poly(vinyl acetate) (Meares, 1954) and in Fig. 2.4 for various gases in poly(ethyl methacrylate) (Stannett and Williams, 1965). The plots in Fig. 2.3 show a distinct change in slope with a lower activation energy below T_g. However, the corresponding plots for poly(ethyl methacrylate) in Fig. 2.4 do not show any inflection. The explanation lies in the size of the diffusing species relative to the average void size in the polymer (Frisch, 1965). For small diffusing molecules, much smaller than the average void size, diffusion occurs by localized activated jumps from one pre-existing cavity to another; only a few contiguous monomer segments are involved. With larger diffusing species, pre-existing cavities may be unable to accommodate the diffusing molecules. Therefore, larger numbers of monomer segments must be rearranged to allow the molecules to diffuse. The process thus becomes dependent on the macroscopic free volume of the polymer. Presumably, poly(ethyl methacrylate) has a larger cavity size, and so involves the first mode of diffusion for the gases studied, while the gaseous molecules in poly(vinyl acetate) appear to follow the second mechanism (Stannett, 1968).

A somewhat similar explanation for diffusivity behavior at and below T_g has been given by Duda (1985) and illustrated in Fig. 2.5. The nature of the observed behavior depends on the amount of free volume trapped in the system as the

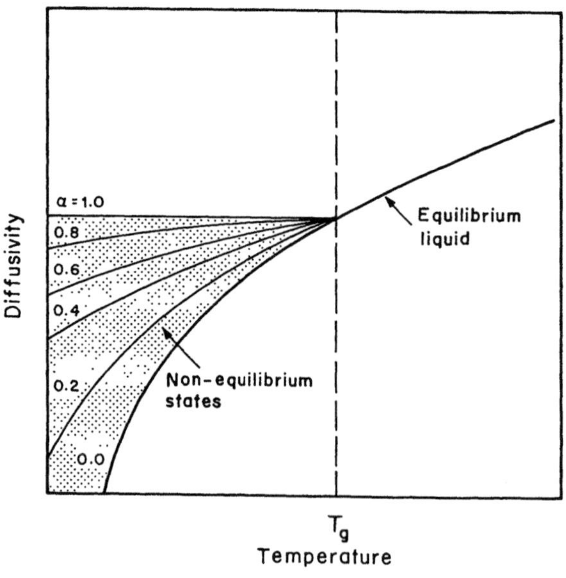

Fig. 2.5. Schematic of the temperature dependence of diffusivity of a solvent molecule in a polymer in the vicinity of the glass transition temperature (T_g) as predicted by the free-volume theory; a is the fraction of free volume at T_g which is available for molecular transport in the glassy polymer (Duda, 1985)

temperature is lowered below T_g. If the amount of hole free volume is maintained at the same level as at T_g (e.g., by rapid cooling), the diffusion coefficient would be a constant. At the other extreme, if no extra free volume exists below T_g (equilibrium cooling), the diffusion coefficient would follow the relationship describing diffusion in an equilibrium liquid. Thus the free-volume theory predicts that a break will occur in the diffusivity versus temperature relationship in the vicinity of T_g (Duda, 1985), depending on the nature of the glassy state. Furthermore, it shows that molecular motion does not cease below T_g.

While the temperature during application of a controlled release device is generally predetermined, a more practical method of adjusting the diffusivity of a given polymer system is via crosslinking. Increasing the degree of crosslinking restricts the freedom of movement of the polymer segments, thus decreasing diffusivity. This effect is illustrated in Fig. 2.6 through different degrees of vulcanization of rubber (Barrer and Skirrow, 1948). The diffusion coefficients decrease sharply with increasing sulfur content. A parallel increase in the glass transition temperature has also been observed (Boyer and Spencer 1946); this agrees with the free-volume concept of diffusion.

The addition of plasticizers to the polymer has an influence opposite to that of crosslinking; it serves to loosen up the polymer matrix, thereby increasing chain mobility. Plasticizer, at low concentration, influences the free volume by reducing interchain interactions. At a higher concentration, it can be considered

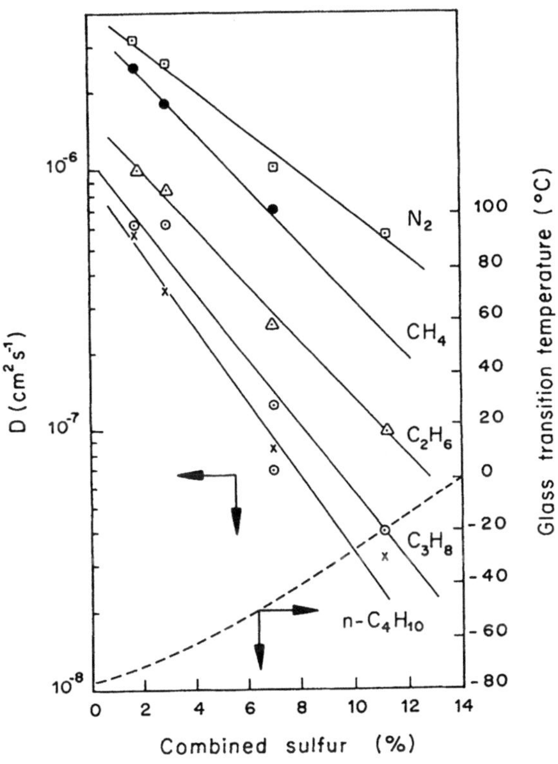

Fig. 2.6. Effect of combined sulfur content of vulcanized natural rubber upon the diffusivity of various gases at 50 °C (Barrer and Skirrow, 1948) and upon the glass transition temperature (Boyer and Spencer 1946; van Amerongen, 1964)

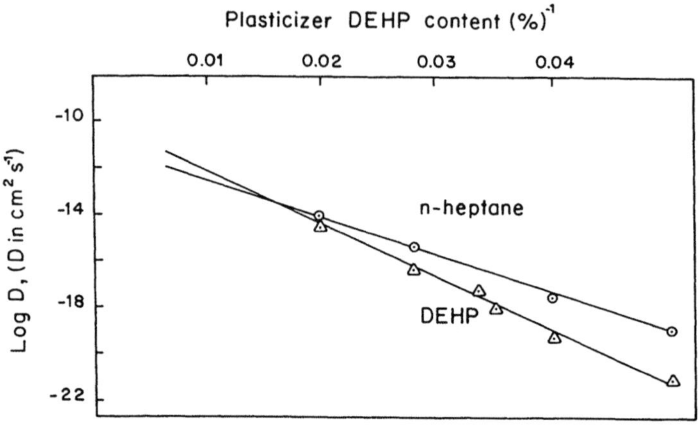

Fig. 2.7. Diffusivity of plasticizer (dioctylphthalate) and *n*-heptane in poly(vinyl chloride) sheets as a function of plasticizer content (Taverdet and Vergnaud, 1984)

a diluent, thereby contributing its free volume to that of the polymer. The net effect is an increase in the free volume with a corresponding increase in the diffusivity (Flynn, 1974). This is illustrated in Fig. 2.7, which shows the diffusivities of plasticizer, dioctylphthalate, and solute, *n*-heptane in poly(vinyl chloride) for various concentrations of the plasticizer (Taverdet and Vergnaud, 1984; also see Hellwege et al., 1961). An increase in T_g corresponding to an increase in plasticizer content can be seen in Fig. 2.8 for diethylphthalate in poly(methyl methacrylate) (Kelley and Bueche, 1961).

In contrast to plasticizers, fillers incorporated into polymers reduce the diffusivity by serving as geometric obstructions and increasing tortuosity of the diffusion path. Furthermore, finely dispersed filler particles can act as effective crosslinks by adsorbing and immobilizing polymer segments (Flynn, 1974). Observed permeation rates can be further decreased by the stronger adsorption of the diffusing species on the fillers. However, this effect should not be significant in a controlled release device where the polymer (and filler) is saturated with the active agent. The influence of filler content on the diffusion is illustrated in Fig. 2.9 (Barrer et al., 1963). While the general influence of fillers is to reduce the diffusivity, the exact nature of the effect is dependent on the type of filler, porosity, geometry, orientation, and spatial and particle-size distribution of filler particles. Van Ameron-gen (1964) has discussed some of these factors in detail.

Crystallites in polymer exert an influence similar to that of fillers in that they serve to increase the tortuosity of the paths and also serve as crosslinks. A randomly dispersed copolymer can disrupt the formation of crystallites, thus reducing T_g and increasing D. However, block copolymers can have an opposite effect depending on the length of the blocks. In addition, the hydrophilic/hydrophobic nature of a comonomer influences the release with hydrophilic comonomers increasing the release rate (Halpern et al., 1976).

Controlled graft copolymerization procedures can also alter the polymer permeation characteristics. The dominant effect of grafting is to impede the migration of the

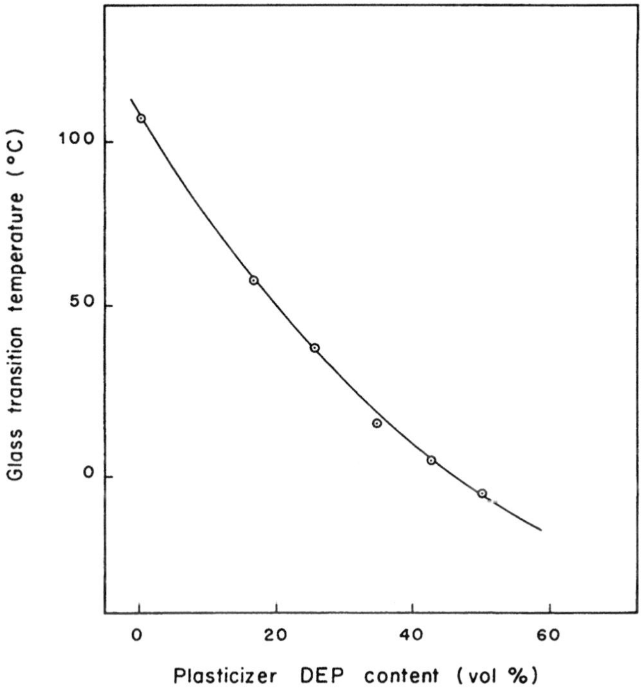

Fig. 2.8. Effect of plasticizer (diethylphthalate) on glass transition of poly(methyl methacrylate) (Kelley and Bueche, 1961)

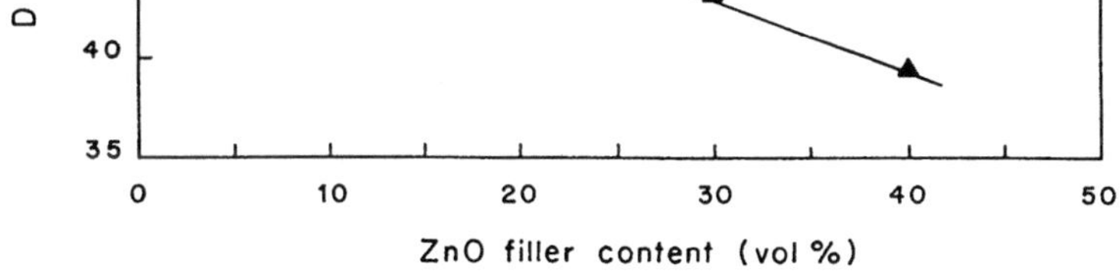

Fig. 2.9. Diffusivity of propane at 40 °C in natural rubber containing zinc oxide filler as a function of filler content (Barrer et al., 1963)

diffusing molecules; it does not cause a variation in the equilibrium content or distribution of these molecules in the polymer. Short graft chains act as more effective inert fillers than longer graft chains since their distribution along the backbone chain allows more efficient packing and structural densification than the relatively isolated long chains (Rogers, 1976).

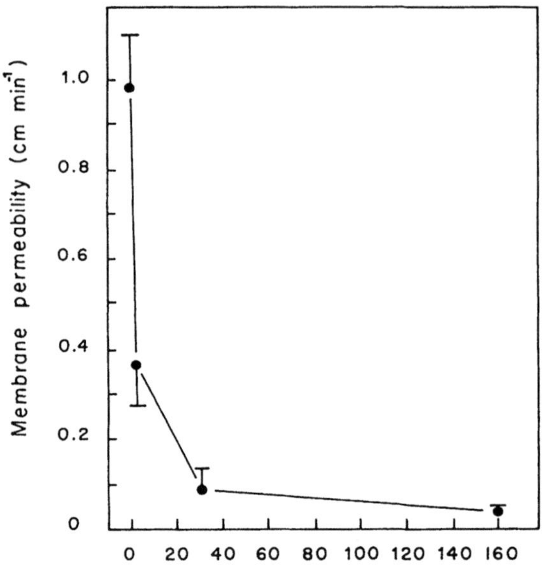

a

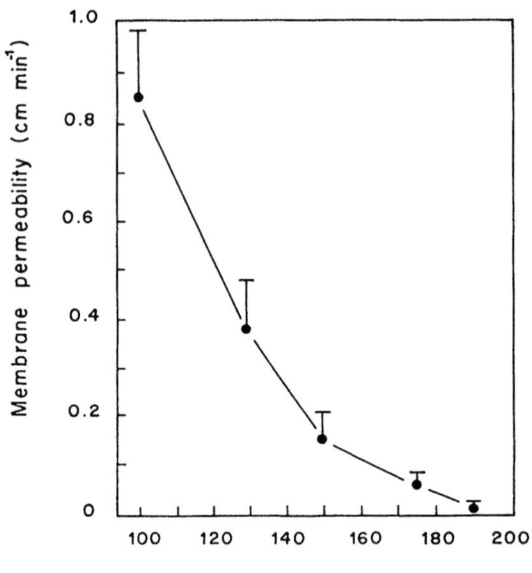

b

Fig. 2.10. **a** Effect of time of heating (at 100 °C) on permeability of poly(vinyl alcohol) membrane to methylene blue at 37 °C (Byron and Dalby, 1987), **b** effect of pretreatment temperature on permeability of poly(vinyl alcohol) membrane to methylene blue at 37 °C (Byron and Dalby, 1987)

The thermal history of polymer affects its transport properties. An example is portrayed in Fig. 2.10 (a, b) (Byron and Dalby, 1987), for poly(vinyl alcohol) films preheated at various temperatures for different durations. While the permeability (= product of diffusivity and partition coefficient) may be partially reduced as a result of changes in the partition coefficient, the diffusivity is also expected to decrease due to geometric obstruction caused by crystallites; heat treatment of poly(vinyl alcohol) has been reported to increase its degree of crystallinity (Kojima et al., 1983, 1984).

Another pretreatment effect besides thermal history is the molecular orientation of polymer chains and their directions relative to the direction of diffusive flow. The rate of penetration parallel to the orientation axis is less than the rate in a randomly oriented sample, which is, in turn, less than the rate perpendicular to the orientation axis. Orientation by stretching of a polymer, well above its T_g, does not appreciably reduce the diffusivity until such time as elongation results in crystallization. The magnitude of this effect is, therefore, dependent on the initial degree of crystallinity (Rogers, 1965).

The factors discussed so far have mainly been characteristics of the polymer. However, the diffusivity of the active agent is also a function of active agent characteristics. The most significant of these is the molecular weight (and thus, size) of the diffusing species, with bigger molecules diffusing slowly (Fig. 2.11) (Baker, 1987). The effects of shape and size of the molecule on the diffusivity are illustrated in Fig. 2.12 for a homologous series of *n*- and iso-paraffins in polyisobutylene (van Amerongen, 1964). The diffusivity of linear hydrocarbons seems to level off at about 5 carbons whereas the branched and cyclic compounds have lower values. The leveling-off may be a result of the diffusion of the

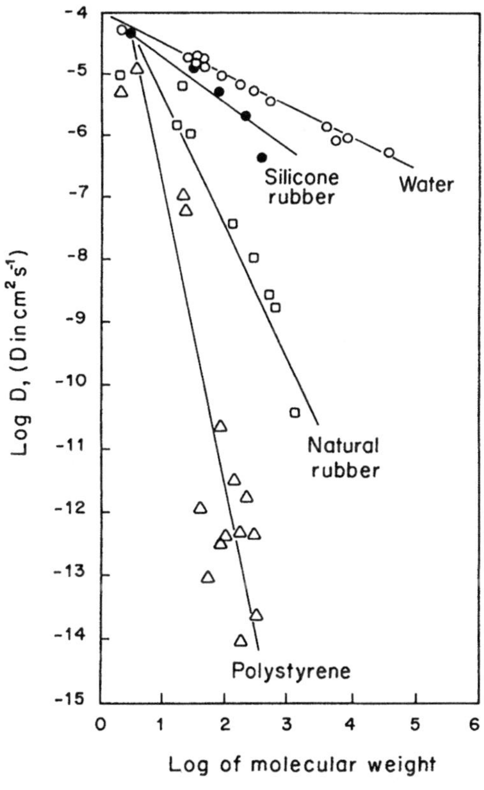

Fig. 2.11. Diffusivity of various solutes in water, silicone rubber, natural rubber, and polystyrene as a function of the molecular weight of the solute (Baker and Lonsdale, 1974; Baker, 1987)

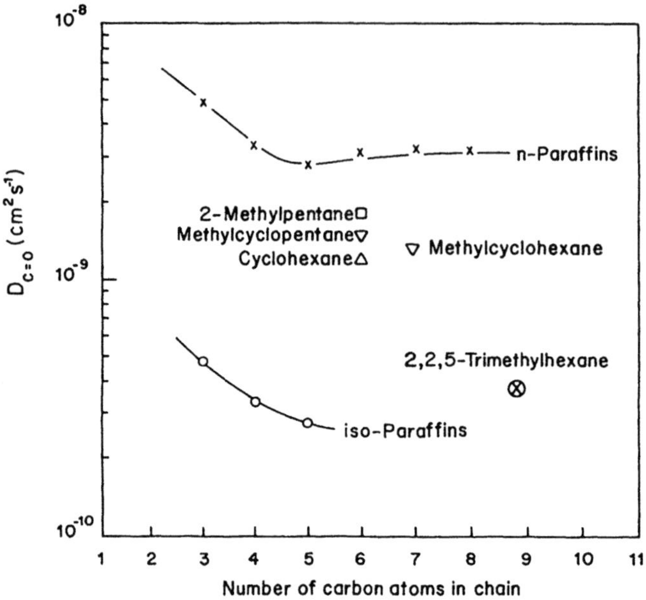

Fig. 2.12. Dependence of diffusivity at 35 °C for simple paraffins in polyisobutylene on the number of carbon atoms in the paraffin chain (van Amerongen, 1964)

linear molecules occurring preferentially along the direction of their length: a linear molecule penetrates the polymer like a needle (van Amerongen, 1964).

The importance of structure has been demonstrated in another study (Lacey and Cowsar, 1974). In this study, the diffusivity of steroids through poly(dimethyl siloxane) membrane was determined. Small modifications in the structure of the steroids have altered the observed diffusion coefficients significantly; this is caused by alteration in their steric requirements.

b) Solubility of active agent in polymer. The rate of transport of the active agent within a polymer is proportional to the concentration gradient within the polymer. This concentration gradient in the amorphous phase of the polymer is dependent on the solubility of the active agent in the void spaces among the polymer chains. Two facets of solubility are salient for controlled release. In a reservoir device, it determines the partition of the active agent between the reservoir fluid and polymer and between the elution medium fluid and polymer, which moreover applies to monolithic devices. In a monolithic device, where the active agent is dispersed within the polymer, it is the absolute solubility (of the active agent) that is important. We shall concern ourselves with the latter case here; the partition coefficients will be examined in the next subsection.

The process of dissolution of the active agent from its matrix-dispersed phase into the polymer phase requires the active-agent molecule to dissociate itself from its crystal cell. Solubility is, therefore, dependent on the active-agent crystal energetics, balanced against the free energy of mixing and other enthalpic and entropic processes occurring during the solution process (Flynn, 1974).

The process of dissociation requires a dissociation energy that is dependent on the melting point of the crystals. The subsequent process of solubilization of the molecule into the polymer is more complicated and is dependent on the chemical natures of the polymer and active agent, as well as on the temperature of the system. A general rule is that compounds similar in polarity to the polymer phase will solubilize more readily (Chien, 1976).

Among the theories employed to predict solubilities is the "solubility parameter" concept (Hildebrand and Scott, 1964), relating the solubility parameter to the cohesive energy density of the material. A polymer will most efficiently dissolve a material having a solubility parameter close to its own. Solubility parameters are a convenient measure of compatibility between a chemical and a polymer, and extensive compilations are available (see, e.g., Burrell, 1975). However, the solubility parameter approach is not useful in a system with polar phases involving hydrogen bonding or hydrophobic association (Flynn, 1974).

Chien (1976) has correlated the solubilities of closely related molecules to their melting point through

$$\ln \bar{C}_p = k - \frac{\Delta H_m}{RT_m} \tag{2.23}$$

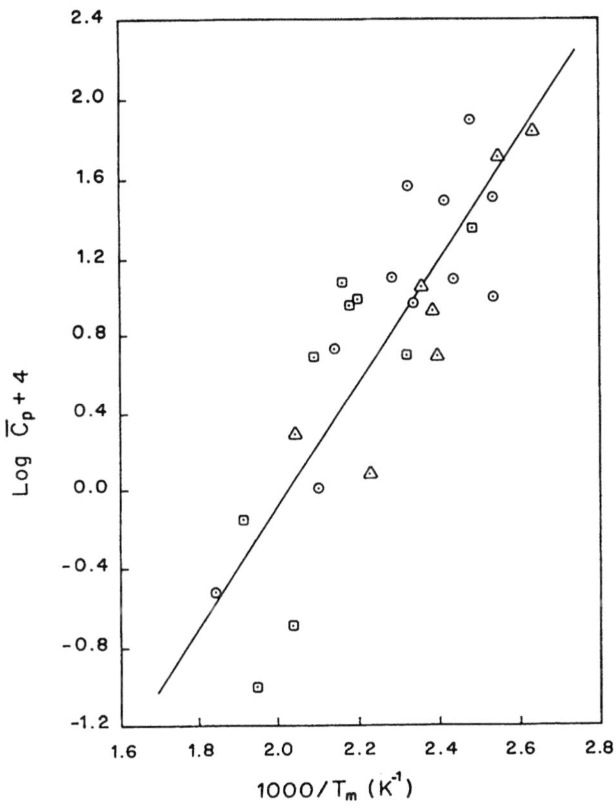

Fig. 2.13. Relationship between mole fraction solubility of testosterone, progesterone, and estradiol derivatives in silicone polymer and the reciprocal of the melting point temperature of the derivative compound (Chien, 1976)

where $\bar{C}_p$ is the mole fraction solubility, ΔH_m the enthalpy of melting and T_m the melting point of the active-agent crystal. This is demonstrated in Fig. 2.13 for a set of steroid derivatives in silicone rubber.

Michaels et al. (1975) have applied the solubility theories of Hildebrand and of Flory-Huggins to the solubility of steroids in polymers. A predictive correlation has been derived between the polymer permeability (= product of diffusivity and solubility) and steroid crystalline melting temperature. This correlation is illustrated in Fig. 2.14 for a set of eleven steroids in six rubbery polymers. Other correlating parameters are entropy of fusion of the steroid and computed solubility parameters of steroid and polymer.

Studies on the solubility of gases in rubber has indicated that the solubility depends on critical temperature or boiling point of gas and on the temperature of the system. In addition, it varies with the nature of gas and rubber, polar gases being more soluble in rubber containing polar groups, non-polar gases in non-polar rubber. The presence of fillers has an added effect because of adsorption of gas in the filler material (van Amerongen, 1964).

In summary, the similarity-dissimilarity in the chemical and physical nature of the polymer and active agent, and the melting enthalpy and temperature of the active agent, are the principal factors determining the solubility. Thus, solubility can be manipulated by utilizing metastable (high energy) or polymorphic crystalline forms of the active agent. Furthermore, analogs or homologs of the original active agent can exhibit altered solubility characteristics and may be substituted if they

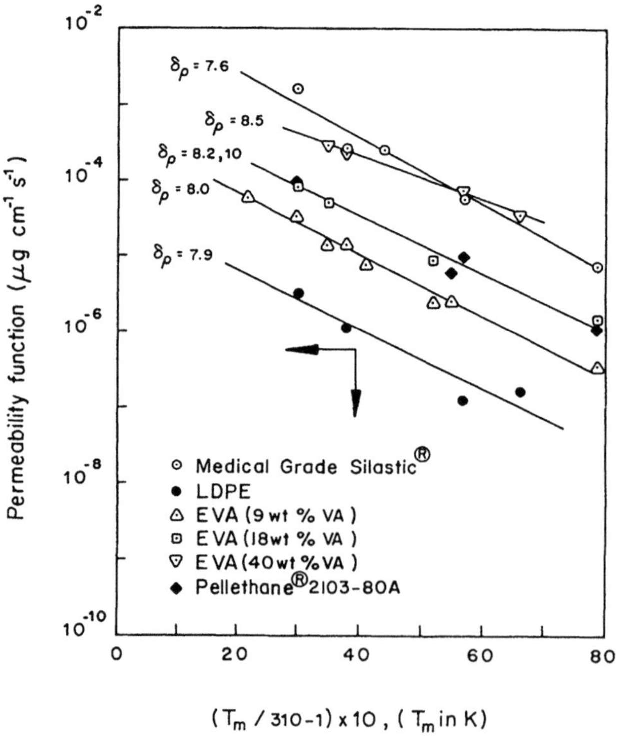

Fig. 2.14. Correlation of permeabilities of some steroids in various polymers with steroid melting point (Michaels et al., 1975)

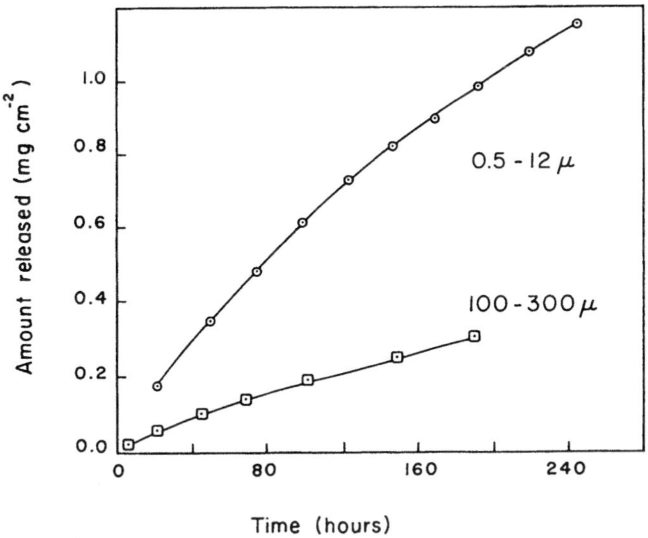

Fig. 2.15. Release of chlormadinone acetate from 10 wt% dispersions in silicone elastomer matrices prepared with micro-sized and macro-sized particles (Haleblian et al., 1971)

exhibit the desired biological activity. Solubilization rates, however, can be more simply manipulated by utilizing particles of different sizes, thus altering performance characteristics of a monolithic device. This is illustrated in Fig. 2.15 for release of chlormadinone acetate from silicone elastomer matrices loaded to 10% with micro-sized (0.5–12 μ) or macro-sized (100–300 μ) particles (Haleblian et al., 1971). The higher surface area of micro-particles leads to a higher rate of availability of chlormadinone acetate for solubilization and diffusion; this leads to a higher release rate. Similar particle-size effects have been demonstrated by a number of investigators for various drugs (see, e.g., Roseman and Cardarelli, 1980).

c) Partition coefficient between medium and polymer. The second important aspect of the solubility of active agent is its distribution between the carrier/elution fluid and polymer phases. This distribution is determined by the requirement for equilibrium between the two phases, given by

$$\alpha_i\big|_{\text{polymer}} = \alpha_i\big|_{\text{elution medium}} \tag{2.24}$$

where α_i is the activity of the active agent in the indicated phases. The activity may be replaced by the product of activity coefficient (v) and concentration (C) so that

$$(v_i C_i)\big|_{\text{polymer}} = (v_i C_i)\big|_{\text{elution medium}} \tag{2.25}$$

or

$$\frac{C_i\big|_{\text{polymer}}}{C_i\big|_{\text{elution medium}}} = \frac{v_i\big|_{\text{elution medium}}}{v_i\big|_{\text{polymer}}} \equiv K \tag{2.26}$$

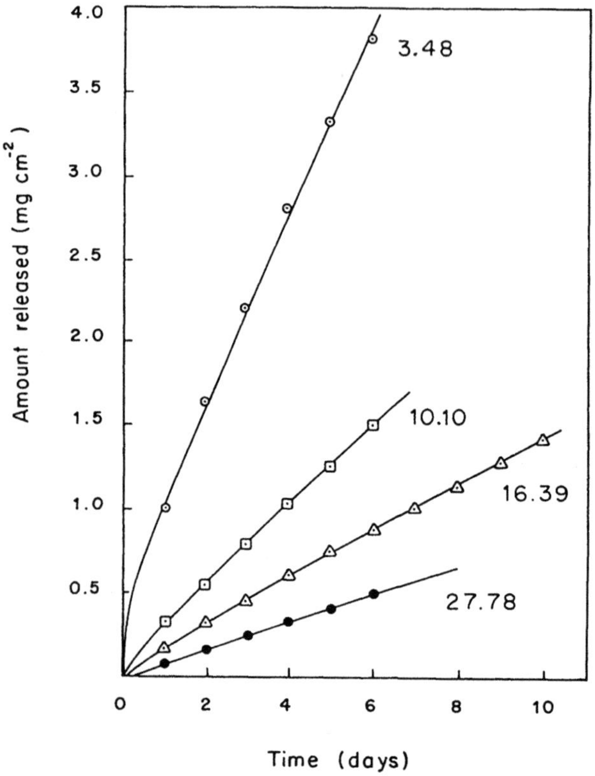

Fig. 2.16. Release of ethynodiol diacetate from a unit area of silicone elastomer matrices into poly(ethylene glycol) (PEG) 400–water media containing varying amounts of PEG 400; partition coefficient of ethynodiol diacetate between polymer and elution media varies with PEG 400 content in the media (Chien and Lambert, 1974)

where K is the partition coefficient and defined as the ratio of solubility of the active agent in the polymer to its solubility in the medium.

The most general approach to the distribution phenomena is to treat the partition law as an extension of Henry's law. In the ideal case, the concentrations of any single molecular species in two phases in equilibrium with each other will bear a constant ratio to each other, as long as the activity coefficients remain relatively constant. This definition assumes no significant solute-solute interactions and also no strong specific solute-solvent interactions (Leo et al., 1971).

An example of the influence of the partition coefficient on release characteristics is given in Fig. 2.16 for release of micro-sized ethynodiol diacetate from silicone elastomer matrix devices to an elution medium containing poly(ethylene glycol) (PEG) 400. The partition coefficient is seen to significantly affect the rate of release. The partition coefficients themselves were varied by altering the solubility of ethynodiol diacetate in water with the addition of the PEG 400 (Chien and Lambert, 1974; also, see Tojo et al., 1985b). Variation of the partition coefficient over a relatively small range (0.1 to 1) also has resulted in a change in the rate controlling mechanism from the matrix-diffusion controlled to the elution medium-polymer partition controlled (Chien and Lambert, 1974).

As with absolute solubility, the partition coefficient is a function of the chemical nature of the active agent and polymer. In one study, Tojo et al. (1987) examined

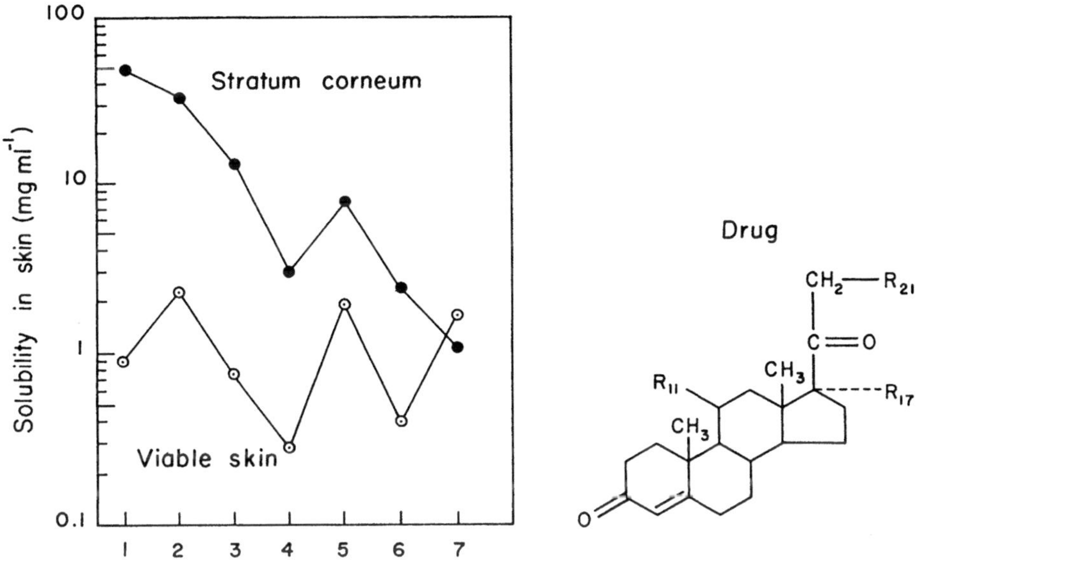

	Drug	R_{21}	R_{11}	R_{17}
1	Progesterone	H	H	H
2	Desoxycorticosterone	OH	H	H
3	11α-Hydroxyprogesterone	H	OH	H
4	17α-Hydroxyprogesterone	H	H	OH
5	Corticosterone	OH	OH	H
6	17α-Hydroxydesoxycorticosterone	OH	H	OH
7	Hydrocortisone	OH	OH	OH

Fig. 2.17. Effect of the number and position of OH groups in progesterone derivatives on the solubility of drug in the skin (Tojo et al., 1987)

the influence of varying the number and position of OH groups in progesterone on its partitioning between a saline-PEG 400 medium and skin. The results are shown in Fig. 2.17 as solubility in the skin versus the progesterone derivatives, and the effect is significant.

The importance of the size of a solute molecule is evident in the plot of solubility against the molar volume of the condensed vapor for sorption in polyethylene (Fig. 2.18) (Rogers et al., 1960). Moreover, two general classes of behavior can be distinguished on the basis of shape of the molecules: the straight chain paraffins in one group, and the aromatic and halogenated compounds in the other. The pH of the medium has been shown to influence the partition due to the effect of pH on the integrity of the polymer, as demonstrated in Fig. 2.19 for distribution of *p*-toluene sulfonic acid between an aqueous solution containing the acid and a nylon film immersed in the solution (Wirbrant and Sundelöf, 1983).

An extensive compilation of partition coefficients has been provided by Leo et al. (1971).

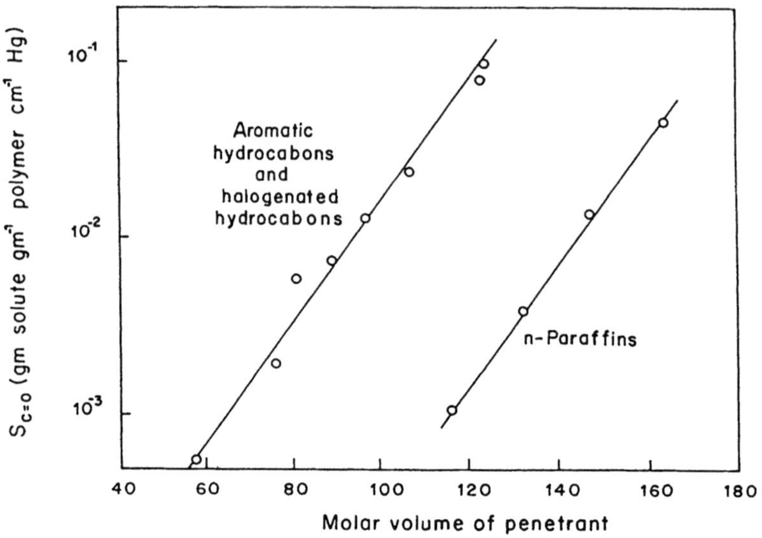

Fig. 2.18. Dependence of the solubility of various permeants in polyethylene at 25 °C on the molar volume of the condensed vapor of the permeant (Rogers et al., 1960)

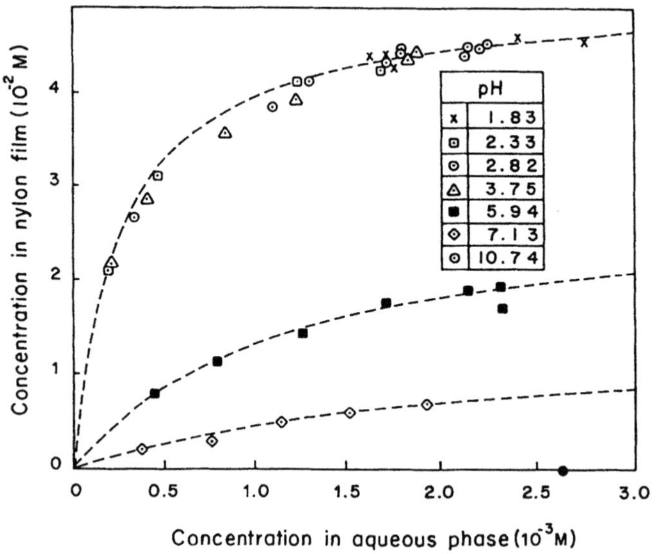

Fig. 2.19. Dependence of the solubility of *p*-toluene sulfonic acid in nylon films on the concentration of acid in, and pH of, the aqueous phase (Wibrant and Sundelöf, 1983)

d) Active agent loading and other design factors. Apart from the factors discussed in the preceding subsections, which are in the most part fixed, once the choice of the polymer-active agent system has been made, there exist some additional parameters that can affect the release characteristics but which can be altered independent of the choice of the system. The first of these is the active agent

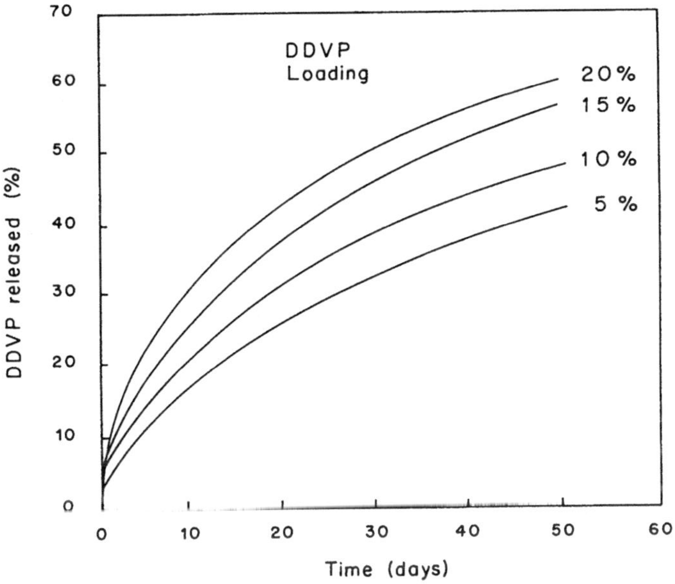

Fig. 2.20.a. Effect of initial loading on amount of 0-,0-dimethyl-2,2-dichlorovinyl phosphate (DDVP) pesticide released from poly(vinyl chloride) matrices containing the dissolved pesticide (Gillett et al., 1972)

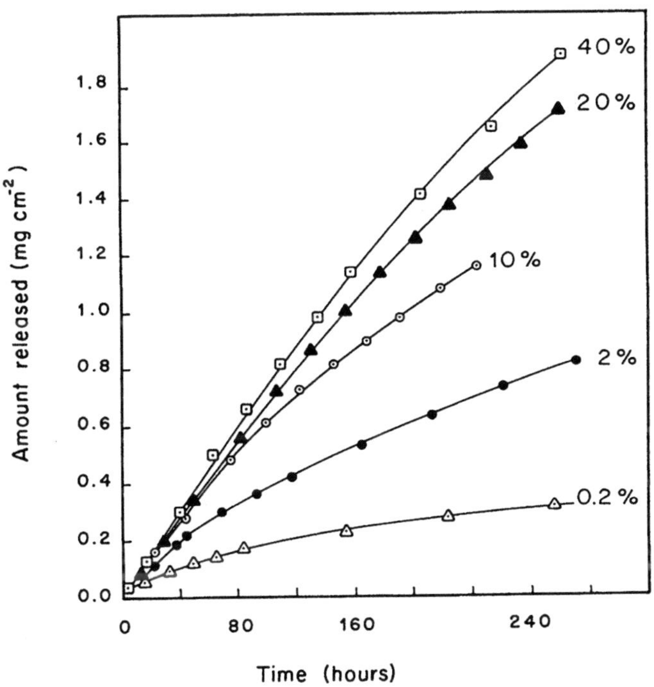

Fig. 2.20.b. Effect of initial loading on amount of chlormadinone acetate released from silicone elastomer matrices containing micro-sized dispersions of the drug (Haleblian et al., 1971)

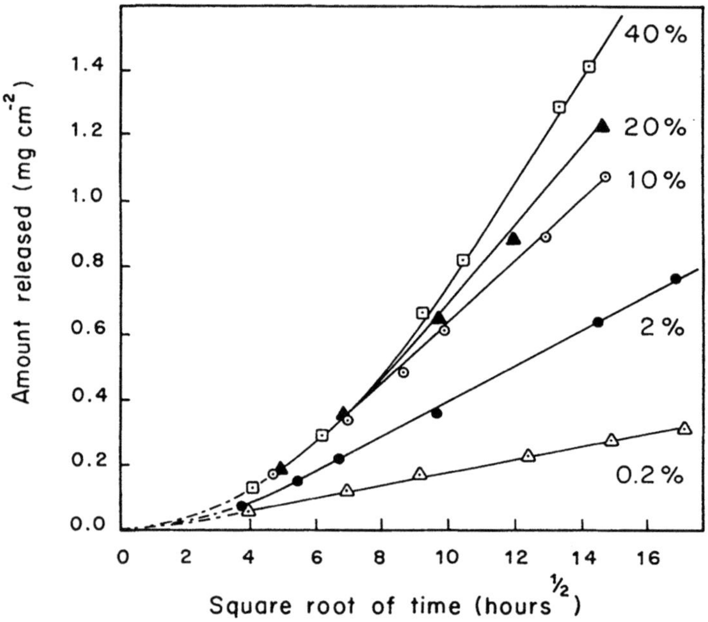

Fig. 2.20.c. Release of chlormadinone acetate as a function of square root of time from silicone elastomer matrices containing micro-sized dispersion of the drug in various initial concentrations (Haleblian et al., 1971)

loading, signifying the amount of active agent incorporated in the reservoir or dissolved/dispersed in the matrix. For a reservoir device, the loading primarily determines the life-time of the device and the length of the zero-order release period, since the diffusion barrier layer is of a fixed thickness. A reservoir loaded with an excess of the active agent will keep the inner surface of the membrane saturated and lead to a steady state concentration profile and a zero-order release. Once the concentration in the reservoir falls below the level required for saturation of the inner surface of the membrane, the release rate will drop continuously.

In contrast to reservoir devices, the average distance the active agent molecules must diffuse before release from a monolithic device changes with time as the active agent is depleted. For monolithic devices containing only dissolved active agent, the release rate is initially proportional to $t^{-1/2}$ and falls exponentially after approximately 60% of the active agent has been released. Figure 2.20a shows the release of pesticide from one such device (Gillett et al., 1972). If increasing amounts of the active agent are incorporated into the polymer, a dispersion of the active agent is obtained after the polymer is saturated. The release rate is still proportional to $t^{-1/2}$; however, the release rate changes over a smaller range than in the case of dissolved active agent devices since the dispersions serve as reservoirs of the active agent. The performance of a dispersed active-agent device is illustrated in Fig. 2.20b for release of micro-sized chlormadinone acetate from silicone elastomer matrices (Haleblian et al., 1971). The dispersions slow the increase in diffusional path by dissolving, thus providing more diffusible active agent at all points in the matrix than would be possible if the polymer were simply saturated with the active agent. However, the loading level in a dispersed active agent monolithic device has another more complex effect on the release mechanism.

At a low (volumetric) loading level (0–5 vol. %), the active agent is widely dispersed in the polymer. The mechanism of release is dissolution into the polymer continuum followed by diffusion to the surface. A pseudo-steady state analysis of this type of system yields a release rate proportional to $t^{-1/2}$, and devices of this type have been called "simple monolithic dispersions". At slightly higher loading levels (5–10 vol.%), the release mechanism becomes more complex. The cavities arising from loss of the active agent near the surface are filled with the elution medium fluid and provide short-cuts for the remaining active agent to diffuse. The cavities do not form continuous pathways at these loading levels; however, they enhance the effective release rate. These devices have been called "complex mono-

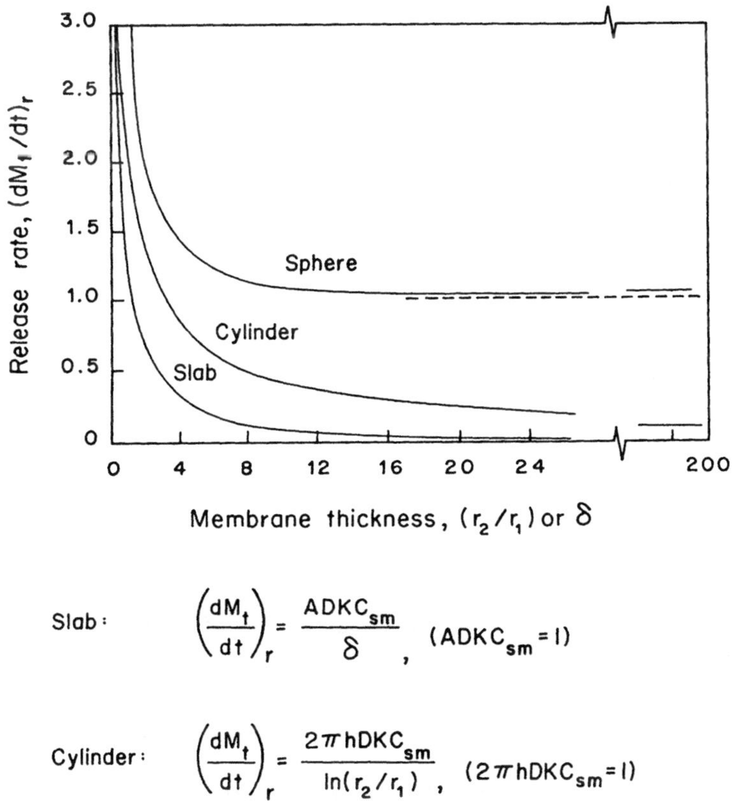

Slab:

$$\left(\frac{dM_t}{dt}\right)_r = \frac{ADKC_{sm}}{\delta}, \quad (ADKC_{sm} = 1)$$

Cylinder:

$$\left(\frac{dM_t}{dt}\right)_r = \frac{2\pi hDKC_{sm}}{\ln(r_2/r_1)}, \quad (2\pi hDKC_{sm} = 1)$$

Sphere:

$$\left(\frac{dM_t}{dt}\right)_r = \frac{4\pi DKC_{sm}\, r_2 r_1}{r_2 - r_1}, \quad (4\pi DKC_{sm} = 1\,;\ r_1 = 1)$$

Fig. 2.21. Steady state release rate as a function of membrane thickness for reservoir devices of various geometries (Baker and Lonsdale, 1974)

lithic dispersions". Finally, at loading levels greater than 20 vol.%, the cavities become sufficiently numerous to allow formation of continuous channels where release can then occur by a leaching mechanism. These devices have been called "monolithic matrix systems" (Baker, 1987).

The effects discussed in the preceding paragraph have been observed in various experimental systems. In Fig. 2.20c, the data for 0.2 and 2 wt% loading appear to follow the behavior typical for "simple monolithic dispersions". However, the mechanism for the 10, 20, and 40 wt% loadings is different since they exhibit somewhat higher release rates. Studies on release of proteins from polymeric matrices have shown high release rates that have been explained by the concepts of percolation theory (Rhine et al., 1980a; Saltzman and Langer, 1987). An indication is that the percolation thresholds (creation of continuous channels) for macromolecular drugs are possibly approached at relatively low loadings.

To counter the decrease in the release rate due to an increase in the diffusional path, monolithic devices have been covered by a rate-controlling barrier, making them analogous to reservoir devices (see, e.g., Lee et al., 1980; Zimmerman, 1986). The thickness of the rate-controlling membrane is crucial in determining the steady-state release rate in a reservoir device, although the dependence is a function of the geometry of the device. For slab geometries, the release rate is inversely proportional to the membrane thickness, while for cylindrical and spherical geometries, the increase in the membrane thickness beyond a certain point does not influence the release rate significantly (Fig. 2.21; Baker, 1987). An example of the influence of this membrane thickness is displayed in Fig. 2.22 for release of testosterone from silicone elastomer capsules of cylindrical geometry (Shippy et al., 1973).

The three most common geometries are slab, cylinder, and sphere. The effect of reservoir device geometry for the same active agent load can be seen in Fig. 2.23 for cylinders and spheres. For monolithic devices, the effect of geometry is

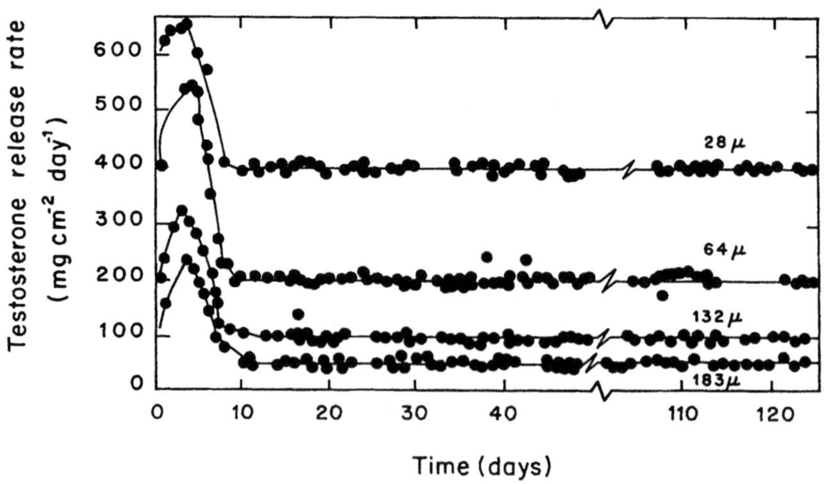

Fig. 2.22. Release rate of testosterone from cylindrical reservoir devices of different membrane thickness made from silicone rubber (Shippy et al., 1973)

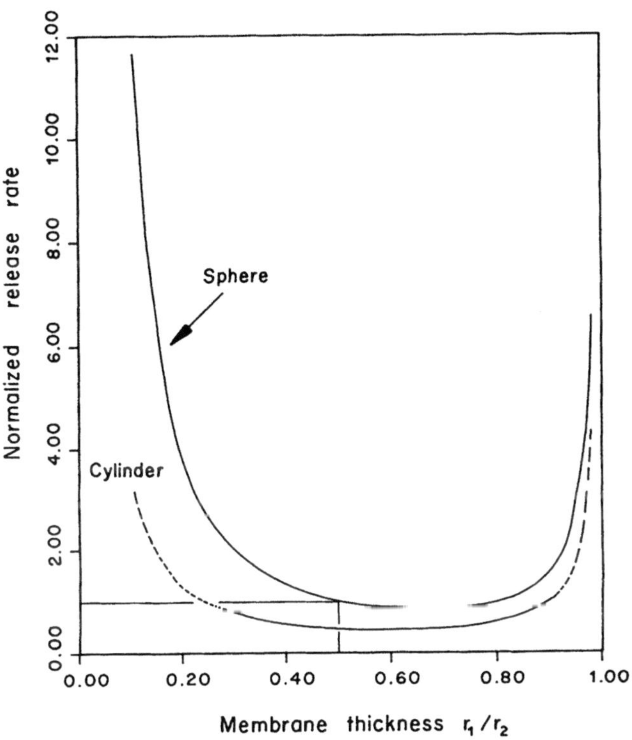

Sphere: $$\left(\frac{dM_t}{dt}\right)_r = 4\pi DKC_{sm}\frac{r_2 r_1}{r_2 - r_1} \quad ; \quad 4\pi DK = 1$$

Cylinder: $$\left(\frac{dM_t}{dt}\right)_r = 2\pi DKC_{sm}\frac{h}{\ln(r_2/r_1)} \quad ; \quad 2\pi DK = 1/2$$

Fig. 2.23. Normalized steady-state release rate as a function of membrane thickness for spherical and cylindrical reservoir devices containing the same amount of active agent; normalization has been performed with respect to release rate from a spherical device with $(r_1/r_2) = 0.5$

dependent on whether the active agent is simply dissolved (Fig. 2.24a) or dispersed in the matrix (Fig. 2.24b; simple monolithic dispersions only). In addition to the three common geometries, special geometries have been devised to obtain constant release rates. These include hemispherical (Rhine et al., 1980b) and sectioned-cylinder (Brooke and Washkuhn, 1977) geometries. The objective of the special geometries is to create an increasing area over which the active agent dissolves into the polymer as the diffusion distance increases (Fig. 2.25a, b).

A final factor important for reservoir devices is the nature of the reservoir fluid. If the active agent is in the form of dry particulate matter, the lack of a good contact with the membrane will lead to non-constant and possibly irreproducible release rates. The active agent, therefore, is frequently suspended in a carrier fluid in which it is highly permeable. Liquid carriers can provide for good contact with the membrane due to bulk movement of the liquid. However, care need be taken that the carrier liquid itself does not diffuse out of the membrane along with the active agent. An example of the improvement with a carrier fluid is demonstrated in Fig. 2.26, in which is displayed the release rate of a series of devices loaded with different amounts of estriol (Baker et al., 1979). Devices loaded with dry powdered drug showed release rates proportional to their loading level. When the device was loaded with cocoa butter as the carrier, the release rate remained unchanged until almost 95% of the original drug load had been released. This problem of lack of contact is significantly smaller for liquid active agents.

e) Hydrodynamic diffusion layer. A factor not controlled by the designer of a controlled release system, but which nevertheless may have a significant influence on the release rate is the presence of a boundary layer on the surface of the

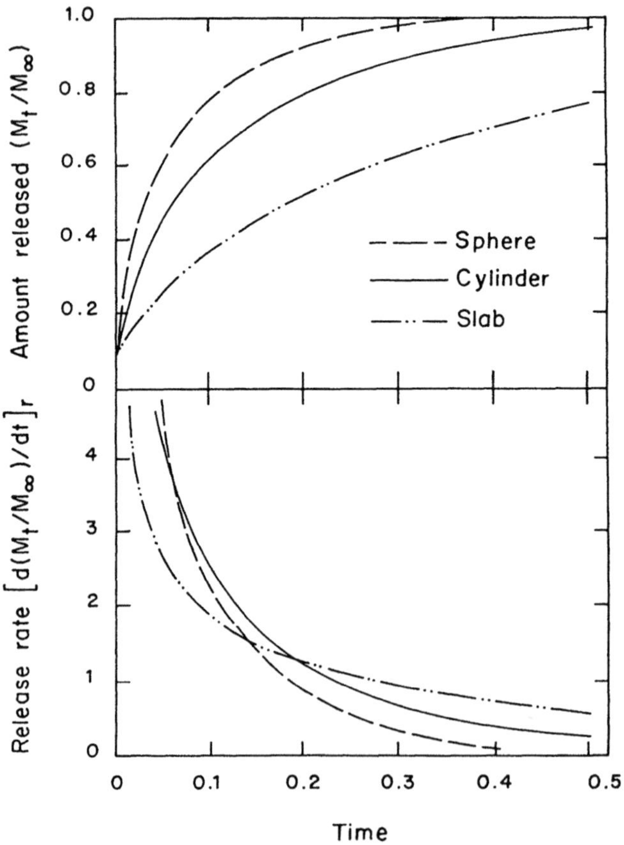

Fig. 2.24. a Theoretical fractional release and release rate as a function of time from monolithic devices of various geometries containing active agent in the dissolved state (Baker and Lonsdale, 1974)

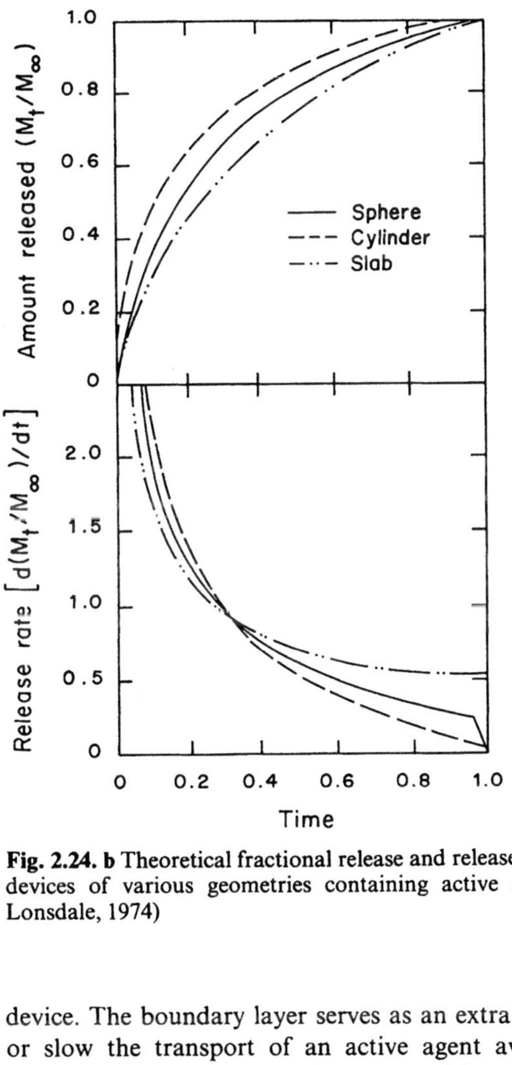

Fig. 2.24. b Theoretical fractional release and release rate as a function of time from monolithic devices of various geometries containing active agent in the dispersed state (Baker and Lonsdale, 1974)

device. The boundary layer serves as an extra diffusional resistance and can restrict or slow the transport of an active agent away from the device. In an extreme case, especially with an active agent of low solubility, a stagnant boundary layer can become saturated with the active agent and stop its release from the device (Baker and Lonsdale, 1974).

While experimental studies in the laboratory are frequently performed under well-stirred infinite sink conditions, such conditions often do not reflect the *in vivo* situation. An example is shown in Fig. 2.27 for release of medroxy progesterone from vaginal rings containing excess drug in suspension, both *in vitro* and *in vivo* (Roseman and Yalkowsky, 1976). An initial period of linear release is obtained during the first 21 days and then a gradual decrease occurs. A high initial rate of release from the ring will shift the controlling resistance to the boundary layer and lead to a constant effective delivery rate. However, as the release rate from the ring drops, the matrix diffusion resistance becomes controlling and the release rate drops. Similar results were observed by Wood et al. (1984) who found that increasing the

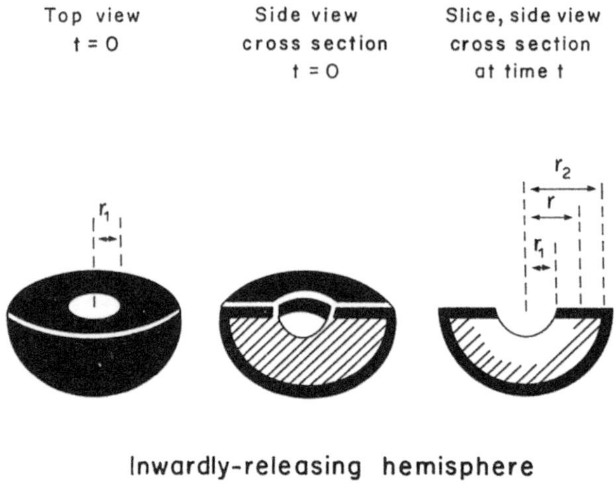

a

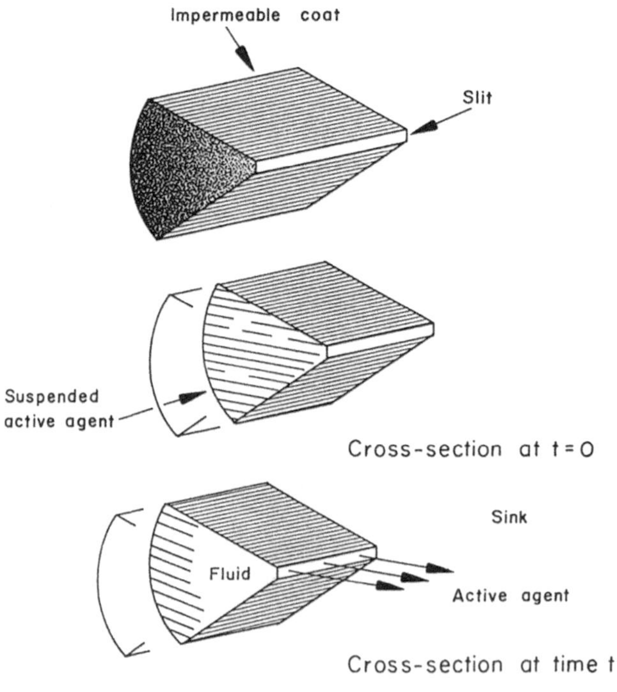

b

Fig. 2.25. a Inwardly releasing hemispherical monolithic device (Rhine et al., 1980b), **b** Inwardly releasing sectioned cylindrical monolithic device (Brooke and Washkuhn, 1977)

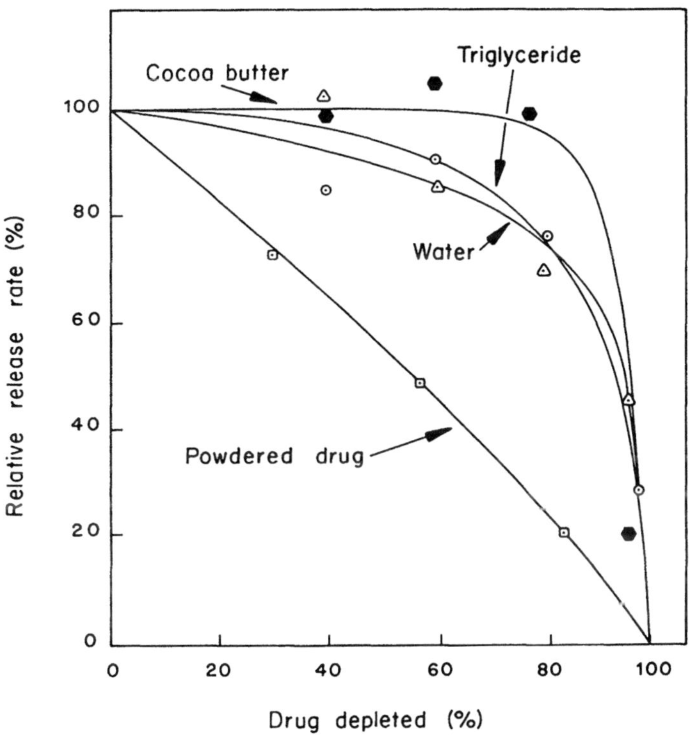

Fig. 2.26. Release rate of estriol from various partially loaded polyurethane devices as influenced by carrier fluid for the drug (Baker et al., 1979)

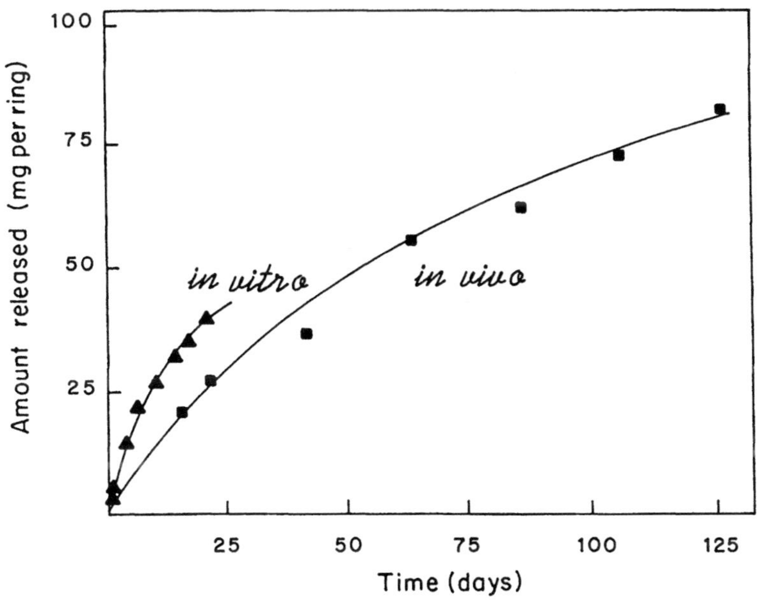

Fig. 2.27. *In vitro* and *in vivo* release of medroxyprogesterone acetate from vaginal rings (Roseman and Yalkowsky, 1976)

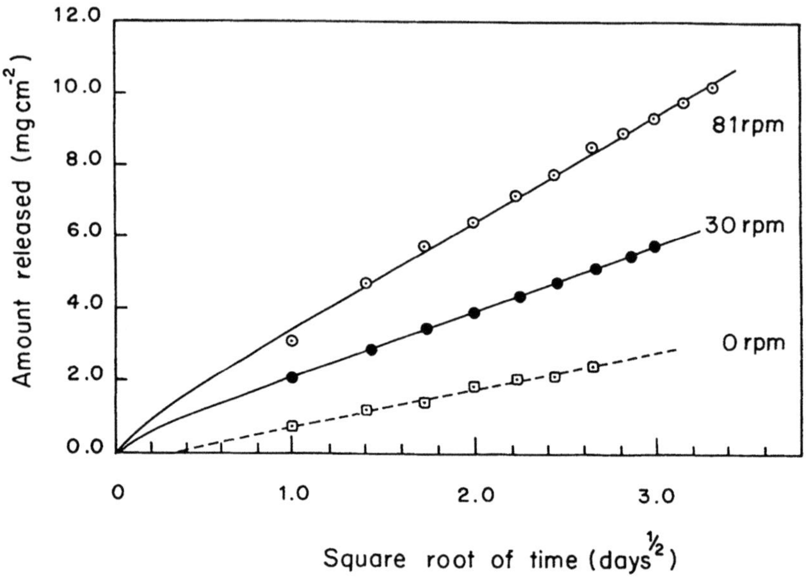

Fig. 2.28. Release of ethynodiol diacetate into a poly(ethylene glycol) 400-water medium as a function of square root of time and agitation speed (Chien et al., 1975)

crosslinker content (ethylene glycol dimethacrylate) in a poly(2-hydroxyethylmethacrylate) gel resulted in a closer agreement between *in vivo* and *in vitro* release rates.

In *in vitro* studies, reservoir devices that show the release rate to be independent of the thickness of the membrane but proportional to the membrane surface area, indicate the existence of a controlling hydrodynamic diffusion layer. An *in vitro* technique for estimating the significance of this resistance is to change the agitation rate in release studies. Chien et al. (1975) observed that the release rate of ethynodiol diacetate from silicone matrices was strongly dependent on the agitation rate (Fig. 2.28).

The boundary-layer diffusional resistance can alter the release characteristics of a device, as evident from Figs. 2.27 and 2.28. It can upset the performance of an otherwise well-designed device and therefore must be considered during the design process itself.

2.2 Models for Diffusion-Controlled Release Systems

A device is categorized as diffusion controlled when the release rate is determined only by the rate of diffusion of the active agent. As mentioned earlier, the basic controlled-release mechanism in any of these devices is diffusion through one or more of the structural components of the device, such as the membrane, matrix and pores or the surface boundary layers surrounding the device. The basic governing equation of a model for this type of device, whether reservoir or monolithic, is

$$\frac{\partial C_{im}}{\partial t} = \nabla \cdot (D_{im} \nabla C_{im}) \tag{2.27}$$

where the subscripts "*im*" refer to species *i* in the membrane/matrix. The subscript *i* has been dropped in the subsequent discussion for simplicity since a single active agent species will be considered. However, in the absence of cross-diffusion effects, the discussion is valid for the case of multiple active agents also. Equation (2.27) has been solved, subject to various boundary conditions depending on the characteristics of the device and environmental conditions. The release rate $(dM_t/dt)_r$ at any instant t, is given by

$$\left(\frac{\mathrm{d}M_t}{\mathrm{d}t}\right)_r = (\text{external surface area}) \times \text{flux}|_{\text{external surface}} .$$

This general category of devices can be divided into reservoir, monolithic and porous devices according to their physical structure or construction, as discussed earlier. The models for these systems discussed here have also been summarized in Tables 2.1, 2.2, and 2.3. (In these tables, PSSA = pseudo steady-state analysis.)

Table 2.1. Non-erodible reservoir devices

Controlling resistances	System description	Model description and solution type	Reference
Membrane mass transfer resistance	Infinite source, Infinite well-agitated sink	Rectangular reservoir no end effects, time-lag and burst effects, analytic	Baker and Lonsdale (1974), Flynn et al. (1974), Crank (1975)
		Cylindrical reservoir no end effects, time-lag and burst effects, analytic	Crank (1975)
		Spherical reservoir time-lag and burst effects, analytic	Crank (1975)
	Infinite source, finite well-agitated activated agent	Rectangular reservoir PSSA, analytic	Flynn et al. (1974)
		Spherical reservoir time-lag, effect of a number of pellets of same size in the sink, analytic	Christensen et al. (1982)

Table 2.1 (continued)

Controlling resistances	System description	Model description and solution type	Reference
Membrane mass transfer resistance	Finite source, finite well-agitated sink	Rectangular reservoir PSSA, analytic	Baker and Lonsdale (1974) Flynn et al. (1974)
		Unsteady state, time-lag analytic	Spacek and Kubin (1967)
	Finite source, finite well-agitated sink	Spherical reservoir time-lag, effect of a number of pellets of same/different sizes in the sink, analytic	Christensen et al. (1982)
	Finite source, infinite well-agitated sink	Rectangular reservoir PSSA, analytic	Flynn et al. (1974)
		Spherical reservoir time-lag, effect of a number of pellets of same size in the sink, analytic	Christensen et al. (1982)
Membrane and laminates mass transfer resistance	Infinite source, infinite sink	Rectangular reservoir PSSA, analytic	Flynn et al. (1974)
		Unsteady state, analytic	Barrie et al. (1963)
		Rectangular reservoir PSSA, analytic	Flynn et al. (1972)
Membrane and boundary layer mass transfer resistance	Infinite source, infinite sink	Cylindrical reservoir unsteady state, analytic	Carslaw and Jaeger (1959)
		Spherical reservoir unsteady state, analytic	Carslaw and Jaeger (1959)
Reservoir, membrane and boundary layer mass transfer resistance	Finite source, infinite sink	Spherical reservoir unsteady state, time-lag, burst, and general initial conditions, active agent decay, membrane properties and temperature effects, numerical	Fan et al. (1978), Fan et al. (1979), Tojo and Fan (1981), Tojo et al. (1983)

Table 2.2. Non-erodible monolithic devices

Controlling resistances	System description	Model description and solution type	Reference
Diffusive mass transfer in polymer matrix	Infinite well-agitated sink, active agent dissolved in polymer ($C_{a0} \leqq C_{as}$)	Rectangular geometry short/long time approx, analytic; semi-infinite slab, analytic; numerical	Crank (1975) Paul and McSpadden (1976) Jaeger and Clarke (1947)
		Cylindrical geometry analytic; no end effects, short/long time approximation, analytic; numerical	Fu et al. (1976) Crank (1975) Jaeger (1942)
	Finite well-agitated sink, active agent dissolved in polymer ($C_{a0} \leqq C_{as}$)	Rectangular geometry analytic; approximate analytic	Crank (1975) Lee (1980a)
		Cylindrical geometry analytic; approximate analytic	Crank (1975) Lee (1980b)
		Spherical geometry analytic; approximate analytic	Crank (1975) Lee (1980b)
Diffusive mass transfer in polymer matrix and boundary layer resistance	Infinite sink with boundary layer, active agent dissolved in polymer ($C_{a0} \leqq C_{as}$)	Rectangular geometry analytic;	Crank (1975)
		phase transfer resistance at source-sink interface, analytic	Somasekharan and Subramaniam (1980)
		Cylindrical geometry analytic solution	Crank (1975)
		Spherical geometry analytic solution	Crank (1975)
Diffusive mass transfer in polymer matrix and sink	Infinite unagitated sink, active agent dissolved in polymer ($C_{a0} \leqq C_{as}$)	Rectangular geometry analytic solution	Somasekharan and Subramaniam (1980)
Diffusive mass transfer in polymer matrix, microporous membrane, adhesive layer, and skin	Infinite well-agitated sink, active agent dissolved in polymer ($C_{a0} \leqq C_{as}$)	Rectangular geometry diffusivity in matrix and partition coefficient across matrix and membrane first-order rate functions of time, numerical	Gienger et al. (1984)

Table 2.2 (continued)

Controlling resistances	System description	Model description and solution type	Reference
Diffusive mass transfer resistance in polymer matrix	Infinite well-agitated sink, active agent dispersed in polymer ($C_{a0} > C_{as}$)	Rectangular geometry PSSA; no end effects, analytic; exact analytic; approximate analytic	Higuchi (1961), Roseman and Higuchi (1970) Paul andMcSpadden (1976) Lee (1980a)
		Cylindrical geometry PSSA with end effects	Cobby (1974)
		Spherical geometry PSSA, analytic; approximate analytic	Cobby (1974) Lee (1980a)
		General geometry, PSSA, Biconvex Inwardly releasing hemispheres; Sector of a cylinder Multiple holes in rectangular matrix	Cobby (1974) Rhine et al. (1980) Brooke and Washkuhn (1977) Kuu and Yalkowsky (1985)
Diffusive mass transfer resistance in polymer matrix, and boundary-layer resistance	Infinite agitated sink, active agent dispersed in polymer ($C_{a0} > C_{as}$)	Rectangular geometry PSSA, analytic; exact analytic	Roseman and Higuchi (1970) Tojo (1985) Paul and McSpadden (1976)
		Cylindrical geometry PSSA, analytic	Roseman and Higuchi (1970)
	Finite agitated sink, active agent dispersed in polymer ($C_{a0} > C_{as}$)	Spherical geometry PSSA, analytic	Gupta and Sparks (1980)
Diffusive mass transfer resistance in polymer, and laminate layer	Infinite well-agitated sink, active agent dispersed in polymer ($C_{a0} > C_{as}$)	Cylindrical geometry PSSA, analytic	Lee et al. (1980)
Dissolution of active agent, and diffusive mass transfer resistance in polymer and barrier layer	Infinite well-agitated sink, active agent dispersed in polymer ($C_{a0} > C_{as}$)	Rectangular geometry analytic, numerical; semi-infinite geometry, analytic solution, valid while concentration in matrix > C_{as} everywhere.	Ayers and Lindstrom (1977) Lindstrom and Ayers (1977) Chandrasekaran and Paul (1982)

Table 2.3. Porous Devices

Controlling resistances	System description	Model description and solution type	Reference
Pore diffusion resistance	Reservoir device, infinite source and sink	Hollow fiber PSSA, analytic	Ashare et al. (1976)
	Monolithic device, infinite well-agitated sink, dispersed active agent ($C_{a0} > \epsilon C_{bs}$)	Rectangular geometry PSSA, analytic;	Higuchi (1963)
		diffusion-dissolution mechanism, analytic;	Peppas (1983)
		diffusion only in continuously forming pores, analytic	Miller and Peppas (1983)
		Cylindrical geometry PSSA, analytic	Roseman and Higuchi (1970)
		Spherical geometry PSSA, analytic	Higuchi (1963)
	Monolithic device, infinite well-agitated sink, dissolved active agent ($C_{a0} \leqq C_{bs}$)	Rectangular geometry PSSA, analytic	Fessi et al. (1982)

2.2.1 Reservoir Devices

By assuming a constant level of activity of the active agent throughout the release period, a simple model can be derived for a reservoir device. This assumption essentially corresponds to an infinite source and can theoretically be achieved by loading the reservoir with the active agent such that its concentration is much greater than its solubility in the polymeric membrane. The steady-state behavior of this device can be characterized by the expression for flux of the active agent, i.e.

$$J = -D_m \frac{\partial C_m}{\partial x} \tag{2.28}$$

where C_m is the active agent concentration in the membrane.

For such an infinite source in a rectangular device with negligible end effects and membrane thickness, δ, integration of Eq. (2.28) yields

$$J_2 = D_m \frac{C_{m1} - C_{m2}}{\delta}. \tag{2.29}$$

The subscript "1" refers to the reservoir side of the membrane, while "2" refers to the sink side. For the device with total external area (both surfaces), A, placed in an infinite well-agitated sink, the steady-state release rate can thus be written as

$$\left(\frac{\mathrm{d}M_t}{\mathrm{d}t}\right)_r = AD_m \frac{C_{m1} - C_m}{\delta} \tag{2.30}$$

or

$$\left(\frac{\mathrm{d}M_t}{\mathrm{d}t}\right)_r = AD_m K \frac{C_1 - C_2}{\delta}\,. \tag{2.31}$$

The total amount released up to time t is then

$$M_t = AD_m K \frac{C_1 - C_2}{\delta} t \tag{2.32}$$

where C_1 and C_2 are the concentrations in the reservoir and sink (assumed constant and uniform), respectively, and the product of D_m and K, i.e., D_mK, is the permeability.

A similar expression can be obtained for an infinite source in a cylindrial device of length, h, with negligible end effects and with internal and external radii, r_1 and r_2, respectively. The steady-state release rate is

$$\left(\frac{\mathrm{d}M_t}{\mathrm{d}t}\right)_r = \frac{2\pi h D_m K(C_1 - C_2)}{\ln (r_2/r_1)}\,. \tag{2.33}$$

The corresponding expression for a spherical device with internal and external radii, r_1 and r_2, respectively, is

$$\left(\frac{\mathrm{d}M_t}{\mathrm{d}t}\right)_r = \frac{4\pi D_m K(C_1 - C_2)}{(r_2 - r_1)/r_1 r_2} \tag{2.34}$$

Apart from the assumptions of infinite well-agitated source (which implies a constant C_1) and constant D_m, the above constant activity source model also assumes that the distribution coefficient is independent of concentration (constant K); that the medium on both sides of membrane (reservoir and sink) is the same (K same on both sides); and that the sink is infinite and well-agitated (no boundary layer, unchanging C_2). For a simple model, none of these assumptions is particularly restrictive except for the assumption of similar reservoir and sink media, which would appear to be rare in actual situations. By resorting to the use of the parameter "transference $\bar{T}$" instead of "permeability D_mK", the restriction about the medium can be lifted (Theeuwes et al., 1976). The transference, $\bar{T}$, is independent of the medium, unless there is medium-membrane interaction, and is a function only of the active agent and membrane combination. The permeability, on the other hand, depends on all three, irrespective of whether medium-membrane interaction exists or not. $\bar{T}$ is defined as

$$\bar{T} = J_{sat}\delta = D_m K C_s = D_m C_{sm} \tag{2.35}$$

where

J_{sat} = steady-state flux through the membrane from a saturated source to a pure sink (zero concentration), both comprised of the same medium,

δ = thickness of the membrane,

C_s = saturation active agent concentration in the source medium, and

C_{sm} = saturation active agent concentration in the membrane at equilibrium.

When saturated active agent solutions in different media are in equilibrium with the same membrane material, the following equation holds in the absence of medium–membrane interactions.

$$C_{sm} = K_1 C_{s1} = K_2 C_{s2} = \dots = K_n C_{sn} \tag{2.36}$$

where the subscripts 1, 2, ... , denote different media. Thus, when K is independent of concentration in the membrane, Eqs. (2.35) and (2.36) can be used in rewriting Eqs. (2.31), (2.33), and (2.34) for the case when the source and sink media are not the same (and in the absence of medium–membrane interactions). This gives, respectively,

$$\left(\frac{dM_t}{dt}\right)_r = \frac{A\bar{T}}{\delta}\left(\frac{C_1}{C_{s1}} - \frac{C_2}{C_{s2}}\right) \tag{2.37}$$

for a rectangular device,

$$\left(\frac{dM_t}{dt}\right)_r = \frac{2\pi h\bar{T}(C_1/C_{s1} - C_2/C_{s2})}{\ln(r_2/r_1)} \tag{2.38}$$

for a cylindrical reservoir device, and

$$\left(\frac{dM_t}{dt}\right)_r = \frac{4\pi\bar{T}(C_1/C_{s1} - C_2/C_{s2})}{(r_2 - r_1)/r_1 r_2} \tag{2.39}$$

for a spherical reservoir device.

These constant activity sources modeled above have the characteristic that after an initial transition period, the release rate reaches the steady state values given by Eqs. (2.31), (2.33), (2.34) or (2.37), (2.38), and (2.39). This very desirable feature is commonly termed the "zero-order release".

In actual devices, many of the assumptions in the above model would be violated, leading to less-than-ideal release characteristics. The model presented assumes a steady-state profile in the membrane but does not account for the time it may take to attain the steady state. The performance of a real device would depend on, among other factors, its age. If the device is applied immediately after its reservoir is loaded with the active agent, it will require time to establish a concentration profile in the membrane. This is called the "time-lag". On the other hand, if the device is stored before usage, the membrane will become saturated with the active agent during storage and produce a "burst effect" on application. Solutions for these two initial conditions are available for an infinite source and infinite, well-agitated sink (Baker and Lonsdale, 1974; Crank, 1975).

When the initial concentration in a rectangular membrane is C_m^*, the resultant concentration profile in it is

$$C_m = C_{m1} + (C_{m2} - C_{m1})\frac{x}{\delta} + \frac{2}{\pi}\sum_{n=1}^{\infty}\frac{C_{m2}\cos n\pi - C_{m1}}{n}\sin\frac{n\pi x}{\delta}\exp(-D_m n^2\pi^2 t/\delta^2) + \frac{4C_m^*}{\pi}\sum_{m=0}^{\infty}\frac{1}{2m+1}\sin\frac{(2m+1)\pi x}{\delta}\exp[-D_m(2m+1)^2\pi^2 t/\delta^2] \quad (2.40\,\mathrm{a})$$

and the amount released by time t is

$$\frac{M_t}{A} = \frac{D_m(C_{m1} - C_{m2})t}{\delta} + \frac{2\delta}{\pi^2}\sum_{n=1}^{\infty}\frac{C_{m1}\cos n\pi - C_{m2}}{n^2}[1 - \exp(-D_m n^2\pi^2 t/\delta^2)] + \frac{4C_m^*\delta}{\pi^2}\sum_{m=0}^{\infty}\frac{1}{(2m+1)^2}\{1 - \exp[-D_m(2m+1)^2\pi^2 t/\delta^2]\}. \quad (2.40\,\mathrm{b})$$

A is the total area of exchange surface. For the "time-lag" case,

$$C_m^* = 0,\ 0^+ \leqq x \leqq \delta^-$$

and thus the amount released by time t with $C_{m2} = 0$ is given by (Crank, 1975),

$$\frac{M_t}{A} = \frac{D_m C_{m1} t}{\delta} - \frac{\delta C_{m1}}{6} - \frac{2\delta C_{m1}}{\pi^2}\sum_{n=1}^{\infty}\frac{\cos n\pi}{n^2}\exp\left[\frac{-D_m n^2\pi^2 t}{\delta^2}\right]. \quad (2.41)$$

After a long period, the transient term vanishes, giving the steady-state release rate of

$$\left(\frac{dM_t}{dt}\right)_r = \frac{AD_m C_{m1}}{\delta}. \quad (2.42)$$

The corresponding release equation for the "burst case" with

$$C_m^* = C_{m1},\ 0 \leqq x \leqq \delta^-$$

is (Baker and Lonsdale, 1974),

$$\frac{M_t}{A} = \frac{D_m C_{m1} t}{\delta} + \frac{\delta C_{m1}}{3} - \frac{2\delta C_{m1}}{\pi^2}\sum_{n=1}^{\infty}\frac{\cos n\pi}{n^2}\exp\left[\frac{-D_m n^2\pi^2 t}{\delta^2}\right] - \frac{4C_{m1}\delta}{\pi^2}\sum_{m=0}^{\infty}\frac{1}{(2m+1)^2}\exp\left[\frac{-D_m(2m+1)^2\pi^2 t}{\delta^2}\right]. \quad (2.43)$$

The steady-state release rate, after the transient term vanishes is

$$\left(\frac{\mathrm{d}M_t}{\mathrm{d}t}\right)_r = \frac{AD_m C_{m1}}{\delta} . \tag{2.44}$$

Equations (2.42) and (2.44) are equivalent to the solution obtained by the steady state analysis, Eq. (2.30) for $C_{m2} = 0$; however, a similar reduction is possible with $C_{m2} \neq 0$.

The solution for release rates from cylindrical and spherical devices for the "time lag" and "burst case" can be found in numerous treatises on diffusion (see, e.g., Crank, 1975). The more general case where the initial concentration profile is a function of position in the membrane has also been briefly considered by Crank (1975). Numerical solutions for some of the general initial condition cases have been presented by Tojo et al. (1983).

The preceding analysis still has some restricting assumptions including those of an infinite source (constant activity); infinite sink; and well-agitated sink and well-mixed source (no boundary layers and no concentration profiles in either source or sink).

Relaxing the assumptions of infinite source and sink would result in a non-constant concentration of the active agent in each of them. Even if a reservoir is initially loaded with a high concentration of the active agent resulting in a period of zero-order-release (such as from constant activity sources), depletion of the active agent will eventually result in the reservoir concentration declining with time. The concentration of active agent in a finite sink will also increase. A simple analysis of this case involves the assumption of a linear concentration gradient in the membrane at all times (Baker and Lonsdale, 1974; Flynn et al., 1974). Based on this pseudo steady-state assumption, the flux in a rectangular device of exchange area A, may be written as

$$J = -D_m \frac{C_{m2} - C_{m1}}{\delta} \tag{2.45}$$

where active agent concentrations in membrane, (C_{m1} and C_{m2}), are related to the reservoir and sink conditions (C_1 and C_2 respectively) through the distribution coefficient K. These reservoir and sink conditions at time t can be represented as

$$C_1 = \frac{M_1 - M_t}{V_1} \tag{2.46a}$$

$$C_2 = \frac{M_2 + M_t}{V_2} \tag{2.46b}$$

where

M_1, M_2 = total initial amounts of active agent in the reservoir and sink, respectively,
V_1, V_2 = volumes of the reservoir and sink, respectively,
M_t = net amount of the active agent exhanged in time t.

Equations (2.46a) and (2.46b) hold only for cases where the amount of active agent hold-up in the membrane is negligibly small compared to that in the reservoir and sink. Combining Eqs. (2.45), (2.46a), and (2.46b) and integrating yield (Flynn et al., 1974)

$$M_t = \frac{M_1 V_2 - M_2 V_1}{V_1 + V_2}\left[1 - \exp\left[-\frac{(V_1 + V_2)}{V_1 V_2}\frac{A D_m K}{\delta}\,t\right]\right]. \tag{2.47}$$

Variations of this analysis for the case of finite source and infinite sink and for the case of infinite source and finite sink can be easily accomplished. In the first case (finite source and infinite sink), V_2 may be assumed infinite (very large with respect to V_1), thereby reducing Eq. (2.47) to (Flynn et al., 1974)

$$M_t = M_1\left[1 - \exp\left[-\frac{A D_m K}{\delta V_1}\,t\right]\right]. \tag{2.48}$$

Similarly, for the second case (infinite source and finite sink), Eq. (2.47) may be rewritten as (Flynn et al., 1974)

$$M_t = \frac{M_1 V_2 - M_2 V_1}{V_1}\left[1 - \exp\left[-\frac{A D_m K}{V_2 \delta}\,t\right]\right]. \tag{2.49}$$

It should be noted that one of the assumptions of this analysis is the instantaneous equilibriation of the membrane with the changing external conditions, thus instantaneously establishing a linear (pseudo steady-state) concentration profile in membrane. This and the earlier stated assumption of negligible membrane hold-up, require that the membrane be sufficiently thin.

Spacek and Kubin (1967) have treated the same problem in describing a method to determine the diffusion and partition coefficients of solutes in membranes. Their analysis has yielded a solution for the unsteady state diffusion from a finite well-stirred source into another well-stirred finite sink, through area A, subject to the conditions

$$t = 0\,; \quad C_m = 0\,, \quad 0^+ \leqq x \leqq \delta^- \quad \text{(time lag)} \tag{2.50}$$

$$t \geqq 0\,; \quad C_m = KC_1\,, \quad x = \delta^- \quad \text{(reservoir)} \tag{2.51 a}$$

$$t \geqq 0\,; \quad C_m = KC_2\,, \quad x = 0^+ \quad \text{(sink)} \tag{2.51 b}$$

$$C_2 = \left(\frac{D_m A}{V_2}\right)\int_0^t \left.\frac{\partial C_m}{\partial x}\right|_{x=0} dt \tag{2.51 c}$$

$$C_1 = C_0 - \left(\frac{D_m A}{V_1}\right)\int_0^t \left.\frac{\partial C_m}{\partial x}\right|_{x=\delta} dt \tag{2.51 d}$$

where C_0 is the initial concentration of the active agent in the reservoir. The resultant concentration profiles are

$$S(x, p) = C \sinh [x(p/D_m)^{1/2}] + D \cosh [x(p/D_m)^{1/2}] \tag{2.52a}$$

$$s_1(q) = \frac{(a_2 \cos q - q \sin q)\, C_0}{(\alpha_1\alpha_2/q^2 - 1)\, q \sin q + (\alpha_1 + \alpha_2) \cos q} \tag{2.52b}$$

$$s_2(q) = \frac{\alpha_2 C_0}{(\alpha_1\alpha_2/q^2 - 1)\, q \sin q + (\alpha_1 + \alpha_2) \cos q} \tag{2.52c}$$

where

$$C = \frac{iqKC_0}{(\alpha_1\alpha_2/q^2 - 1)\, q \sin q + (\alpha_1 + \alpha_2) \cos q} \tag{2.53a}$$

and

$$D = -\frac{i\alpha_2 C}{q}\,. \tag{2.53b}$$

In these expressions,

$$S(x, p) = \mathscr{L}\{C_m(x, t)\} = p \int_0^\infty C_m(x, t) \exp(-pt)\, dt \tag{2.54a}$$

$$s_1(p) = \mathscr{L}\{C_1(t)\} = p \int_0^\infty C_1(t) \exp(-pt)\, dt \tag{2.54b}$$

$$s_2(p) = \mathscr{L}\{C_2(t)\} = p \int_0^\infty C_2(t) \exp(-pt)\, dt \tag{2.54c}$$

$$iq = \delta \left(\frac{p}{D_m}\right)^{1/2} \tag{2.55a}$$

$$\alpha_1 = \frac{K\delta A}{V_1} \tag{2.55b}$$

and

$$\alpha_2 = \frac{K\delta A}{V_2}\,. \tag{2.55c}$$

The above system of equations can be used to evaluate the unsteady state concentration of active agent in finite volume source and sink and also the concentration profile in the membrane. Under the condition that the source and sink volumes, V_1 and V_2, are almost the same, the total amount of active agent released by time t is given by (Spacek and Kubin, 1967)

$$M_t = \frac{VC_0}{2}\left[1 - \sum_{i=1}^{\infty} 4\alpha[q_i^2 + \alpha(2+\alpha)]^{-1} \exp(-q_i^2 D_m t/\delta^2)\right] \tag{2.56}$$

where

$$V = \frac{2V_1V_2}{V_1 + V_2}, \tag{2.57a}$$

$$\alpha = \frac{K\delta A}{V}, \tag{2.57b}$$

and q_i are the positive roots of the characteristic equation

$$\tan q = \frac{2\alpha q}{(q^2 - \alpha^2)}. \tag{2.57c}$$

The next assumption to be removed is that of a well-agitated sink and a well-mixed source, leading to the consideration of boundary layer resistances on the surfaces of the membrane. Boundary layers will exist even in well-agitated real systems, giving rise to a small but appreciable resistance. The resistances can be treated by resorting to film mass transfer coefficients, k_1 and k_2, for the two surfaces of the membrane; one at the reservoir and the other at the sink.

Introduction of an extra resistance renders the consideration of relative rates important. If the characteristic time of diffusion through the membrane is larger than that of mass transfer across the boundary layers, the system is essentially membrane-diffusion controlled. If, on the other hand, the boundary layer mass-transfer controls the release rate, the concentration profiles in the membrane are assumed to be always at steady state (quasi steady-state assumption).

Boundary conditions for the general boundary-layer problem for an infinite activity reservoir and sink are:

$$t = 0; \quad C_m = f(x), \quad -\delta/2 \leqq x \leqq +\delta/2 \tag{2.58a}$$

$$t \geqq 0; \quad C_{m1} = KC_1, \quad x = -\delta/2 \quad \text{(reservoir)} \tag{2.58b}$$

$$t \geqq 0; \quad C_{m2} = KC_2, \quad x = +\delta/2 \quad \text{(sink)} \tag{2.58c}$$

$$t \geqq 0; \quad k_1[C_{1\infty} - C_1] = -D_m \frac{\partial C_m}{\partial x}, \quad x = -\frac{\delta}{2} \quad \text{(reservoir)} \tag{2.58d}$$

$$t \geqq 0; \quad k_2[C_2 - 0] = -D_m \frac{\partial C_m}{\partial x}, \quad x = +\frac{\delta}{2} \quad \text{(sink)} \tag{2.58e}$$

where $C_{1\infty}$ is the concentration of the active agent in the bulk of the reservoir; the active agent concentration in the bulk of the sink remains zero for all times. Solution to the above problem can be found by analogy with the corresponding heat transfer problem for $k_1 = k_2 = k_a$ and $f(x) = 0$ (Carslaw and Jaeger, 1959). Their solution in terms of the concentration profile in the membrane is

$$C_m = -\frac{hKC_{1\infty}x}{2\left(1+\frac{\delta}{2}h\right)} + hKC_{1\infty}\sum_{n=1}^{\infty}\exp(-D_m\beta_n^2 t)\frac{\sin\beta_n x}{\left[h+\frac{\delta}{2}(\beta_n^2+h^2)\right]\sin(\beta_n\delta/2)}$$
$$+\frac{KC_{1\infty}}{2} - hKC_{1\infty}\sum_{n=1}^{\infty}\exp(-D_m\alpha_n^2 t)\frac{\cos\alpha_n x}{\left[h+\frac{\delta}{2}(\alpha_n^2+h^2)\right]\cos(\alpha_n\delta/2)} \tag{2.59}$$

where

$$h = \frac{k_a}{KD_m}. \tag{2.60}$$

In Eq. (2.59), α_n and β_n are the roots of

$$\alpha\tan\frac{\alpha\delta}{2} - h = 0 \tag{2.61 a}$$

and

$$\beta\cot\frac{\beta\delta}{2} + h = 0\,, \tag{2.61 b}$$

respectively. With the magnitude of C_m at $x = \pm\delta/2$ known, the flux or release rate can be determined from

$$\left(\frac{dM_t}{dt}\right)_r = \frac{Ak_a}{K}(KC_{1\infty} - C_{m1})\bigg|_{x=-\delta/2} = \frac{Ak_a}{K}C_{m2}\bigg|_{x=+\delta/2} \tag{2.62}$$

The steady-state release rate is

$$\left(\frac{dM_t}{dt}\right)_r = \frac{1}{2}Ak_aC_{1\infty}\left[1 - \frac{k_a(\delta/2)}{KD_m + k_a(\delta/2)}\right] \tag{2.63}$$

where A is the total area of the exchange surface. When the membrane diffusion controls the process, i.e., $(\delta/2)\,k_a \gg KD_m$, the steady-state release rate becomes

$$\left(\frac{dM_t}{dt}\right)_r = \frac{AKD_mC_{1\infty}}{\delta} \tag{2.64}$$

which is similar to Eq. (2.31) for $C_2 = 0$. When the boundary-layer diffusion controls the process, i.e., when $(\delta/2)\, k_a \ll KD_m$, the corresponding expression is

$$\left(\frac{\mathrm{d}M_t}{\mathrm{d}t}\right)_r = \frac{1}{2}\, Ak_a C_{1\infty} \,. \tag{2.65}$$

The expression of Eq. (2.65) is identical to that obtained by Flynn et al. (1972) by a pseudo steady-state analysis for the same case assuming a constant concentration (negligible resistance) in the membrane. They have also given a pseudo steady-state solution for the boundary-layer diffusion-controlled case when the partition and mass transfer coefficients differed for the two sides of the membrane. Thus for partition coefficients, K_1 and K_2, and the mass transfer coefficients, k_1 and k_2, on the reservoir and sink side respectively, the expression for M_t is (under the assumptions that the concentration profiles in the boundary layers are always at steady state with respect to the membrane interface concentrations, and that there exists no concentration gradient across the membrane)

$$M_t = \frac{k_2 A}{K_2} \cdot \frac{1}{B_1} \left\{ t - \frac{1}{B_1 B_0} \left[1 - \exp\left(-B_1 B_0 t\right)\right] \right\} \tag{2.66}$$

where

$$B_1 = \left(\frac{k_1}{K_1} + \frac{k_2}{K_2}\right) \Big/ (k_1 C_{1\infty}) \tag{2.67 a}$$

and

$$B_0 = \frac{k_1 C_{1\infty}}{\delta} \,. \tag{2.67 b}$$

For large t, Eq. (2.66) reduces to the well-known form in terms of resistance

$$M_t = \frac{AC_{1\infty} t}{\dfrac{1}{k_1} + \dfrac{K_2}{K_1 k_2}} \,. \tag{2.67 c}$$

Introduction of a steady-state membrane diffusion resistance into the above equation leads to another familiar form (see, e.g., Tojo et al., 1985 a)

$$M_t = \frac{AC_{1\infty} t}{\dfrac{1}{k_1} + \dfrac{\delta}{K_1 D_m} + \dfrac{K_2}{K_1 k_2}} \,. \tag{2.67 d}$$

A variation to the case of boundary-layer diffusion control is the case of multiple resistances-in-series in a laminated membrane device formed by using two or more different membranes in layers. For laminated devices, where the resistances in each lamina is independent of the concentration of the active agent, the diffusional resistance R_i in the ith lamina can be expressed as (Flynn et al., 1974)

$$R_i = \frac{\delta_i}{P_i} = \frac{\delta_i}{D_{mi}K_i} \tag{2.68}$$

where P_i is the permeability of the ith lamina, and δ_i and D_{mi} are the thickness and diffusivity for this lamina, respectively. K_i is the partition coefficient for the ith lamina based on the reservoir and sink media which are assumed to be of the same composition. The overall diffusional resistance for n laminae is then given by

$$R_T = \frac{\delta_T}{P_T} = \frac{\delta_1}{D_{m1}K_1} + \frac{\delta_2}{D_{m2}K_2} + \dots + \frac{\delta_n}{D_{mn}K_n}\,. \tag{2.69}$$

The composite permeability coefficient P_T can thus be obtained from knowledge of the thickness of each lamina, the diffusivity of active agent in each lamina, and the distribution coefficient between each lamina and the external medium. The release rate of active agent is obtained as

$$\left(\frac{dM_t}{dt}\right)_r = \frac{A(C_1 - C_2)}{\dfrac{\delta_1}{D_{m1}K_1} + \dfrac{\delta_2}{D_{m2}K_2} + \dots + \dfrac{\delta_n}{D_{mn}K_n}} = \frac{A(C_1 - C_2)}{\left(\dfrac{\delta_T}{P_T}\right)}\,. \tag{2.70}$$

Equation (2.70) gives the steady-state release rate from an infinite source at C_1 to an infinite sink at C_2 when both the reservoir and sink are comprised of the same medium and are well mixed. This expression reduces to Eq. (2.31) when the resistance in any one layer becomes controlling. It is evident from Eq. (2.70) that the order of arrangement of the laminates is immaterial in determining the performance — a consequence of the assumption of independence of resistances from the concentration. An analysis of the type made above has been utilized by Berner (1985) to determine the pharmacokinetics of transdermal drug delivery (the steady-state flux and the pharmacokinetic time lag). The skin serves as an extra lamina in the transdermal drug delivery from a reservoir-type (patch) device.

While the steady-state flux can be determined by evaluating the compound permeability coefficient, P_T, from Eq. (2.69), the unsteady-state flux is more difficult to compute. Barrie et al. (1963) have presented an expression which shows that the unsteady-state flux is dependent on the order of placement of the laminae.

The models considered so far are limited mainly to rectangular geometries with no edge effects. This represents the simplest geometry for which analytic solutions can be given. Solutions for the corresponding cylindrical and spherical systems, for all but the simplest cases, are often substantially more difficult. Solutions for some analogous heat transfer problems can be found in Carslaw and Jaeger (1959), including the following cases:

a. Finite hollow cylinder $a \leqq r \leqq b$, $0 \leqq z \leqq L$; the reservoir-membrane interphase at $r = a$ maintained at $f(z)$; sink at zero concentration (constant activity source, well-agitated infinite sink).

b. Finite hollow cylinder $a \leqq r \leqq b$, $0 \leqq z \leqq L$; the reservoir at $r = a$ kept at $f(z)$; boundary layer at external surface; sink at zero concentration (well-agitated constant activity source, boundary layer at infinite sink).
c. Hollow sphere $a \leqq r \leqq b$; the face at $r = a$ maintained at C_1; boundary layer at $r = b$; sink at C_2 (well-agitated constant activity source, infinite sink with boundary layer).
d. Hollow sphere $a \leqq r \leqq b$; boundary layers at $r = a$ and b; source at C_1 and sink at C_2 (constant activity source, infinite sink, boundary layers at both source and sink).

Christensen et al. (1982) have analyzed finite and infinite reservoir spherical devices (pellets) under both finite and infinite well-agitated sink conditions. The governing differential equation for transport in the membrane has been solved by the method of Laplace transform. They have included the effect of having a number of pellets of different sizes releasing into the same sink. This is trivial only if the sink is infinite and well agitated, since under that condition, the release from each pellet is independent of others. In such a case, the total release is equal to the sum of the individual releases. The equations for the amount released are quite complicated, especially for the case of pellets of different size with variable reservoir and sink concentrations. Numerical evaluation of matrix equations is required.

Models in which fewer assumptions are made can be very difficult to solve analytically; numerical solution may be the only approach to these. A case in point is a spherical geometry, non-constant activity source, finite sink, boundary layer in sink, and an unstirred source leading to concentration profiles in the reservoir. This may well represent a more realistic model and may be further complicated by reaction/temperature effects. The only approach to solution of these coupled partial differential equations may be to evaluate them numerically.

A fairly comprehensive model for reservoir devices has been presented in a series of papers including Fan et al. (1978), Fan et al., (1979), Tojo and Fan (1981), and Tojo et al. (1983). The model presented in these papers involves mass and heat balances around both the reservoir and membrane. The governing equations of the model are

$$\frac{\partial(\varepsilon C)}{\partial t} = \frac{1}{x^\lambda}\frac{\partial}{\partial x}\left(D_m x^\lambda \frac{\partial C}{\partial x}\right) - \bar{\gamma}_A - \bar{\gamma}_s \tag{2.71}$$

and

$$\varrho C_p \frac{\partial T}{\partial t} = \frac{1}{x^\lambda}\frac{\partial}{\partial x}\left(k x^\lambda \frac{\partial T}{\partial x}\right) + (-\Delta H_1)\,\bar{\gamma}_A + (-\Delta H_2)\,\bar{\gamma}_s \tag{2.72}$$

where
λ = 0, 1 or 2 for a slab, cylindrical or spherical geometry, respectively,
$\bar{\gamma}_A$ = rate of spontaneous decay of active agent having a hyperbolic rate form,
$\bar{\gamma}_s$ = rate of sublimation or dissolution of active agent existing as a solid phase,
ΔH_1 = heat of reaction for decay of active agent, and
ΔH_2 = heat of dissolution of active agent.

The boundary conditions involve an infinite sink with boundary-layer transport at the surface; solutions are given for the "time lag", "burst effect" and more general initial conditions. Variations of the different physical parameters in the model, such as the mass and heat transport properties, heat capacity, and temperature dependence of the specific rate of decay of active agent have been considered by the authors. They have resorted to the method of lines to solve the coupled non-linear partial differential equations for the spherical geometry case.

Analysis is also given for the effect of heat and mass transfer coefficients, diffusivity, thermal conductivity, membrane thickness, void fraction of the membrane, perturbations in surrounding temperature, and heat of reaction on the concentration profile, temperature profile, and release rate (Fan et al., 1978). Dynamic characteristics of a device, which is exposed to a sinusoidal perturbation in the surrounding temperature, are presented in Fan et al. (1979). Both isothermal and non-isothermal systems have been explored in Tojo and Fan (1981). Effects of the void fraction, membrane thickness, concentration dependence of diffusivity, and step change in environment temperature on the release rate are studied. Results are compared to those obtained by making the quasi steady-state assumption. Optimization of the controlled release in terms of the membrane thickness and void fraction is also discussed in their works. The effects of the storage period of the product and of the environment temperature on both the transient and steady-state release rate are elucidated in the work by Tojo et al. (1983). The model has also been employed to describe an experimentally determined release pattern of an active agent through a bioerodible membrane. The effect of temperature on the release rate are also considered.

In conclusion, it seems that a whole spectrum of models is available for the reservoir devices in diffusion-controlled systems. The models range from the simple to the very complicated. Analytic solutions are available in many cases and the coverage is fairly comprehensive. While the majority of the models consider only the isothermal cases, non-isothermal cases have also been solved.

2.2.2 Monolithic Devices

Monolithic devices are manufactured by blending the active agent in the polymeric matrix such that the active agent forms a solution with the polymer. Depending on the concentration of the active agent, it may be simply dissolved in the matrix or it may also exist as a dispersion (supersaturation). The performance and release characteristics are different in these two cases. With appropriate boundary conditions, the governing differential equation for both cases is

$$\frac{\partial C_m}{\partial t} = \nabla \cdot (D_m \nabla C_m) \, . \tag{2.73}$$

a) Dissolved active agent case. This type of device is created when the active agent is dissolved in the polymer at or below saturation level ($C_{a0} \leq C_{as}$). For such a planar matrix of thickness δ and negligible end effects, immersed in a well-agitated infinite medium, the amount of active agent released by time t is given by (Crank, 1975)

$$\frac{M_t}{M_\infty} = 1 - \sum_{n=0}^{\infty} \frac{8}{(2n+1)^2 \pi^2} \exp\left[\frac{-D_m(2n+1)^2 \pi^2 t}{\delta^2}\right] \tag{2.74}$$

where M_∞ is the amount released at infinite time; it is equal to the initial active agent loading less the amount in the device which is in equilibrium with the medium at infinite time. The corresponding result for a cylindrical device with negligible end effects is

$$\frac{M_t}{M_\infty} = 1 - \sum_{n=1}^{\infty} \frac{4}{a^2 \alpha_n^2} \exp\left[-D_m \alpha_n^2 t\right] \tag{2.75}$$

where a is the radius of the cylinder and α_n are the positive roots of

$$J_0(a\alpha) = 0 \tag{2.76}$$

where $J_0(x)$ is the Bessel function of the first kind of order zero. Similarly, for a sphere of radius a in a well-stirred infinite medium,

$$\frac{M_t}{M_\infty} = 1 - \frac{6}{\pi^2} \sum_{n=1}^{\infty} \frac{1}{n^2} \exp\left[-\frac{D_m n^2 \pi^2 t}{a^2}\right]. \tag{2.77}$$

Simplified expressions for short- and long-time approximations in the above cases are also available. For the slab, the expression for the initial stages of release is (Baker and Lonsdale, 1974; Crank, 1975)

$$\frac{M_t}{M_\infty} = 4\left(\frac{D_m t}{\pi \delta^2}\right)^{1/2} \quad \text{for} \quad 0 \leqq \frac{M_t}{M_\infty} \leqq 0.6 \tag{2.78a}$$

and the long-time approximation is

$$\frac{M_t}{M_\infty} = 1 - \frac{8}{\pi^2} \exp\left(-\frac{\pi^2 D_m t}{\delta^2}\right) \quad \text{for} \quad 0.4 \leqq \frac{M_t}{M_\infty} \leqq 1.0\,. \tag{2.78b}$$

The release rate thus varies as $t^{-1/2}$ during the early stages of release and then decays exponentially. The same variation of release rate with time of release is observed for cylindrical and spherical geometries also (Baker and Lonsdale, 1974).

Solutions for many other cases can be found through analogy to the corresponding heat-transfer problems (see e.g., Jaeger, 1942; Carslaw and Jaeger, 1959). Some of them are:

a. A system with a general initial distribution $C_{a0}(x)$ of the active agent in a rectangular slab. The sink is infinite at zero concentration and a boundary layer exists at the surface.
b. A rectangular slab with only one face in contact with a finite sink at zero initial concentration.

Solution for the active agent release from a long cylinder into an unstirred infinite medium can be obtained from the work of Tranter (1947). Numerical solutions for release from rectangular slabs for various surface conditions can be obtained in Jaeger and Clarke (1947).

In order to have a model applicable to devices that range from a flat disk (radius $\gg$ thickness) to a cylindrical rod (radius $\ll$ thickness), the diffusion equation must be solved in cylindrical coordinates in three dimensions. This also removes the restriction of negligible end effects. For such a polymer-active agent composite in a well-stirred infinite medium of zero initial concentration, the fraction of active agent released by time t is given by (Fu et al., 1976)

$$\frac{M_t}{M_\infty} = 1 - \frac{8}{l^2 a^2} \sum_{k=1}^{\infty} \exp\left(-D_m \alpha_k^2 t\right)\left(\alpha_k^{-2}\right) \cdot \sum_{j=0}^{\infty} \exp\left(-D_m \beta_j^2 t\right)\left(\beta_j^{-2}\right) \tag{2.79}$$

where l is the half-length of the cylindrical tablet and a is its radius; α_k are the zeroes of

$$J_0(a\alpha) = 0\,, \tag{2.80a}$$

and β_k is defined as

$$\beta_k = \frac{(2k+1)\pi}{2l}\,. \tag{2.80b}$$

The authors have suggested that for most purposes, 10 terms in each series in Eq. (2.79), is sufficient.

When an infinitely long planar matrix of thickness δ (no end effects) is immersed in an agitated infinite medium, a boundary-layer mass-transfer resistance may exist at the surface. The boundary condition at the surfaces ($x = \pm\delta/2$) is then given by

$$-D_m \frac{\partial C_m}{\partial x} = h(C_m - KC_\infty) \tag{2.81}$$

where h is the mass transfer coefficient in the boundary layer. The total amount of active agent, M_t, diffusing out of the matrix is given by (Crank, 1975)

$$\frac{M_t}{M_\infty} = 1 - \sum_{n=1}^{\infty} \frac{2L^2 \exp\left(-\dfrac{4\beta_n^2 D_m t}{\delta^2}\right)}{\beta_n^2(\beta_n^2 + L^2 + L)} \tag{2.82}$$

where β_n are the positive roots of

$$\beta \tan \beta = L \tag{2.83a}$$

and

$$L = \frac{\delta}{2}\frac{h}{D_m}\,. \tag{2.83b}$$

The corresponding solution for an infinite cylinder of radius a is (Crank, 1975)

$$\frac{M_t}{M_\infty} = 1 - \sum_{n=1}^{\infty} \frac{4L^2 \exp(-\beta_n^2 D_m t/a^2)}{\beta_n^2(\beta_n^2 + L^2)} \tag{2.84}$$

where β_n are the roots of

$$\beta J_1(\beta) - LJ_0(\beta) = 0 \tag{2.85a}$$

and

$$L = \frac{ah}{D_m} \tag{2.85b}$$

where $J_1(x)$ and $J_0(x)$ are Bessel functions of the first kind of first- and zero-order, respectively.

For a sphere of radius a with a boundary layer at its surface, the amount of active agent released at time t is (Crank, 1975)

$$\frac{M_t}{M_\infty} = 1 - \sum_{n=1}^{\infty} \frac{6L^2 \exp(-\beta_n^2 D_m t/a^2)}{\beta_n^2\{\beta_n^2 + L(L-1)\}} \tag{2.86}$$

where β_n are the roots of

$$\beta_n \cot \beta_n + L - 1 = 0 \tag{2.87a}$$

and

$$L = \frac{ah}{D_m}. \tag{2.87b}$$

Application of monolithic devices often involves their introduction into an external medium or sink with a limited volume and thus the assumption of a constant medium concentration, made in the previous analyses, does not hold. The concentration of the medium rises as an increasing amount of the active agent is released, resulting in a lowering of the release rate; the system eventually approaches an equilibrium. The final uptake of the active agent by the medium (M_∞) is obtained by a mass balance on the system and is given by (Crank, 1975)

$$M_\infty = \frac{M_1}{1 + \left(\frac{V_1 K}{V_2}\right)} \tag{2.88}$$

where

M_1 = initial active agent loading in the device,
V_1 = volume of the device, and
V_2 = volume of the sink (medium).

For a planar device of thickness δ (neglecting end effects) placed symmetrically in the center of an agitated sink (Crank, 1975), the boundary condition representing the changing condition of the medium for a unit area of exchange at each surface, is given by

$$\frac{V_2}{2K}\frac{\partial C_m}{\partial t} = \mp D_m \frac{\partial C_m}{\partial x}; \qquad x = \pm\delta/2, \qquad t > 0. \tag{2.89a}$$

The amount of active agent released by time t is then given by

$$\frac{M_t}{M_\infty} = 1 - \sum_{n=1}^{\infty} \frac{2\alpha(1+\alpha)}{1+\alpha+\alpha^2 q_n^2} \exp\left(-\frac{4D_m q_n^2 t}{\delta^2}\right) \tag{2.89b}$$

where q_n are the non-zero positive roots of

$$\tan q = -\alpha q \tag{2.90a}$$

and

$$\alpha = \frac{V_2}{V_1 K}. \tag{2.90b}$$

Similarly for a cylindrical device of radius a placed in a finite sink with negligible end effects, (Crank, 1975)

$$\frac{M_t}{M} = 1 - \sum_{n=1}^{\infty} \frac{4\alpha(1+\alpha)}{4+4\alpha+\alpha^2 q_n^2} \exp\left(-D_m q_n^2 t/a^2\right) \tag{2.91}$$

where q_n are the non-zero positive roots of

$$\alpha q J_0(q) + 2J_1(q) = 0 \tag{2.92}$$

and α is defined as in Eq. (2.90b). For a spherical device of radius a immersed in a limited volume agitated bath, the amount released by time t is (Crank, 1975)

$$\frac{M_t}{M_\infty} = 1 - \sum_{n=1}^{\infty} \frac{6\alpha(\alpha+1)}{9+9\alpha+q_n^2\alpha^2} \exp\left(-D_m q_n^2 t/a^2\right) \tag{2.93}$$

where q_n are the non-zero roots of

$$\tan q = \frac{3q}{3+\alpha q^2} \tag{2.94}$$

and α is defined as in Eq. (2.90b).

The solution for the case of a semi-infinite matrix, of exposed surface area $2A$, immersed in an unagitated finite medium is also available (Reid et al., 1980).

When the mass transfer in the medium is diffusive, the amount of active agent released by time t is given by

$$\frac{M_t}{2A} = \frac{2C_{a0}\beta}{1+\beta}\left(\frac{D_m t}{\pi}\right)^{1/2} \times \left\{1 - \frac{2}{1+\beta}\sum_{n=1}^{\infty}\left[-\frac{(1-\beta)}{(1+\beta)}\right]^{n-1} \exp(-n^2W^2)\left[1-E(nW)\right]\right\} \quad (2.95)$$

where

$$\beta = \frac{(D_b/D_m)^{1/2}}{K} \quad (2.96\,\text{a})$$

$$W = \frac{\delta_b}{(D_b t)^{1/2}} = \frac{\delta_b}{\beta K (D_m t)^{1/2}} \quad (2.96\,\text{b})$$

and

$$E(W) = \pi^{1/2} W \exp(W^2)\,\mathrm{erfc}(W)\,. \quad (2.96\,\text{c})$$

In the above equations, D_b is the diffusivity in the unagitated medium, and δ_b the volume of medium per unit area of transfer surface.

It is evident that the exact solutions of even the simple cases considered so far, are quite cumbersome; however, approximate analytic solutions have been achieved for some systems, by means of a refined heat balance integral method (Lee, 1980a). The resultant solutions are easier to evaluate than the exact solutions presented earlier. The method involves double integration of the transient diffusion equation with respect to the space coordinate which eliminates spatial derivatives in the integral equation. The concentration distribution is then approximated by a quadratic function. Resorting to this technique, the amount of active agent released from a planar matrix of thickness δ placed in a well-agitated finite sink of volume V_2 (neglecting end effects), is given by (Lee, 1980a)

$$\frac{M_t}{M_\infty} = (1+\lambda)(1-\Theta_b) \quad (2.97)$$

where λ is defined as

$$\lambda = \frac{V_2}{\delta A K} \quad (2.98\,\text{a})$$

and Θ_b is defined implicitly through the expression

$$\frac{4D_m t}{\delta^2} = \frac{3\lambda^2}{4}\left[\ln\Theta_b + \frac{1}{2\Theta_b^2} - \frac{1}{2}\right]. \quad (2.98\,\text{b})$$

In Eq. (2.98a), A is the surface area of each face of the planar matrix. The solution agrees well with the exact numerical solution especially for small λ and small t. Deviations became appreciable at the tail end of the release process.

For a cylindrical device of radius a and length h (neglecting end effects) in a well-agitated finite sink of volume V_2 (Lee, 1980b)

$$\frac{M_t}{M_\infty} = (1 + \lambda)(1 - \Theta_b) \tag{2.99a}$$

where λ is defined as

$$\lambda = \frac{V_2}{\pi a^2 h K} \tag{2.99b}$$

and Θ_b is defined by the expression

$$\frac{D_m t}{a^2} = \frac{3}{40}\lambda^3\left(\frac{1}{\Theta_b} - 1\right)^3 + \frac{3(3\lambda + 5)}{160}\lambda^2\left(\frac{1}{\Theta_b} - 1\right)^2 + \frac{3\lambda^2(5 - 3\lambda)}{80}\left(\frac{1}{\Theta_b} - 1\right) - \frac{3\lambda^2(5 - 3\lambda)}{80}\ln\left(\frac{1}{\Theta_b}\right). \tag{2.99c}$$

The corresponding solution for a device of spherical geometry with radius a (in a well-agitated finite sink of volume V_2) is (Lee, 1980a)

$$\frac{M_t}{M_\infty} = (1 + \lambda)\left[1 - \frac{4\lambda}{4(1 + \lambda) + (2 - \beta)^2}\right] \tag{2.100}$$

where β is defined through the expression

$$\frac{D_m t}{a^2} = -\frac{\beta}{3} - \left(\frac{\lambda + 2}{3}\right)\ln\left[\frac{4\lambda + 4 - (\beta - 2)^2}{4\lambda}\right] + \frac{2}{3}(1 + \lambda)^{0.5}\ln\left\{\frac{[2(1 + \lambda)^{0.5} + (\beta - 2)][(1 + \lambda)^{0.5} + 1]}{[2(1 + \lambda)^{0.5} - (\beta - 2)][(1 + \lambda)^{0.5} - 1]}\right\} \tag{2.101a}$$

and λ is given by

$$\lambda = \frac{3V_2}{4\pi a^3 K}. \tag{2.101b}$$

The solutions for the cylindrical and spherical cases are in better agreement with the corresponding exact solutions than are the solutions of the planar case. As λ becomes larger, deviation appears for the tail end of the release. It is recommended that the solution for the infinite sink condition be used when $\lambda \gg 10$, for all geometries (Lee, 1980a, 1980b).

Apart from the diffusional resistances in the matrix and the boundary layer, a third matrix-medium interface mass transfer resistance has also been incorporated into models presented by Somasekharan and Subramanian (1980). The interface mass transfer resistance is accounted for by a first-order rate expression

$$\text{flux at interface} = k_i C_m|_{\text{interface}}$$

where k_i is the interface mass transfer coefficient. Models representing three different combinations of conditions have been presented for a semi-infinite matrix of cross-sectional area, $2A$, immersed in an infinite medium.

The first model (N-R-S) considers unsteady-state diffusion in the matrix, a matrix-medium interface mass transfer resistance, and a steady-state boundary-layer resistance. Combining these three resistances in series, the release rate is obtained as (Somasekharan and Subramanian, 1980)

$$\frac{1}{2A}\left(\frac{dM_t}{dt}\right)_r = \frac{C_{a0} D_m^{1/2} k_i D_b}{k_i D_b \pi^{1/2} t^{1/2} + D_m^{1/2} D_b + D_m^{1/2} k_i \delta_b} \tag{2.102}$$

where D_b and δ_b are the diffusivity in, and thickness of, the boundary layer, respectively.

The second model (N-E-S) considers unsteady state diffusion in the matrix, equilibrium at the matrix-medium interface, and a steady-state boundary-layer resistance at the surface. The expression for release rate is

$$\frac{1}{2A}\left(\frac{dM_t}{dt}\right)_r = \left(\frac{C_{a0}}{K} - C_{ab}\right)\frac{D_b}{\delta_b}\exp\,(k_2^2 D_m t)\,\text{erfc}\,(k_2 D_m^{1/2} t^{1/2}) \tag{2.103}$$

where

$$k_2 = \frac{D_b/\delta_b}{D_m K} \tag{2.104}$$

and C_{ab} is the bulk medium concentration of the active agent.

The third model (N-E-N) considers unsteady-state diffusion in the matrix, equilibrium at the matrix-medium interface, and unsteady-state diffusion in the medium (unagitated medium). The resultant release rate is

$$\frac{1}{2A}\left(\frac{dM_t}{dt}\right)_r = \frac{D_m^{1/2}}{(\pi t)^{1/2}}\,\frac{KC_{ab} - C_{a0}}{1 + K(D_m/D_b)^{1/2}}\,. \tag{2.105}$$

When a monolithic device is employed to deliver active agent through the skin (transdermal drug delivery), the skin layer also acts as a laminate. Gienger et al. (1984) have modeled and studied such a system involving transport through a series of resistances including: the matrix, the microporous membrane covering the matrix, the adhesive layer, and the skin. The geometry considered is rectangular (slab); the sink (blood capillary) is considered infinite and well stirred. The model comprises unsteady-state Fickian diffusion equations for the four layers. Equilibrium, characterized by partition coefficients, is assumed at each interface. The diffusivity of active agent in the matrix, and its partition coefficient between the matrix and the microporous membrane, are taken to be first-order rate functions of time. The system of four governing partial differential equations has been solved numerically, and the concentration profiles within each layer are presented as functions of time. For the values of parameters assumed by the authors, significant mass transfer resistances are found only in the matrix and skin. The effect of time-dependent changes of the diffusivity and partition coefficient is such that the concentration gradient within the skin remains essentially unchanged with time. The system, therefore, comes close to the desired zero-order release (Gienger et al., 1984).

b) Dispersed active agent case. When the initial active agent loading in the polymer (C_{a0}) exceeds its solubility in the polymer (C_{as}), the excess, ($C_{a0} - C_{as}$), gives rise to a separate phase at equilibrium. This phase is dispersed in the polymer and causes the release characteristics to be different from the dissolved active agent ($C_{a0} \leqq C_{as}$) case.

Higuchi (1961) offered the first conceptual model for this type of device. This model postulates that the active agent is first extracted from the surface layer; after this layer has been depleted to the saturation level, extraction is initiated at the next layer, leading to a concentration discontinuity at a well-defined, extraction boundary (see Fig. 2.2b). The mechanism of extraction is the "solution-diffusion" process discussed earlier. [A more rigorous description of this mechanism has been given by Chein (1976). He visualizes the process as consisting of three energy-activated steps: (a) dissociation of active agent molecules from their crystal lattice at the extraction boundary, (b) solvation of these dissociated molecules into the polymer matrix at the extraction boundary, and (c) diffusion of these dissolved active agent molecules through the polymer continuum. The majority of the models negelect the influence of the first two steps on the release rate; the release is assumed to be "diffusion controlled"]. Higuchi's analysis is based on the pseudo steady-state assumption that the concentration profile in the extracted region is always at steady state with respect to the movement of the extraction front. He also assumes that dissolution of active agent in the polymer does not influence the rate, i.e., the release is diffusion controlled with dissolution being relatively rapid.

For an infinite rectangular slab (or a rectangular slab with end effects neglected) of thickness δ, in a well-stirred infinite medium, the release rate at the surfaces of the polymer can be written as (under the pseudo steady-state assumption)

$$\left(\frac{dM_t}{dt}\right)_r = \frac{2AD_m[C_{as} - C_{ai}]}{\left(\frac{\delta}{2} - x\right)} \tag{2.106}$$

when the extraction front is at x; C_{ai} is the concentration of the active agent in the polymer at the matrix-medium interface and A is the cross-sectional area of each face of the matrix. A mass balance on the active agent gives

$$\frac{M_t + 2A\left(\frac{\delta}{2} - x\right)\frac{C_{as} + C_{ai}}{2}}{M_\infty} = \frac{(\delta/2 - x)}{(\delta/2)} \; . \tag{2.107}$$

Combining the above two equations, we have

$$\left(\frac{dM_t}{dt}\right)_r = \frac{2AD_m(C_{as} - C_{ai})\left[1 - \frac{2A(\delta/2)\left(\frac{C_{as} + C_{ai}}{2}\right)}{M_\infty}\right]}{\left(\frac{\delta}{2}\right)\left(\frac{M_t}{M_\infty}\right)} . \tag{2.108}$$

Integrating this expression,

$$\frac{M_t^2}{2} = 2AD_m(C_{as} - C_{ai})\left(1 - \frac{A\delta\,\frac{C_{as} + C_{ai}}{2}}{M_\infty}\right)\frac{M_\infty}{(\delta/2)}\,t . \tag{2.109}$$

This, in turn, leads to

$$M_t = 2A\left[2D_m(C_{as} - C_{ai})\left(C_{a0} - \frac{C_{as} + C_{ai}}{2}\right)t\right]^{1/2} \tag{2.110}$$

since

$$M_\infty = A\delta C_{a0} . \tag{2.111}$$

For an infinite medium with a negligible concentration of the active agent ($C_{ab} = 0$, i.e., $C_{ai} = 0$), the above expression reduces to

$$M_t = 2A\left[2D_m C_{as}\left(C_{a0} - \frac{C_{as}}{2}\right)t\right]^{1/2} \tag{2.112}$$

and for $C_{a0} \gg C_{as}$, it further simplifies to

$$M_t = 2A(2D_m C_{a0} C_{as} t)^{1/2} . \tag{2.113}$$

Comparing the above expression with that for the corresponding dissolved active agent case ($C_{a0} \leqq C_{as}$, rectangular slab in infinite well-agitated medium), it is apparent that the initial release in both systems is a function of $t^{1/2}$ [Eqs. (2.78a) and (2.113)]. For the dissolved active agent case, this dependence changes at long extraction times [Eq. (2.78b)], but in the dispersed active agent case, Eq. (2.113) holds until the concentration at the center of the matrix falls below saturation level.

This pseudo steady-state approach can also be applied to other geometries. For $C_{a0} \gg C_{as}$ and $C_{ab} = 0$, the fractional active agent loss from a cylinder of radius a with negligible end effects, can be obtained from (Roseman and Higuchi, 1970; Baker and Lonsdale, 1974)

$$\left(1 - \frac{M_t}{M_\infty}\right)\ln\left(1 - \frac{M_t}{M_\infty}\right) + \frac{M_t}{M_\infty} = \frac{4C_{as}D_m}{C_{a0}a^2}\,t . \tag{2.114}$$

For a sphere of radius a, the corresponding expression is (Baker and Lonsdale, 1974)

$$\frac{3}{2}\left[1-\left(1-\frac{M_t}{M_\infty}\right)^{2/3}\right]-\frac{M_t}{M_\infty}=\frac{3D_mC_{as}}{C_{a0}a^2}\,t\,. \qquad (2.115)$$

It must be noted that apart from requiring the process to be diffusion controlled, the pseudo steady-state approach also assumes that $C_{a0} \gg C_{as}$; the rate of movement of the extraction front is then much slower than the rate of removal of the active agent, thus enabling a steady-state (linear) concentration profile to be established with respect to the (moving) extraction front. This analysis must, therefore, be restricted to devices in which C_{a0} exceeds C_{as} by a factor of at least three or four.

By combining Eqs. (2.111) and (2.112), the fractional amount released from a slab with $C_{a0} > C_{as}$, under the pseudo steady-state assumption, is obtained as

$$\frac{M_t}{M_\infty}=2\left[\frac{2D_mC_{as}\left(C_{a0}-\frac{C_{as}}{2}\right)t}{\delta^2C_{a0}^2}\right]^{1/2} \qquad (2.116)$$

For $C_{a0} \rightarrow C_{as}$, this equation reduces to

$$\frac{M_t}{M_\infty}=2\left(\frac{D_mt}{\delta^2}\right)^{1/2}. \qquad (2.117)$$

For the dissolved active agent case ($C_{a0} \leqq C_{as}$, rectangular slab in infinite agitated medium) Eq. (2.78a) holds in the limit $C_{a0} \rightarrow C_{as}$. However, it can be seen that Eq. (2.78a) for the case of $C_{a0} \leqq C_{as}$, and Eq. (2.117) for the case of $C_{a0} > C_{as}$, do not lead to the same expression in the limit of $C_{a0} \rightarrow C_{as}$. This discrepancy can be resolved by not resorting to the pseudo steady-state assumption (Paul and McSpadden, 1976). The exact expression for the amount of the active agent released by time t from a semi-infinite slab of exposed surface area, $2A$, is

$$\frac{M_t}{2A}=\frac{2C_{as}}{\operatorname{erf}(\eta^*)}\left(\frac{D_mt}{\pi}\right)^{1/2} \qquad (2.118)$$

where η^* is a dimensionless parameter for the position of the extraction front, defined through the following relationship;

$$\pi^{1/2}\eta^*\exp(\eta^{*2})\operatorname{erf}(\eta^*)=\frac{C_{as}}{C_{a0}-C_{as}}\,. \qquad (2.119)$$

Equation (2.118) indicates that M_t is linear in $t^{1/2}$ as is the case for the other geometries. When $C_{a0} \rightarrow C_{as}$, the right-hand side of Eq. (2.119) approaches ∞, implying that η^* becomes very large and thus $\operatorname{erf}(\eta^*)$ approaches unity; hence,

$$\frac{M_t}{2A}=2C_{as}\left[\frac{D_mt}{\pi}\right]^{1/2} \qquad (2.120)$$

which is the same as the exact solution for semi-infinite slab with $C_{a0} \leqq C_{as}$ in the limit $C_{a0} \to C_{as}$ (Crank, 1975). Similarly, when $C_{a0} \gg C_{as}$, Eq. (2.118) (obtained from exact analysis) reduces to Eq. (2.113) obtained from pseudo steady-state analysis, thus defining the region of applicability of the pseudo steady-state approach.

An approximate analytic solution is also available for the diffusion from a polymeric matrix for $C_{a0} > C_{as}$ (Lee, 1980a). This solution is not only more accurate than that obtained from the pseudo steady-state analysis, but is also easier to evaluate than the exact solution due to Paul and McSpadden (1976), discussed above. For the planar matrix with $C_{a0} > C_{as}$, immersed in an infinite agitated sink of negligible active agent concentration (Lee, 1980a)

$$M_t = (1 + H)(3H)^{-1/2} C_{as}(D_m t)^{1/2} \tag{2.121}$$

where

$$H = 5\frac{C_{a0}}{C_{as}} - 4 + \left[\left(\frac{C_{a0}}{C_{as}}\right)^2 - 1\right]^{1/2}. \tag{2.122}$$

The linear relation between M_t at $t^{1/2}$ is observed in this solution too. In the limit $C_{a0} \to C_{as}$, this solution differs by only 2.3% from the exact solution in Eq. (2.118), as compared to Higuchi's solution [Eq. (2.113)] which is 11.4% too small. The solution is also applicable for (C_{a0}/C_{as}) ratios near unity, unlike Eq. (2.113).

The approximate analytic solution for a spherical matrix (with $C_{a0} > C_{as}$) of radius a in an infinite agitated sink is (Lee, 1980a)

$$\frac{M_t}{M_\infty} = [1 - (1-\beta)^3]\left(1 - \frac{C_{as}}{C_{a0}}\right) + 3\beta\frac{C_{as}}{C_{a0}}\left[\left(a_1 + \frac{a_2}{2} + \frac{a_3}{3}\right) - \left(\frac{a_1}{2} + \frac{a_2}{3} - \frac{a_3}{4}\right)\beta\right] \tag{2.123}$$

where β is defined from

$$\frac{D_m t}{a^2} = \frac{1}{12}\left[6\frac{C_{a0}}{C_{as}} - 4 - a_3\right]\beta^2 - \frac{1}{3}\left(\frac{C_{a0}}{C_{as}} - 1\right)\beta^3 \tag{2.124a}$$

and

$$a_1 = 1 \tag{2.124b}$$

$$a_2 = -a_3 - 1 \tag{2.124c}$$

and

$$a_3 = 1 - \left(1 - \frac{C_{a0}}{C_{as}}\right)(1-\beta) - \left[\left[1 - \left(1 - \frac{C_{a0}}{C_{as}}\right)(1-\beta)\right]^2 - 1\right]^{1/2}. \tag{2.124d}$$

The situation where an external mass transfer resistance exists in the form of a boundary layer has also been subjected to a pseudo steady-state analysis (Roseman and Higuchi, 1970). For a planar matrix of total exposed surface area, $2A$, in an infinite sink with a bulk concentration of zero and $C_{a0} \gg C_{as}$

$$\frac{M_t}{2A} = C_{a0} l \tag{2.125}$$

where l is the distance between the extraction boundary and the surface; it is determined from

$$l^2 + \frac{2D_m K l}{(D_b/\delta_b)} = \frac{2D_m C_{as} t}{C_{a0}}. \tag{2.126}$$

D_b and δ_b are the diffusion coefficient of active agent in, and thickness of, the boundary layer, respectively. The pseudo steady-state assumption has been applied to both the matrix and the boundary layer in this analysis.

The above solution is valid only when $C_{a0} \gg C_{as}$. The more complete solution for cases when $C_{a0} > C_{as}$, under the pseudo steady-state assumption (applied to both the matrix and the boundary layer) is (Tojo, 1985)

$$\frac{M_t}{2A} = \left[\left(\frac{\beta D_m}{\alpha}\right)^2 - \left(C_{a0} - \frac{C_{as}}{2}\right)^2 + 2D_m\left(C_{a0} - \frac{C_{as}}{2}\right) C_{as} t\right]^{1/2}$$
$$+ \left[\frac{D_m(\beta - 1)}{\alpha} \log\left[\left(\beta^2 + \frac{2C_{as}\alpha^2 t}{D_m\left(C_{a0} - \frac{C_{as}}{2}\right)}\right)^{1/2} + 1 - \beta\right] - \frac{\beta D_m}{\alpha}\right]$$
$$\times \left(C_{a0} - \frac{C_{as}}{2}\right) \tag{2.127}$$

where

$$\alpha = \frac{D_b/\delta_b}{K} \tag{2.128a}$$

and

$$\beta = \frac{C_{a0} - C_{as}}{C_{a0} - \frac{C_{as}}{2}}. \tag{2.128d}$$

For $C_{a0} \gg C_{as}$, the above expression reduces to

$$\frac{M_t}{2A} = \left[\left(\frac{\beta D_m}{\alpha}\right)^2 \left(C_{a0} - \frac{C_{as}}{2}\right)^2 + 2D_m\left(C_{a0} - \frac{C_{as}}{2}\right) C_{as} t\right]^{1/2}$$
$$- \frac{D_m \beta}{\alpha}\left(C_{a0} - \frac{C_{as}}{2}\right). \tag{2.129}$$

In the limit of $C_{a0} \gg (C_{as}/2)$, Eqs. (2.125) and (2.129) can be shown to be identical.

For a cylindrical matrix with $C_{a0} \gg C_{as}$ (radius a, height h, no end effects), the total amount released by time t is (Roseman and Higuchi, 1970)

$$M_t = \pi h(a^2 - r^2)\, C_{a0} \tag{2.130}$$

where r is the position of the extraction front and is given by

$$\frac{r^2}{2} \ln \frac{r}{a} + \frac{1}{4}(a^2 - r^2) + \frac{D_m K(a^2 - r^2)}{2(D_b/\delta_b)\, a} = \frac{C_{as} D_m}{C_{a0}}\, t\,. \tag{2.131}$$

A discrepancy exists between the exact solution for $C_{a0} < C_{as}$ with appreciable boundary-layer resistance, and the solution based on the pseudo steady-state approach for $C_{a0} > C_{as}$ (with boundary-layer resistance), in the limit of $C_{a0} \to C_{as}$. The amount released from a semi-infinite slab of cross-sectional area, $2A$, in an infinite sink ($C_{bulk} = 0$) with $C_{a0} \leqq C_{as}$ for large values of t is (Crank, 1975)

$$\frac{M_t}{2A} = 2C_{a0}\left[\frac{D_m t}{\pi}\right]^{1/2} - \frac{C_{a0} D_m}{(h_b/K)} \tag{2.132}$$

where h_b is the mass transfer coefficient in the boundary layer. The above expression also holds in the limit as $C_{a0} \to C_{as}$. For $C_{a0} > C_{as}$, under the pseudo steady-state assumption, the solution at large t is (Paul and McSpadden, 1976)

$$\frac{M_t}{2A} = \left[2D_m C_{as}\left(C_{a0} - \frac{C_{as}}{2}\right) t\right]^{1/2} - \frac{C_{a0} D_m}{(h_b/K)}\,, \tag{2.133}$$

which, in the limit $C_{a0} \to C_{as}$, reduces to

$$\frac{M_t}{2A} = C_{a0}(D_m t)^{1/2} - \frac{C_{a0} D_m}{(h_b/K)}\,. \tag{2.134}$$

Thus, the solutions in Eqs. (2.132) and (2.134) do not agree with each other. This discrepancy can again be eliminated by employing the exact solution for $C_{a0} > C_{as}$. For a semi-infinite slab of cross section, $2A$, immersed in an infinite sink ($C_{bulk} = 0$) with a boundary layer on the surface (Paul and McSpadden, 1976)

$$\frac{M_t}{2A} = \frac{2C_{as}}{\operatorname{erf}(\eta^*)}\left[\frac{D_m t}{\pi}\right]^{1/2} - \frac{C_{a0} D_m}{(h_b/K)} \tag{2.135}$$

where

$$\pi^{1/2}\eta^* \exp(\eta^{*2}) \operatorname{erf}(\eta^*) = \frac{C_{as}}{C_{a0} - C_{as}}\,. \tag{2.136}$$

As $C_{a0} \to C_{as}$, $\operatorname{erf}(\eta^*) \to 1$ and Eq. (2.135) becomes equivalent to Eq. (2.132).

An additional complication is introduced in the analysis when the sink is considered to be finite. Under the assumption that pseudo steady-states exists in both the boundary layer and the extraction zone, the following solution is obtained for a spherical matrix of radius a (Gupta and Sparks, 1980)

$$\left(\frac{dM_t}{dt}\right)_r = \frac{4\pi C_{a0} r^2 \left[D_m a^2 \left(\frac{C_{as}}{K} - \frac{LC_{a0}}{\varrho V_2} \right) + \frac{D_m L C_{a0}}{\varrho V_2 a} r^3 \right]}{C_{a0} r^2 \left[\frac{a}{K} - \frac{D_m}{(D_b/\delta_b)} \right] - \frac{a^2}{K} C_{a0} r} \tag{2.137}$$

where L is the weight of the device, ϱ is the density, and r the position of the extraction front given through the relation

$$\begin{aligned} t = {} & \frac{\varrho V_2 a}{3 D_m L} \left(\frac{a}{K} - \frac{D_m}{D_b/\delta_b} \right) \ln \left[\frac{K L C_{a0}}{C_{as} \varrho V_2} \left(\frac{r^3}{a^3} - 1 \right) + 1 \right] \\ & + \frac{\varrho V_2 a^3}{1.732 K D_m L y} \left[\tan^{-1} \left(\frac{2a + y}{1.732 y} \right) - \tan^{-1} \left(\frac{2r + y}{1.732 y} \right) \right] \\ & + \frac{\varrho V_2 a^3}{6 K D_m L y} \ln \left[\frac{(r^2 + ry + y^2)(a - y)^2}{(a^2 + ay + y^2)(r - y)^2} \right] \end{aligned} \tag{2.138 a}$$

and y by

$$y^3 = \frac{a^3 (1 - C_{as} \varrho V_2 / K)}{L C_{a0}} \,. \tag{2.138 b}$$

When matrix diffusion controls the release rate (in case of well-agitated external medium)

$$\frac{D_m}{(D_b/\delta_b)} \approx 0 \,,$$

and Eq. (2.137) reduces to

$$\left(\frac{dM_t}{dt}\right)_r = \frac{4\pi r}{(a^2 - ar)} \left[C_{as} D_m a^2 + \frac{D_m K L C_{a0}}{\varrho V_2 a} (a^3 - r^3) \right] . \tag{2.139}$$

It is apparent that the mathematics for describing the behavior of a device, of even the simplest geometry, becomes quite complicated as various assumptions are dropped. When finite-size devices and sinks are considered instead of semi-infinite or infinite geometries, the analysis of anything but the simplest cases is quite involved. A kinetic expression, having cubic form, that relates the fraction of active agent released to time elapsed, has been derived by a pseudo steady-state approach:

$$\frac{M_t}{M_\infty} = a_1 K_r t^{1/2} - a_2 (K_r t^{1/2})^2 + a_3 (K_r t^{1/2})^3 \,. \tag{2.140}$$

a_1, a_2, a_3 are shape factors, values for which are available for spherical cylindrical, and biconvex shaped matrices (Cobby et al., 1974; Cobby, 1978). K_r is the release rate constant dependent only on characteristics of the matrix and is given by

$$K_r = \frac{1}{C_{a0} r_0} [D_m (2C_{a0} - C_{as}) \, C_{as}]^{1/2} \tag{2.141}$$

where r_0 is a characteristic radius of the device.

One of the motives for the mathematical modeling of the controlled-release devices is to provide guidelines for design of better devices. A preferred release pattern is the zero-order release where the rate remains a constant over the major portion of the release period. Many ingenious shapes have been studied for this purpose. Most of the common shapes have release rates that are linear with respect to $t^{-1/2}$. This is due to the fact that the distance through which the active agent has to diffuse increases with time, and the release rate is inversely proportional to this distance. When the area of the active agent core within the matrix decreases, a further complication is added. The main requirement for obtaining zero-order release is therefore to keep the ratio between diffusion distance and core area a constant.

One shape satisfying the above criterion is a section from a cylinder with an opening at the center to release the drug (an inwardly-releasing sectioned-cylinder; Fig. 2.25 b) (Brooke and Washkuhn, 1977). With the pseudo steady-state assumption, it has been shown that an ideal device of such a geometry would produce a zero-order release rate. The ideal device has an infinitely small opening but one which presents no barrier to diffusion. Thus, in a device of finite dimensions, the zero-order phase develops only after the length of the diffusion path becomes substantially larger than the discharge slit width. A small portion of the active agent within the device thus would be released quickly but the major part would diffuse out at a rate approximating zero-order (Lipper and Higuchi, 1977).

Another geometry which keeps the ratio between diffusion distance and core area constant is an inwardly releasing hemisphere shown in Fig. 2.25 a (Rhine et al., 1980 b). An active-agent-loaded hemispherical matrix is covered with an impermeable membrane. The active agent escapes only through a concave opening in the membrane on the flat surface. Comparison of the performances of inwardly releasing hemisphere and the inwardly releasing cylindrical section shows that the hemisphere approaches the zero-order more closely. Experimental evidence for zero-order release from such a device has been obtained, for both micro- and macromolecules, by Hsieh et al. (1983).

The above inwardly-releasing-hemisphere concept has been extended by Kuu and Yalkowsky (1985) to one of multiple holes in a rectangular matrix covered by an impermeable membrane. Three basic designs have been proposed based on the thickness of matrix being comparable to, much less than, or much greater than, the distance between the holes. The device may have perforations on one or both faces. The rate of active agent release depends on the number of holes, the distance between the holes, the size of the holes, the hole configuration, and the initial active agent loading. The device has been analyzed by applying the pseudo steady-state assumption to the extracted region of the matrix. Zero-order release rates are expected from devices where the thickness of the matrix is approximately equal to or much less than the distance between holes (Kuu and Yalkowsky, 1985).

The zero-order release rate can also be produced by surrounding monolithic device with a rate limiting barrier (Lee et al., 1980). The diffusion path controlling the process (thickness of barrier) is thus a constant and so is the area of release. A constant concentration gradient is produced in the barrier at steady state. Based on the pseudo steady-state analysis of Roseman and Higuchi (1970), the release rate from a cylindrical device with a rate limiting barrier into an infinite well-stirred medium is

$$\left(\frac{dM_t}{dt}\right)_r = \frac{2\pi ahC_B D_B}{C_{as}\delta_B}\left[\frac{D_m C_{as}\delta_B}{-\dfrac{C_B D_B a}{C_{as}}\ln\dfrac{r}{a} + D_m\delta_B}\right]. \tag{2.142a}$$

where a and h are radius and height of cylindrical core, and r is the position of the extraction front. C_B is the solubility of the active agent (corresponding to C_{as}), and D_B its diffusivity, in the barrier layer of thickness δ_B. When barrier diffusion controls, i.e., $D_B \ll D_m$, the above equation reduces to (Lee et al., 1980)

$$\left(\frac{dM_t}{dt}\right)_r = 2\pi ahC_B D_B/\delta_B\,. \tag{2.142b}$$

Zero-order release is thus obtained when the external barrier controls the diffusion process. [A condition, implicity assumed in the derivation of Eq. (2.142a) by Roseman and Higuchi (1970) and therefore required Eq. (2.142b), is that $\delta_B \ll a$]. Such devices can be manufactured by covering a monolithic matrix with a membrane. Such a device has been tested experimentally by Lee et al. (1980) and the result indicated that a zero-order release rate was obtainable from this design. Another interesting approach to attaining zero-order release may be obtained from the transdermal transport model of Gienger et al. (1984). According to their model, time varying diffusivity and partition coefficients may render the concentration gradient essentially constant in the external laminate. This can lead to near zero-order rates; however, this approach may be difficult to implement.

As stated earlier, all the models reviewed in this subsection have been derived under the assumption that the active agent dissolution step at the extraction boundary does not influence the rate. These "diffusion-controlled" models have been found to be quite suitable for a number of experimental situations. However, when the dissolution rate is inadequate, both the dissolution and diffusion steps must be accounted for (Chandrasekaran and Paul, 1982). In such a situation, a sharp well-defined extraction boundary is not observed; the dissolution process occurs at all points over the region being extracted where the concentration is greater than C_{as}. The dissolution step then appears as a source term in the governing differential equation. Approximating this step by a first-order rate term, the differential equation describing transport of active agent in the matrix becomes

$$\frac{\partial C_m}{\partial t} = D_m \frac{\partial^2 C_m}{\partial x^2} + k'(C_{as} - C_m) \tag{2.143}$$

where k' is the dissolution constant. Chandrasekaran and Paul (1982) have solved the above equation for a semi-infinite matrix immersed in a well-stirred infinite medium ($C_{bulk} = 0$). The amount of active agent released, while there is undissolved active agent at all points in the matrix, is obtained as

$$\frac{M_t}{2A} = C_{as} \left(\frac{D_m}{k'}\right)^{1/2} \left[\left(k't + \frac{1}{2}\right) \operatorname{erf}(\sqrt{k't}) + \left(\frac{k't}{\pi}\right)^{1/2} \exp(-k't)\right] \tag{2.144}$$

where $2A$ is the cross-sectional area of exposed face. The above equation is only valid for time t such that

$$C_{as}k't \leqq C_{a0} - C_{as}\,. \tag{2.145}$$

Approximating the matrix by a slab of thickness δ (for large δ) and a cross-sectional area of A, the fraction release may be written as

$$\frac{M_t}{M_\infty} = 2\frac{C_{as}}{C_{a0}} \left(\frac{D_m}{k'\delta^2}\right)^{1/2} \left[\left(k't + \frac{1}{2}\right) \operatorname{erf}(\sqrt{k't}) + \left(\frac{k't}{\pi}\right)^{1/2} \exp(-k't)\right]. \tag{2.146}$$

In the dissolution-controlled regime (large $k't$), the above equation reduces to

$$\frac{M_t}{M_\infty} = 2\frac{C_{as}}{C_{a0}} \left(\frac{D_m k'}{\delta^2}\right)^{1/2} \left(\frac{1}{2k'} + t\right) \tag{2.147}$$

indicating a zero-order release rate. When controlled by diffusion (small $k't$), the fractional release is given by

$$\frac{M_t}{M_\infty} = 4\frac{C_{as}}{C_{a0}} \left(\frac{D_m t}{\pi\delta^2}\right)^{1/2}. \tag{2.148}$$

This equation may be compared to Eq. (2.120).

The above described diffusion-dissolution process has also been comprehensively analyzed by Ayers and Lindstrom (1977) and Lindstrom and Ayers (1977). They have considered a rectangular monolithic device releasing active agent from one face to an unagitated sink. The model is quite general; however, numerical techniques are required for the evaluation of their analytic solutions.

Monolithic devices form a very important class of controlled-release devices due to the ease with which they can be manufactured. They have been the subject of a whole spectrum of models. The effect of temperature on the release has also been studied by Chien (1976) based on the activation energies of the dissolution and diffusion steps. However, heat transport effects have not been considered in any model. Temperature may be an important factor for monolithic devices especially when used for pesticide/fertilizer delivery where diurnal or even seasonal temperature changes may influence the performance through the decaying of the active agent.

2.2.3 Porous Devices

In the cases considered earlier, transport of the active agent was assumed to occur through the polymer continuum itself, by the "solution-diffusion" mechanism. In this mechanism, the active agent dissolves in the polymer and then diffuses through it. Diffusion in pores was assumed negligible. This condition may not be true in some formulations where pore transport may play a larger role.

Porous devices may also be divided into the two categories mentioned earlier, the reservoir-membrane systems and the monolithic systems, by virtue of the technique employed for loading the active agent. The major difference from devices considered previously is that polymer and active agent interaction is not quite as strong in porous devices, as it is in devices where the solution-diffusion mechanism is applicable.

Hollow fibers are the simplest types of porous reservoir type devices. Ashare et al. (1976) analyzed such a system for dispensing volatile materials from capillary channels. The mechanism of release has three steps: evaporation at the vapor-liquid interface, diffusion from the interface to the open end of hollow fiber, and convection away from the open end. By taking diffusion to be the rate controlling step, a simple analysis leads to

$$\left(\frac{dM_t}{dt}\right)_r = \pi r^2 \left[-\frac{mC_a D_b \varrho}{2} \ln\left(1 - \frac{p_a^{vap}}{p}\right)\right]^{1/2} t^{-1/2} \tag{2.149}$$

where

m = molecular weight of the volatile liquid,
D_b = diffusivity in the ambient fluid,
p_a^{vap} = vapor pressure of the liquid,
p = ambient pressure,
r = radius of hollow fiber,
ϱ = density of the liquid, and
C_a = molar density of the vapor.

The analysis assumes vapor-liquid equilibrium at interface and a perfect sink (zero bulk concentration at the open end). The liquid being dispensed must be pure active agent.

Other porous devices can be constructed by impregnation of porous plastics, hydrogels, etc. These can be analyzed by methods analogous to those used for monolithic systems. A pseudo steady-state analysis similar to that employed for monolithic systems with $C_{a0} \gg C_{as}$ gives an expression for amount released from a granular matrix as (Higuchi, 1963)

$$M_t = \left[\frac{D_b \varepsilon}{\tau} C_{bs}(2C_{a0} - \varepsilon C_{bs})\, t\right]^{1/2} \tag{2.150}$$

where ε and τ are the porosity and tortuosity of the matrix and C_{bs} is the solubility of active agent in the ambient fluid. D_b is the diffusivity in the ambient fluid. The mechanism of release in such systems is leaching, wherein the ambient fluid enters the pores and dissolves the active agent. Similar expressions can be written by analogy from Higuchi (1963) for spherical pellets and from Roseman and Higuchi (1970) for cylindrical pellets. For cylindrical pellets of radius a (Roseman and Higuchi, 1970)

$$\left(1 - \frac{M_t}{M_\infty}\right) \ln\left[1 - \frac{M_t}{M_\infty}\right] + \frac{M_t}{M_\infty} = \frac{4C_{bs}D_b\varepsilon}{C_{a0}a^2\tau}\, t \tag{2.151}$$

and for spherical pellets of radius a (Higuchi, 1963)

$$\begin{aligned}(1-\alpha) &+ 2\left(1 - \frac{M_t}{M_\infty}\right)(1-\alpha) \\ &- \left(1 - \frac{M_t}{M_\infty}\right)^{2/3}(3 - 4\alpha) - \alpha\left(1 - \frac{M_t}{M_\infty}\right)^{1/3} \\ &- \frac{\alpha}{3}\ln\left(1 - \frac{M_t}{M_\infty}\right) = \frac{6C_{bs}D_b\varepsilon}{C_{a0}a^2\tau}\, t\end{aligned} \tag{2.152}$$

where

$$\alpha = \frac{\varepsilon C_{bs}}{C_{a0}} \tag{2.153}$$

Equations (2.150–152) have been derived on the assumption that the active agent particles are quite small relative to the diffusion distance and are uniformly distributed through the matrix. Futhermore, the pseudo steady-state condition will hold only if $C_{a0} \gg \varepsilon C_{bs}$.

In order to account for the change in porosity as the active agent is leached out, we may substitute

$$\varepsilon \approx C_{a0}/\varrho_a\,,$$

into Eqs. (2.150–152). ϱ_a is the density of the active agent and the above relation assumes negligible initial porosity. The relation may also be used if the fraction of matrix volume occupied by the active agent is large.

A moving boundary approach has been employed by Miller and Peppas (1983) for diffusional release from a polymer with continuously forming porous structure. The active agent is assumed to be loaded in the pores at levels higher than its solubility in the external medium ($C_{a0} > \varepsilon C_{bs}$). The active agent dissolves into the medium at the interface which moves into the pore as the dissolution progresses; however, the kinetics of dissolution are neglected. Defining the position of medium-solid active agent interface by

$$2\alpha(D_b t/\tau)^{1/2} ,$$

the release rate of active agent from the device is obtained as (Miller and Peppas, 1983)

$$\left(\frac{dM_t}{dt}\right)_r = \frac{\varepsilon^{2/3} A C_{bs}}{\operatorname{erf}(\alpha)} \left(\frac{D_b}{\tau \pi t}\right)^{1/2} . \tag{2.154}$$

α is obtained from the transcedental equation

$$\pi^{1/2} \alpha \exp(\alpha^2) \operatorname{erf}(\alpha) = \frac{C_{bs}}{\varrho_a - C_{bs}} \tag{2.155}$$

where ϱ_a is the density of the solid active agent.

The above analysis does not account for the bulk motion of the pore fluid as increasing amounts of pore volume are created. This convection effect may be significant if the end of the pore moves rapidly, i.e., when $\varrho_a \to C_{bs}$. Convection currents created in the pore fluid due to density differences resulting from concentration differences may, however, be neglected due to the small cross-sectional area of the horizontal pore in question.

In all the models for porous monolithic devices reviewed, the active agent loading is assumed to be quite high, i.e., $C_{a0} > \varepsilon C_{bs}$. The release rate is found to be a function of $t^{1/2}$. However, Fessi et al. (1982) showed the same $t^{1/2}$ dependence for a matrix with $C_{a0} < \varepsilon C_{bs}$. The result is based on the pseudo steady-state analysis (linear concentration profile in pore fluid) of transport in a constantly increasing linear pore. Dissolution of active agent at the end of the pore is assumed instantaneous. The release rate per unit cross-sectional is determined to be independent of porosity and solubility, being directly proportional to the initial active agent loading C_{a0}

$$\frac{1}{A}\left(\frac{dM_t}{dt}\right)_r = \frac{D_b C_{a0}}{2t^{1/2}} . \tag{2.156}$$

The validity of the pseudo steady-state assumption made above, depends on the requirement that the end of the pore moves much slower than the rate at which the active agent diffuses in the pore fluid. This enables the (linear) steady-state concentration profile to be established in the pore. When the loading of active agent C_{a0} is much greater than εC_{bs}, this condition is satisfied, since the excess loading ($C_{a0} - \varepsilon C_{bs}$) has to be dissolved at one layer before the extraction from the next layer begins (consequently advancing the pore). If $C_{a0} < \varepsilon C_{bs}$, this assumption cannot be expected to hold. However, the moving-boundary analysis of Miller and Peppas (1983) reviewed above, can be employed in this case, under the assumption that dissolution in the pore fluid is not rate limiting. The rate of release from a single pore may then be given as

$$\frac{1}{A}\left(\frac{dM_t}{dt}\right)_r = \frac{2C_{b0}}{\operatorname{erf}(\alpha)}\left(\frac{D_b}{\pi t}\right)^{1/2} \exp(-\alpha^2) \tag{2.157}$$

where α is obtained from

$$\frac{\pi^{1/2}(\varrho_a - C_{b0})}{C_{b0}} = [\alpha \exp(\alpha^2) \operatorname{erf}(\alpha)]^{-1} . \tag{2.158}$$

ϱ_a is the bulk density of the solid active agent in the pore and C_{b0} is the equilibrium solubility of the active agent in the pore fluid corresponding to the loading $C_{a0}(C_{b0} < C_{bs})$. The release rate is seen to be a function of $t^{1/2}$ again.

Peppas (1983) has presented a model for dissolution-controlled active agent release from porous monolithic systems based on a diffusion-dissolution model. For high active agent loading ($C_{a0} > \varepsilon C_{bs}$), transport in the pore fluid is governed by the following differential equation

$$\frac{\partial C}{\partial t} = D_{\text{eff}} \frac{\partial^2 C}{\partial x^2} + \frac{2k'}{r_p}(C_{bs} - C) \tag{2.159}$$

where k' is a dissolution constant and r_p is the radius of the pore. C_{bs} is the saturation concentration of active agent at the pore wall–dissolution medium interface. The second term on the right-hand side thus represents an active agent dissolution source term; the active agent dissolves in the pore fluid throughout the perimeter of the pore. This term drops out when concentration of active agent in the matrix drops below εC_{bs}. A solution to Eq. (2.159) has been presented (employing $r_p = \bar{r}_p$ as a first approximation); the release rate at long times from a slab of thickness (and therefore pore length) δ, and unit area in a well-stirred infinite medium has been obtained as (Peppas, 1983)

$$\left(\frac{dM_t}{dt}\right)_r = \varepsilon^{2/3} C_{bs}\left(\frac{2k' D_{\text{eff}}}{\bar{r}_p}\right)^{1/2} \tanh\left[\frac{\delta}{2}\left(\frac{2k'}{\bar{r}_p D_{\text{eff}}}\right)^{1/2}\right] \tag{2.160}$$

where $\bar{r}_p$ is the average pore radius over the release period. Evidently, a zero-order release rate may be obtained in the dissolution-controlled regime. Plots have also been presented showing the effect of the dissolution constant, active agent solubility, porosity, and slab thickness on the release rate (also see, Korsmeyer et al., 1983).

To adequately account for pore-size distributions as they relate to diffusivity in the matrix, principles of percolation theory would have to be used (Langer and Peppas, 1983; Peppas, 1984). Specific attempts at modeling controlled release with such theories are limited (see, e.g., Saltzman and Langer, 1987) although the literature on adsorption and diffusion in porous media is extensive. Porous devices are difficult to characterize and their performance difficult to predict due to the wide range of pore sizes possible in a single device.

References

Aris, R.: The Mathematical Theory of Diffusion and Reaction in Permeable Catalysts, Vol. I, The Theory of the Steady State, pp. 18–30, Clarendon Press, Oxford (1975)

Ashare, E., Brooks, T. W., Swenson, D. W.: Controlled Release from Hollow Fibers. In: Preprints of papers presented of the 171st meeting, ACS, Division of Organic Coatings and Plastics Chemistry *36* (1), 453–457 (1976)

Ayres, J. W., Lindstrom, F. T.: Diffusion Model for Drug Release from Suspensions. I: Theoretical Considerations, J. Pharm. Sci. *66*, 654–662 (1977)

Baker, R. W.: Controlled Release of Biologically Active Agents, pp. 1–83, John Wiley and Sons, New York (1987)

Baker, R. W., Lonsdale, H. K.: Controlled Release: Mechanisms and Rates, pp. 15–71. In: Controlled Release of Biologically Active Agents, A. C. Tanquary and R. E. Lacey (Eds.), Plenum Press, New York (1974)

Baker, R. W., Tuttle, M. E., Lonsdale, H. K., Ayres, J. W.: Development of an Estriol-Releasing Intrauterine Device, J. Pharm. Sci. *68*, 20–26 (1979)

Balazs, A. C., Calef, D. F., Deutch, J. M., Siegel, R. A., Langer, R.: The Role of Polymer Matrix Structure and Interparticle Interactions in Diffusion-Limited Drug Release, pp. 87–102. In: Polymeric Materials in Medication, C. G. Gebelein and C. E. Carraher, Jr. (Eds.), Plenum Press, New York (1985)

Barrer, R. M.: Diffusion and Permeation in Heterogeneous Media, pp. 165–217. In: Diffusion in Polymers, J. Crank and G. S. Park (Eds.), Academic Press, London (1968)

Barrer, R. M., Skirrow, G.: Transport and Equilibrium Phenomena in Gas-Elastomer System. II. Equilibrium Phenomena, J. Polym. Sci. *3*, 564–575, (1948)

Barrer, R. M., Barrie, J. A., Rogers, M. G.: Heterogeneous Membranes: Diffusion in Filled Rubbers, J. Polym. Sci. *A1*, 2565–2586 (1963)

Barrie, J. A., Levine, J. D., Michaels, A. S., Wong, P.: Diffusion and Solution of Gases in Composite Rubber Membranes, Trans. Faraday Soc. *59*, 869–878 (1963)

Berner, B.: Pharmacokinetics of Transdermal Drug Delivery, J. Pharm. Sci. *74*, 718–721 (1985)

Bird, R. B., Stewart, W. E., Lightfoot, E. N.: Transport Phenomena, pp. 554–563, John Wiley and Sons, New York (1960)

Boyer, R. F., Spencer, R. S.: Second-Order Transition Effects in Rubber and Other High Polymers, Advances in Colloid Science, Vol. II, pp. 1–55, H. Mark and G. S. Whitby (Eds.), John Wiley and Sons, New York (1946)

Brooke, D., Washkuhn, R. J.: Zero-Order Drug Delivery System: Theory and Preliminary Testing, J. Pharm. Sci. *66*, 159–162 (1977)

Burrell, H.: Solubility Parameter Values, pp. IV.337–IV.360. In: Polymer Handbook, J. Brandrup and E. H. Immergut (Eds.), John Wiley and Sons, New York (1975)

Byron, P. R., Dalby, R. N.: Effects of Heat Treatment on the Permeability of Polyvinyl Alcohol Films to a Hydrophilic Solute, J. Pharm. Sci. *76*, 65–67 (1987)

Carslaw, H. S., Jaeger, J. C.: Conduction of Heat in Solids, Oxford University Press, London (1959)

Chandrasekaran, S. K., Paul, D. R.: Dissolution-Controlled Transport from Dispersed Matrixes, J. Pharm. Sci. *71*, 1399–1402 (1982)

Chien, Y. W.: Thermodynamics of Controlled Drug Release from Polymeric Delivery Devices. In: Controlled Release Polymeric Formulations, D. R. Paul and F. W. Harris (Eds.), ACS Symposium Series *33*, 53–71 (1976)

Chien, Y. W., Lambert, H. J.: Controlled Drug Release from Polymeric Delivery Devices II: Differentiation between Partition-Controlled and Matrix-Controlled Drug Release Mechanisms, J. Pharm. Sci. *63*, 515–519 (1974)

Chien, Y. W., Mares, S. F., Berg, J., Huber, S., Lambert, H. J., King, K. F.: Controlled Drug Release from Polymeric Devices III: *In Vitro-In Vivo* Correlation for Intravaginal Release of Ethynodiol Diacetate from Silicon Devices in Rabbits, J. Pharm. Sci. *64*, 1776–1781 (1975)

Christensen, F. N., Hansen, F. Y., Bechgaard, H.: Mathematical Model for *In-Vitro* Drug Release from Controlled Release Dosage Forms Applied to Propoxyphene Hydrochloride Pellets, J. Pharm. Sci. *71*, 694–699 (1982)

Cobby, J., Mayersohn, M., Walker, G. C.: Influence of Shape Factors on Kinetics of Drug Release from Matrix Tablets. I: Theoretical, J. Pharm. Sci. *63*, 725–732 (1974)

Cobby, J.: Mathematical Models for the Release of Drugs from Matrix Tablets, J. Biomed. Mater. Res. *12*, 627–634 (1978)

Cohen, M. H., Turnbull, D.: Molecular Transport in Liquids and Glasses, J. Chem. Phys. *31*, 1164–1169 (1959)

Crank, J.: The Mathematics of Diffusion, Oxford University Press, London (1975)

Crank, J., Park, G. S.: Methods of Measurement, pp. 1–39. In: Diffusion in Polymers, J. Crank and G. S. Park (Eds.), Academic Press, London (1968)

Currie, J. A.: Gaseous Diffusion in Porous Media. Part II. Dry Granular Media, Brit. J. Appl. Phys. *11*, 318–324 (1960)

DeGroot, S. R.: Thermodynamics of Irreversible Processes, North-Holland, Amsterdam (1959)

DiBenedetto, A. T.: Molecular Properties of Amorphous High Polymers. I. A Cell Theory for Amorphous High Polymers, J. Polym. Sci. *A 1*, 3459–3476 (1963a)

DiBenedetto, A. T.: Molecular Properties of Amorphous High Polymers. II. An Interpretation of Gaseous Diffusion Through Polymers, J. Polym. Sci. *A 1*, 3477–3487 (1963b)

Duda, J. L.: Molecular Diffusion in Polymeric Systems, Pure & Appl. Chem. *57*, 1681–1690 (1985)

Evans, III, R. B., Watson, G. M., Mason, E. A.: Gaseous Diffusion in Porous Media at Uniform Pressure, J. Chem. Phys. *35*, 2076–2083 (1961)

Fan, L. T., Fan, L. S., Tojo, K., Miyanami, K.: Dynamics of a Membrane-Moderated Controlled Release, Mathematical Biosciences *39*, 213–234 (1978)

Fan, L. S., Tojo, K., Fan, L. T.: Cyclic Dynamics of a Nonisothermal Controlled Release, Mathematical Biosciences *45*, 69–94 (1979)

Felder, R. M., Huvard, G. S.: Permeation, Diffusion, and Sorption of Gases and Vapors, pp. 315–376. In: Methods of Experimental Physics, L. Marton and C. Marton, Eds.-in-Chief, Vol. 16c, R. A. Fava (Ed.), Academic Press, New York (1980)

Fessi, H., Marty, J. P., Puisieux, F., Carstensen, J. T.: Square Root of Time Dependence of Matrix Formulations with Low Drug Content, J. Pharm. Sci. *71*, 749–752 (1982)

Flynn, G. L.: Influence of Physico-Chemical Properties of Drug and System on Release of Drugs from Inert Matrices, pp. 73–98. In: Controlled Release of Biologically Active Agents, A. C. Tanquary and R. E. Lacey (Eds.), Plenum Press, New York (1974)

Flynn, G. L., Carpenter, O. S., Yalkowsky, S. H.: Total Mathematical Resolution of Diffusion Layer Control of Barrier Flux, J. Pharm. Sci *61*, 312–314 (1972)

Flynn, G. L., Yalkowsky, S. H., Roseman, T. J.: Mass Transport Phenomena and Models: Theoretical Concepts, J. Pharm. Sci. *63*, 479–510 (1974)

Foster, R. N., Butt, J. B.: A Computational Model for the Structure of Porous Materials Employed in Catalysis, AIChE J. *12*, 180–185 (1968)

Frisch, H. L.: Mechanisms for Fickian Diffusion of Penetrants in Polymers, J. Polym. Sci. Polym. Letters, *3*, 13—16 (1965)

Fu, J. C., Hagemeir, C., Moyer, D. L.: A Unified Mathematical Model for Diffusion from Drug-Polymer Composite Tablets, J. Biomed. Mater. Res. *10*, 743–758 (1976)

Fuhrmann, J.: Diffusion in Semi-Crystalline Polymers and Polymer Networks, Ber. Bunsenges. Phys. Chem. *83*, 303–309 (1979)

Fujita, M.: Diffusion in Polymer-Diluent Systems. Fortschritte Der Hochpolymeren-Forschung (Adv. in Pol. Sci.) *3*, 1–43 (1961)

Furusawa, T., Smith, J. M.: Diffusivities from Dynamic Adsorption Data, AIChE J. *19* 401–403 (1973)

Gavalas, G. R., Kim, S.: Periodic Capillary Diffusion in Porous Solids, Chem. Eng. Sci. *36*, 1111–1122 (1981)

Geddes, A. L., Pontius, R. B.: Determination of Diffusivity, pp. 895–1005. In: Technique of Organic Chemistry, A. Weissberg (Ed.), Vol. I. Physical Methods of Organic Chemistry, Part 2, Interscience, New York (1959)

Gienger, G., Knoch, A., Merkle, H. P.: Modelling and Numerical Computation of Drug Transport in Laminates: Model Case Evaluation of Transdermal Delivery System, preprint no. 283, Sonderforschungsbereich 123, Stochastische Mathematische Modelle, Universität Heidelberg, D-6900 Heidelberg 1 (1984)

Gillett, J. W., Harr, J. R., Lindstrom, F. T., Mount, D. A., St. Clair, A. D., Weber, L. J.: Evaluation of Human Health Hazards on Use of Dichlorovos (DDVP), Especially in Resin Strips, Residue Reviews, *44*, 115–159 (1972)

Gupta, D. V. S., Sparks, R. E.: Mathematical Model for Progesterone Release from Injectible Poly(Lactic Acid) Microcapsules *in vitro*, pp. 189–212. In: Controlled Release of Bioactive Materials, R. Baker (Ed.), Academic Press, New York (1980)

Haleblian, J., Runkel, R., Mueller, N., Christopherson, J., Ng, K.: Steroid Release from Silicone Elastomer Containing Excess Drug in Suspension, J. Pharm. Sci. *60*, 541–545 (1971)

Halpern, B. D., Solomon, O., Kopec, L., Korostoff, E., Ackerman, J. L.: Release of Inorganic Fluoride Ion from Rigid Polymer Matrices. In: Controlled Release Polymeric Formulations, D. R. Paul and F. W. Harris (Eds.), ACS Symposium Series *33*, 135–145 (1976)

Haring, R. E., Greenkorn, R. A.: A Statistical Model of a Porous Medium with Nonuniform Pores, AIChE J. *16*, 477–483 (1970)

Hellwege, Von K.-H., Knappe, W., Lohe, P.: Die Konzentrationsabhängigkeit des Diffusionskoeffizienten von Weichmachern in Polyvinylchlorid, Kolloid-Z. *179*, 40–49 (1961)

Higuchi, T.: Rate of Release of Medicaments from Ointment Bases Containing Drugs in Suspension, J. Pharm. Sci. *50*, 874–875 (1961)

Higuchi, T.: Mechanism of Sustained-Action Medication, Theoretical Analysis of Rate of Release of Solid Drugs Dispersed in Solid Matrices, J. Pharm. Sci. *52*, 1145–1149 (1963)

Hildebrand, J. H., Scott, R. L.: The Solubility of Nonelectrolytes, Dover Publications, New York (1964)

Hsieh, D. S. T., Rhine, W. D., Langer R.: Zero-Order Controlled-Release Polymer Matrices for Micro- and Macromolecules, J. Pharm. Sci. *72*, 17–22 (1983)

Jaeger, J. C.: Heat flow in the Region Bounded Internally by a Circular Cylinder, Proc. R. Soc. Edinb. *61A*, 223–229 (1942).

Jaeger, J. C., Clarke, M.: Numerical Results for Some Problems on Conduction of Heat in Slabs with Various Surface Conditions, Phil. Mag. *38*, 504–515 (1947)

Johnson, M. F. L., Stewart, W. E.: Pore Structure and Gaseous Diffusion in Solid Catalysts, J. Catalysis *4*, 248–252 (1965)

Kelley, F. N., Bueche, F.: Viscosity and Glass Temperature Relations for Polymer-Diluent Systems, J. Polym. Sci. *50*, 549–556 (1961)

Kojima, Y., Furuhata, K., Miyasaka, K.: Hydralic Permeability of Water in Poly(vinyl Alcohol) Membranes as à Function of the Degree of Hydration, J. Appl. Polym. Sci. *28*, 2401–2410 (1983)

Kojima, Y., Furuhata, K., Miyasaka, K.: Diffusive Permeability of Solutes in Poly(vinyl Alcohol) Membranes as a Function of the Degree of Hydration, J. Appl. Polym. Sci. *29*, 533–546 (1984)

Korsmeyer, R. W., Gurny, R., Doelker, E., Buri, P., Peppas, N. A.: Mechanisms of Solute Release from Porous Hydrophilic Polymers, Int. J. Pharm. *15*, 25–35 (1983)

Kuu, W. Y., Yalkowsky, S. H.: Multiple-Hole Approach to Zero-Order Release, J. Pharm. Sci. *74*, 926–933 (1985)

Lacey, R. E., Cowsar, D. R.: Factors Affecting Release of Steroids from Silicones, pp. 117–144. In: Controlled Release of Biologically Active Agents, A. C. Tanquary and R. E. Lacey (Eds.), Plenum Press, New York (1974)

Langer, R., Peppas, N.: Chemical and Physical Structure of Polymers as Carriers for Controlled Release of Bioactive Agents: A Review. J. Macromol. Sci.-Rev. Macromol. Chem. Phys. *C23*, 61–126 (1983)

Lee, P. I.: Diffusional Release of a Solute from a Polymeric Matrix-Approximate Analytical Solutions, J. Membrane Sci. *7*, 255–275 (1980a)

Lee, P. I.: Determination of Diffusion Coefficients by Sorption from a Constant, Finite Volume, pp. 135–153. In: Controlled Release of Bioactive Materials, R. Baker (Ed.), Academic Press, New York (1980b)

Lee, E. S., Kim, S. W., Kim, S. H., Cardinal, J. R., Jacobs, H.: Drug Release from Hydrogel Devices with Rate Controlling Barriers, J. Membrane Sci. *7*, 293–303 (1980)

Leo, A., Hansch, C., Elking, D.: Partition Coefficients and Their Uses, Chem. Rev. *71*, 525–616 (1971)

Leuenberger, H., Haas, Ch., Imanidis, G., Usteri, M., Winzap, S.: Percolation Theory and Fractal Geometry-Important Factors in the Design of Controlled Release Systems, pp. 33–34. In: Proceedings of the 14th International Symposium on Controlled Release of Bioactive Materials, P. I. Lee and B. A. Leonhardt (Eds.), Toronto, Ontario, Canada, August 2–5 (1987)

Lightfoot, E. N.: Transport Phenomena and Living Systems, pp. 236–266, John Wiley and Sons, New York (1974)

Lindstrom, F. T., Ayers, J. W.: Diffusion Model for Drug Release from Suspensions. II: Release to a Perfect Sink, J. Pharm. Sci. *66*, 662–668 (1977)

Lipper, R. A., Higuchi, W. I.: Analysis of Theoretical Behavior of a Proposed Zero-Order Drug Delivery System, J. Pharm. Sci. *66*, 163—164 (1977)

Macedo, P. B., Litovitz, T. A.: On the Relative Roles of Free Volume and Activation Energy in the Viscosity of Liquids, J. Chem. Phys. *42*, 245–256 (1965)

McGregor, R.: Diffusion and Sorption in Fibers and Films, Vol. I, An Introduction with Particular Reference to Dyes, pp. 147–220, Academic Press, London (1974)

Meares, P.: The Diffusion of Gases Through Polyvinyl Acetate, J. Am. Chem. Soc. *76*, 3415–3422 (1954)

Michaels, A. S.: Diffusion in a Pore of Irregular Cross Section — A Simplified Treatment, AIChE J. *5*, 270–271 (1959)

Michaels, A. S., Vieth, W. R., Bixler, H. J.: Gas Permeability of Highly Oriented Dibutyl Maleate-Ethylene Copolymer Films, J. Appl. Polym. Sci. *8*, 2735–2750 (1964)

Michaels, A. S., Wong, P. S. L., Prather, R., Gale, R. M.: A Thermodynamic Method of Predicting the Transport of Steroids in Polymer Matrices, AIChE J. *21*, 1073–1080 (1975)

Mikulecky, D. C., Caplan, S. R.: The Choice of Reference Frame in the Treatment of Membrane Transport by Nonequilibrium Thermodynamics, J. Phys. Chem. *70*, 3049–3056 (1966)

Miller, E. S., Peppas, N. A.: Diffusional Release of Water-Soluble Bioactive Agents from Ethylene-Vinyl Acetate Copolymers, Chem. Eng. Commun. *22*, 303–315 (1983)

Pace, R. J., Datyner, A.: Statistical Mechanical Model for Diffusion of Simple Penetrants in Polymers. I. Theory, J. Polym. Sci., Polym. Phys. Ed. *17*, 437–451 (1979a)

Pace, R. J., Datyner, A.: Statistical Mechanical Model for Diffusion of Complex Penetrants in Polymers. I. Theory, J. Polym. Sci., Polym. Phys. Ed. *17*, 1675–1692 (1979b)

Park, G. S.: The Glassy State and Slow Process Anomalies, pp. 141–163. In: Diffusion in Polymers, J. Crank and G. S. Park (Eds.), Academic Press, London (1968)

Paul, D. R., McSpadden, S. K.: Diffusional Release of a Solute from a Polymer Matrix, J. Membrane Sci. *1*, 33–48 (1976)

Peppas, N. A.: A Model of Dissolution-Controlled Solute Release from Porous Drug Delivery Polymeric Systems, J. Biomed. Mater. Res. *17*, 1079–1087 (1983)

Peppas, N. A.: Mathematical Modeling of Diffusion Processes in Drug Delivery Polymeric Systems, pp. 206–237. In: Controlled Drug Bioavailability, Vol. I, Drug Product Design and Performance, V. F. Smolen and L. A. Ball (Eds.), Wiley, New York (1984)

Peterlin, A.: Transport Phenomena and Polymer Morphology, Makromol. Chem. Suppl. *3*, 215–232 (1979)

Petersen, E. E.: Diffusion in a Pore of Varying Cross Section, AIChE J. *4*, 343–345 (1958)

Pismen, L. M.: Diffusion in Porous Media of a Random Structure, Chem. Eng. Sci. *29*, 1227–1236 (1974)

Reid, R. C., Sidman, K. R., Schwope, A. D., Till, D. E.: Loss of Adjuvants from Polymer Films to Food or Food Simulants. Effect of the External Phase, Ind. Eng. Chem. Prod. Res. Dev. *19*, 580–587 (1980)

Rhine, W. D., Hsieh, D. S. T. Langer, R.: Polymers for Sustained Macromolecule Release: Procedures to Fabricate Reproducible Delivery Systems and Control Release Kinetics, J. Pharm. Sci. *69*, 265–270 (1980a)

Rhine, W. D., Sukhatme, V., Hsieh, D. S. T., Langer, R. S.: A New Approach to Achieve Zero-Order Release Kinetics from Diffusion-Controlled Polymer Matrix Systems,

pp. 177–187. In: Controlled Release of Bioactive Materials, R. Baker (Ed.), Academic Press, New York (1980b)
Rogers, C. E.: Solubility and Diffusivity, pp. 510–635. In: Physics and Chemistry of the Organic Solid State, Vol. II, D. Fox, M. M. Labes and A. Weissberger (Eds.), Interscience Publishers, New York (1965)
Rogers, C. E.: Structural Factors Governing Controlled Release. In: Controlled Release Polymeric Formulations, D. R. Paul and F. W. Harris (Eds.), ACS Symposium Series *33*, 15–25 (1976)
Rogers, C. E., Stannett, V., Szwarc, M.: The Sorption, Diffusion, and Permeation of Organic Vapors in Polyethylene, J. Polym. Sci. *45*, 61–82 (1960)
Roseman, T. J., Higuchi, W. I.: Release of Medroxyprogesterone Acetate from a Silicone Polymer, J. Pharm. Sci. *59*, 353–357 (1970)
Roseman, T. J., Yalkowsky, S. H.: Importance of Solute Partitioning on the Kinetics of Drug Release from Matrix Systems. In: Controlled Release Polymeric Formulations, D. R. Paul and F. W. Harris (Eds.), ACS Symposium Series *33*, 33–52 (1976)
Roseman, T. J., Cardarelli, N. F.: Monolithic Polymer Devices, Section I, pp. 21–54. In: Controlled Release Technologies: Methods, Theory, and Applications, Vol. I, A. F. Kydonieus (Ed.), CRC Press, Boca Raton, Florida (1980)
Rowe, R. C., Elworthy, P. H., Ganderton, D.: An Evaluation of a New Pore Model for Plastic Matrix Tablets, J. Pharm. Pharmac. Suppl., *24*, 137P (1972)
Saltzman, W. M., Langer, R.: Microstructural Measurments and Percolation Theory for Predicting Transport Properties in Porous Polymers, pp. 31–32. In: Proceedings of the 14th International Symposium on Controlled Release of Bioactive Materials, P. I. Lee and B. A. Leonhardt (Eds.), Toronto, Ontario, Canada, August 2–5 (1987)
Satterfield, C. N., Colton, C. K., Pitcher, W. H. Jr.: Restricted Diffusion in Liquids with Fine Pores, AIChE J. *19*, 628–635 (1973)
Shippy, R. L., Hwang, S.-T., Bunge, R. G.: Controlled Release of Testosterone Using Silicone Rubber, J. Biomed. Mater. Res. *7*, 95–110 (1973)
Siegel, R. A., Langer, R.: A New Monte Carlo Approach to Diffusion in Constricted Porous Geometries, J. Coll. Interf. Sci. *109*, 426–440 (1986)
Somasekharan, K. N., Subramanian, R. V.: Mechanism of Release of Organotin from Thermoset Polymers, pp. 415–431. In: Controlled Release of Bioactive Materials, R. Baker (Ed.), Academic Press, New York (1980)
Spacek, P., Kubin, M.: Diffusion in Gels, J. Polym. Sci.: Part C, *16*, 705–714 (1967)
Stannett, V.: Simple Gases, pp. 41–73. In: Diffusion in Polymers, J. Crank and G. S. Park (Eds.), Academic Press, London (1968)
Stannett, V., Williams, J. L.: The Permeability of Poly(ethyl Methacrylate) to Gases and Water Vapor, J. Polym. Sci. *C10*, 45–59 (1965)
Taverdet, J. L., Vergnaud, J. M.: Study of Transfer Process of Liquid into and Plasticizer out of Plasticized PVC by using Short Tests, J. Appl. Polym. Sci. *29*, 3391–3400 (1984)
Theeuwes, F., Gale, R. M., Baker, R. W.: Transference: A Comprehensive Parameter Governing Permeation of Solutes Through Membranes, J. Membrane Sci. *1*, 3–16 (1976)
Tojo, K.: Intrinsic Release Rate from Matrix-Type Drug Delivery Systems, J. Pharm. Sci. *74*, 685–687 (1985)
Tojo, K., Fan, L. T.: Membrane-Moderated Controlled Release, Mathematical Biosciences *57*, 279–300 (1981)
Tojo, K., Miyanami, K., Fan, L. T.: Mathematical Simulation of Membrane-Moderated Controlled Release, Powder Technology *35*, 89–96 (1983)
Tojo, K., Sun, Y., Ghannam, M. M., Chien, Y. W.: Characterization of a Membrane Permeation System for Controlled Drug Delivery Studies, AIChE J. *31*, 741–746 (1985a)
Tojo, K., Valia, K. H., Chien, Y. W.: Skin Permeation of Drugs-Diffusion and Partitioning, J. Chem. Eng. Japan *18*, 174–178 (1985b)
Tojo, K., Chiang, C. C., Chien, Y. W.: Drug Permeation Across the Skin: Effect of Penetrant Hydrophilicity, J. Pharm. Sci. *76*, 123–126 (1987)
Tranter, C. J.: Heat flow in an Infinite Medium Heated by a Cylinder, Phil. Mag. *38*, 131–134 (1947)

van Amerongen, G. J.: Diffusion in Elastomers, Rubber Chemistry and Technology *37*, 1065–1152 (1964)

Vrentas, J. S., Duda, J. L.: Diffusion in Polymer-Solvent Systems. I. Reexamination of the Free-Volume Theory, J. Polym. Sci. Polym. Phys. Ed. *15*, 403–416 (1977)

Vrentas, J. S., Duda, J. L.: Molecular Diffusion in Polymer Solutions, AIChE J. *25*, 1–24 (1979)

Wakad, N., Nardse, Y.: Effective Diffusivities and Deadend Pores, Chem. Eng. Sci. *29*, 1304–1306 (1974)

Wakao, N., Smith, J. M.: Diffusion in Catalyst Pellets, Chem. Eng. Sci. *17*, 825–834 (1962)

Wang, T. T., Kwei, T. K., Frisch, H. L.: Diffusion in Glassy Polymers. III, J. Polym. Sci.: Part A-2, *7*, 2019–2028 (1969)

Wirbrant, A., Sundelöf, L.-O.: Permeant Saturation and Regulation of Membrane Transport, J. Appl. Polym. Sci. *28*, 3281–3287 (1983)

Wood, J. M., Attwood, D., Collett, J. H.: Influence of Gel and Solute Structure on *In Vitro* and *In Vivo* Release Kinetics from Hydrogels, pp. 347–360. In: Polymers as Biomaterials, S. W. Shalaby, A. S. Hoffman, B. D. Ratner and T. A. Horbet (Eds.), Plenum Press, New York (1984)

Zimmermann, I.: Colloid Systems for Depot Dosage Forms, Prog. Colloid & Polymer Sci. *72*, 19–27 (1986)

Zygourakis, K.: Modeling the Controlled Release of Drugs from Hydrophobic Polymers, pp. 29–30. In: Proceedings of the 14th International Symposium on Controlled Release of Bioactive Materials, P. I. Lee and B. A. Leonhardt (Eds.), Toronto, Ontario, Canada, August 2–5 (1987)

3. Chemical Reaction Controlled Release

The polymer utilized in diffusion-controlled release devices or systems plays a relatively passive role. It serves simply as a carrier and retards the rate at which the active agent is delivered to the target. Nevertheless, some polymeric carriers are designed to play a more active role in the release process. These polymers undergo chemical reactions at the target site, thereby enabling the active agent to be delivered. Such chemically activated systems fall into two broad categories (Langer, 1980; Baker, 1987):

i. Physical immobilization systems, also called erodible or (bio)degradable systems in each of which the active agent, physically immobilized within the polymer network, is released by erosion of this network (Fig. 3.1 a)

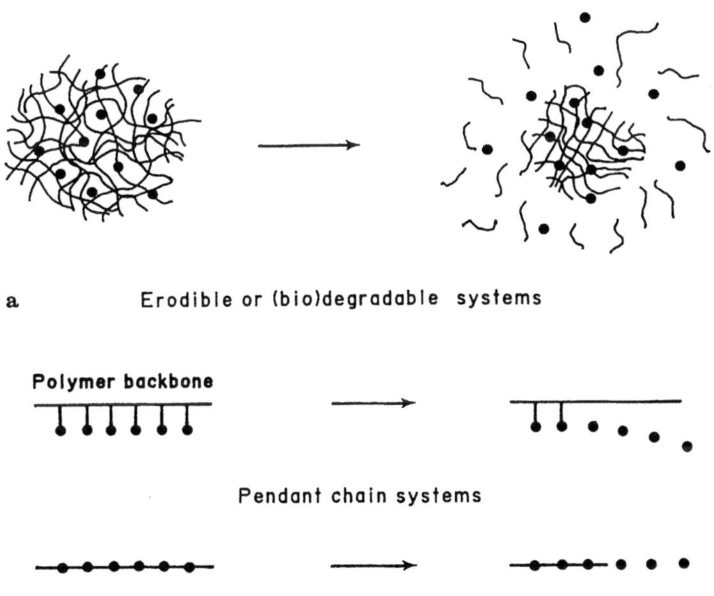

Fig. 3.1 a, b. Chemical reaction-controlled release systems. **a** Physical immobilization, **b** Chemical immobilization (adapted from Baker, 1987)

ii. Chemical immobilization systems, in each of which the active agent is either chemically bonded to the polymer carrier backbone (pendant chain) or is part of the backbone itself (polyagents). Release occurs by hydrolytic or enzymatic degradation of the appropriate bonds (Fig. 3.1 b).

The advantages of an erodible system, especially in pharmaceutical applications, are evident. A controlled-release device of this type will eventually disappear from the target site after having been implanted or orally ingested, thus obviating the need for a surgical removal procedure; however, the degradation products must be benign. Pendant chain and polyagent systems offer the possibility of drug targeting and reduced drug toxicity. They have been receiving increased attention over the past few years. In this chapter, we will examine these two types of reaction-controlled systems. Mathematical models for the first type will be discussed; release from the second type is principally governed by the kinetics of degradation of the appropriate bond and is, therefore, system specific.

3.1 Physically Immobilized Active Agent Systems

An erodible system with physically entrapped active agent(s) can be formulated similarly to the two basic forms of diffusion-controlled systems, namely as a reservoir or as a monolithic device. If significant during the delivery period, erosion of the rate-controlling membrane in reservoir devices can profoundly influence the release rates. This offers a possible mechanism for obtaining a range of delivery rates by coordinating the rates of erosion and depletion. Nevertheless, an unexpected loss of integrity of the membrane through erosion can lead to undesirable "catastrophic release" of the active agent. A more practical erodible reservoir device would be one in which the erosion process is slow during the delivery period and thus does not affect the release rate. However, the slow erosion eventually leads to a disappearance of the device from the target site.

A monolithic device, constructed with erodible polymer, can release the active agent through erosion, diffusion, or a combination of the two, contingent upon the relative rates of these mechanisms. If erosion is slow, diffusion will govern the release rate as described in the preceding chapter; however, the device will eventually disappear. If the active agent is effectively (physically) immobile in the polymer, its release will be governed purely by the erosion mechanism. Finally, a combination of the two mechanisms is possible if the two rates are comparable.

Erosion of a polymeric device can occur either from the surface inwards (surface or heterogeneous erosion) or more or less homogeneously throughout the device (bulk or homogeneous erosion). A polymer that erodes on the surface only is highly desirable. If the diffusion is not significant, such a polymer can lead to a zero-order release rate when the surface area remains constant. Loss of mechanical integrity of the device is less probable, and the release rate becomes independent of the chemical and physical properties of the active agent (Langer and Peppas, 1983). An idealized system such as this has yet to be developed, and most of the available polymers exhibit various degrees of bulk erosion.

The preferred degradation mechanism in such devices is hydrolytic cleavage of a labile bond (Langer and Peppas, 1983), although enzymatic degradation is utilized (see, e.g., Kopecek, 1984); in fact, self-contained enzymatically degradable systems have also been designed (see, e.g., Heller and Trescony, 1979; Yaacobi et al., 1986). The nature of the dominant hydrolytic erosion mechanism can be predicted from the polymer hydrophilicity and morphology. Hydrophobic polymers will exclude water from their interior and, therefore, are more likely to erode from the surface inwards; in contrast, hydrophilic polymers favor homogeneous erosion. Crystalline regions have greater propensity to exclude water than amorphous regions. Thus, the degree of crystallinity of the polymer will influence the mechanism of erosion (Langer and Peppas, 1983).

From the chemical point of view, a variety of labile bonds can be employed in forming (hydrolytically) erodible polymers. The rates of hydrolysis of some of these bonds under neutral conditions estimated from the known hydrolysis rate of low-molecular weight analogs are illustrated in Fig. 3.2, which indicates that

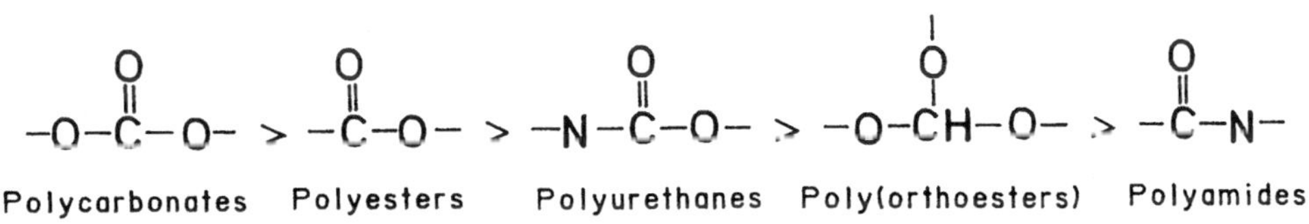

Fig. 3.2. Some common labile bonds and ranking of their estimated rates of hydrolysis under neutral conditions (Baker, 1987)

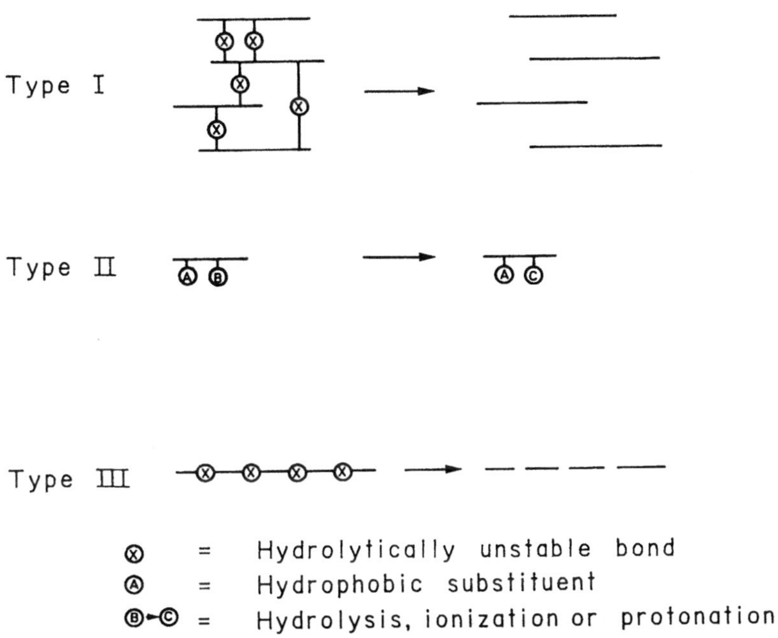

Fig. 3.3. Schematic representation of degradation mechanisms (Heller and Baker, 1980)

polycarbonates are most labile. However, the rates are affected by factors such as morphology, discussed earlier, and the presence and nature of substituent groups. Consequently, the comparison in Fig. 3.2 is only approximate (Baker, 1987).

A polymer, in solid form, can be eroded or solubilized in a number of ways. These have been classified into three categories by Heller and Baker (1980) as exhibited in Fig. 3.3. The first category (type I) consists of water-soluble polymers insolubilized by labile crosslinks. Cleavage of the crosslinks causes these polymers to solubilize, thereby yielding high molecular weight polymers as degradation products. The resultant matrix is completely permeated by water. Since the active agent is in an aqueous environment, its solubility is an important factor; a highly soluble compound will leach out rapidly, independent of the erosion rate. Type I polymers are useful, therefore, for active agents with low solubility or for macromolecular active agents that are entangled in the matrix and cannot be released until a significant number of crosslinks have been cleaved (Heller, 1980). An example of a system of this type is the entrapment of the enzyme, α-chymotrypsin, in a poly(vinyl pyrrolidone) hydrogel, crosslinked with N,N′-methylenebisacrylamide (Torchilin et al., 1977). By varying the ratio of the crosslink to the monomer content from 0.1 to 2.0 w/w % in the polymerization mixture, a set of polymers with dissolution times between several days and virtually insoluble, was obtained. Release of the enzyme, however, did not correlate well with the dissolution rate, presumably due to diffusional release or due to localization of the enzyme in the outer layers of the matrix.

The second category of erodible polymers (type II in Fig. 3.3) consists of polymers initially water-insoluble and rendered soluble by hydrolysis, ionization or protonation of substituent hydrophobic pendant groups. The molecular weight of the polymer does not change significantly during solubilization. Heller et al. (1978) have reported a system of this type, comprising partially esterified copolymers of methyl vinyl ether and maleic anhydride. Solubilization occurs by ionization

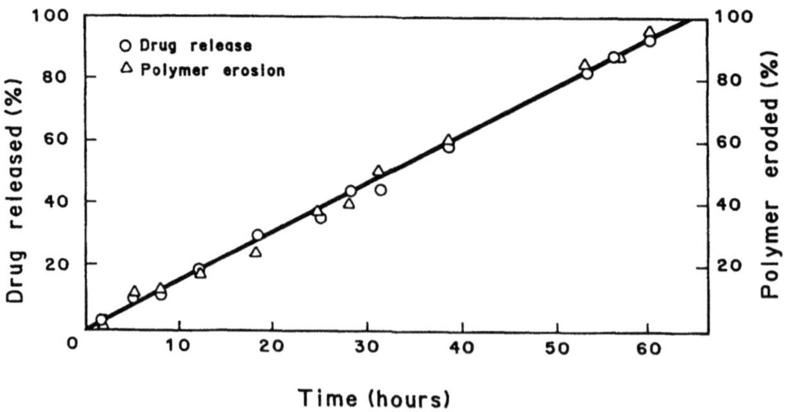

Fig. 3.4. Release of hydrocortisone and polymer erosion for an *n*-butyl half ester or methyl vinyl ether-maleic anhydride copolymer containing a 10 wt% drug dispersion (Heller et al., 1978)

of the carboxyl side groups. Any of these copolymers characteristically exhibits a pH range above which it is soluble and below which, insoluble. This pH range is distinct (~0.25 pH units) and changes with the number of carbon atoms in the ester side groups. Erosion is largely a surface phenomenon, and films of the copolymers containing hydrocortisone have exhibited excellent zero-order kinetics. The release rate correlates well with polymer dissolution rate (Fig. 3.4) (Heller et al., 1978).

The third category of erodible polymers (type III in Fig. 3.3) consists of water insoluble polymers; the polymers are solubilized by conversion to small, water-soluble molecules upon hydrolytic cleavage of their backbone. Thus, the labile bonds are part of the polymer backbone. These polymers undergo a combination of surface and bulk erosion. Examples of systems undergoing this type of degradation mechanism are various (co)polymers of lactic acid and glycolic acid. Yolles et al. (1974) used poly(lactic acid) to deliver cyclazocine. Wise et al. (1978) resorted to copolymers of lactic and glycolic acid to release ^{3}H-Naltrexone base. Altering the monomer ratio of the two acids in the mixture was sufficient to obtain a range of degradation and release rates. Similar results have been reported by Schindler et al. (1977), for releasing progesterone (see Fig. 3.5).

Preparation and properties of various classes of erodible polymers such as polyesters, poly(orthoesters), polyacetals, polyamides, polyanhydrides, etc. have been extensively reviewed by numerous authors (see, e.g., Langer and Peppas, 1983; Pitt and Schindler, 1983; Baker, 1987) and will not be considered here.

Labile bonds in a polymeric system can themselves be arranged in a variety of ways to form a matrix. These arrangements and the resulting modes of active agent immobilization and release have been classified by Pitt and Schindler (1983)

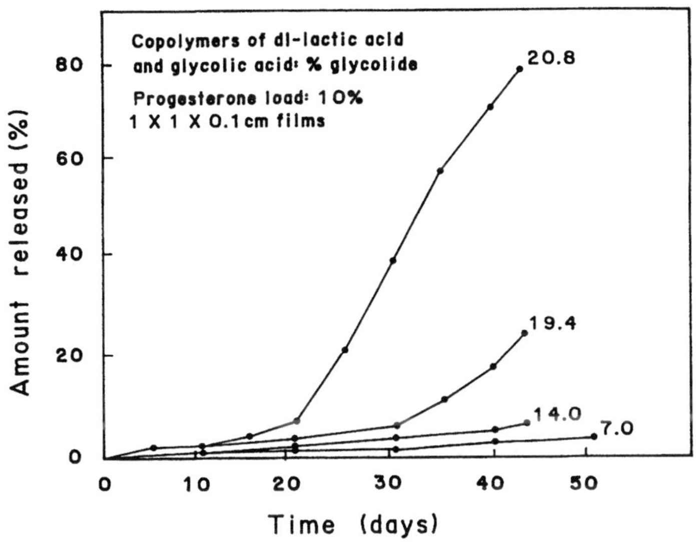

Fig. 3.5. Release of progesterone from poly(glycolide-co-dilactide) films containing 10 wt% steroid (Schindler et al., 1977)

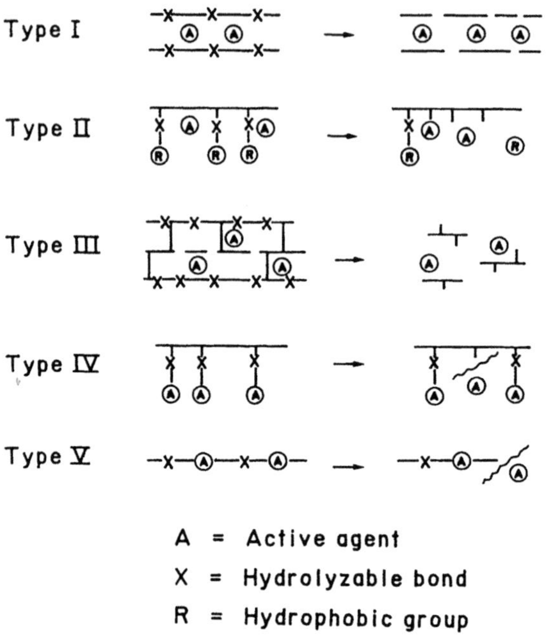

Fig. 3.6. Classification of possible active agent immobilization and release mechanisms from degradable polymers (Pitt and Schindler, 1983)

(Fig. 3.6). The first type involves insoluble polymers containing labile backbones with the active agent trapped between the chains. The second type involves systems in which solubilization of the polymer occurs by the loss of hydrophobic groups and results in active agent release. The third type involves labile bonds along the backbone of soluble polymers that have been insolubilized with crosslinks. The above three types correspond closely with degradation mechanisms given by Heller and Baker (1980) in Fig. 3.3. The final two classes of systems identified by Pitt and Schindler (1983) involve direct chemical bonding of active agent with polymer, either in the backbone or as a side chain (Fig. 3.6). These will be examined below in the category of chemically immobilized systems.

3.2 Chemically Immobilized Active Agent Systems

Active agents can be chemically bonded to a polymeric carrier with hydrolytically or enzymatically erodible bonds, either within the backbone or on side groups (Fig. 3.1b, Fig. 3.6). Sometimes, the polymer itself can be active; however, our consideration here is limited to systems which are biologically active only through release of the active agent. The rate-controlling step is the rate of erosion of the labile bonds. Such carrier systems can usually achieve higher loadings of the active agent as compared to systems in which the active agent is only physically incorporated into the polymer.

It is evident that the mode of attachment to a carrier will be specific for each (class of) active agent. However, some general patterns have been recognized (see, e.g., Allan et al., 1980; Duncan and Kopecek, 1984) and have been systematized by Baker (1987) into the two general categories illustrated in Fig. 3.7. The first

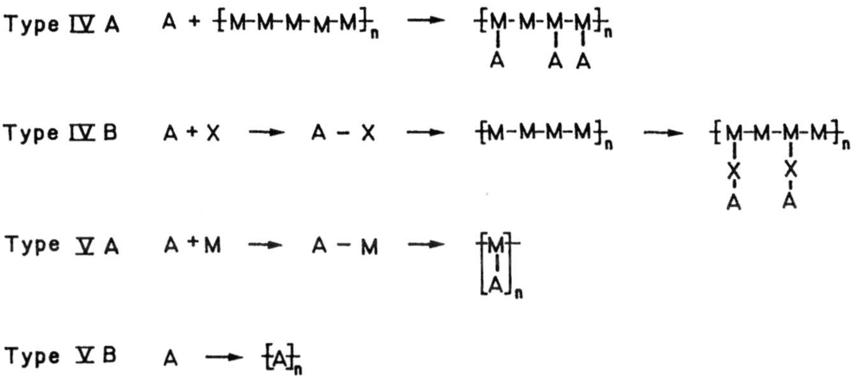

A = Active agent
M = Polymerizable monomer
X = Labile group

Fig. 3.7. Various approaches for synthesis of controlled release polyagents (Baker, 1987)

involves attachment of the active agent to a (pre)formed polymer backbone (type IV), while the second involves co- or homo-polymerization of the active agent or of their derivatives themselves (type V). The two categories have been further subdivided into two schemes, each depending on the actual chemistry of preparation (Baker, 1987).

A device of type IV A (Fig. 3.7) is formed with the active agent containing a reactive group that can be readily linked to a suitable group on a preformed polymer backbone. Some of the common functional moieties employed are (Duncan and Kopecek, 1984; Baker, 1987):

Ester	—CO—O—
Amide	—CO—N<
Anhydride	—O—
Urethane	—O—CO—N<
O-Acylhydroxamic acid	—CO—NH—O—CO—
Hydrazone	—CO—NH—N=C<
and	
Thioether	$—CH_2—S—CH_2—$

The first three types of linkages have been extensively employed by Allan and coworkers (see, e.g., Neogi and Allan, 1974; Allan et al., 1980) and by Harris and coworkers (see, e.g., Harris et al., 1976; Harris, 1980) to link various pesticides and herbicides to polymers. Figure 3.8 presents the relationship between the duration of effectiveness and amount of active agent in different combinations for application of the herbicide 2,4-dichlorophenoxyacetic acid (2,4-D); it is ester-linked to Douglas fir bark and to kraft lignin (Neogi and Allan, 1974). Field tests with herbicide–polymer combinations have demonstrated enhanced growth rates for the treated Douglas fir trees compared to the untreated controls (Allan et al., 1980). Examples of the binding of drugs using these linkages are available.

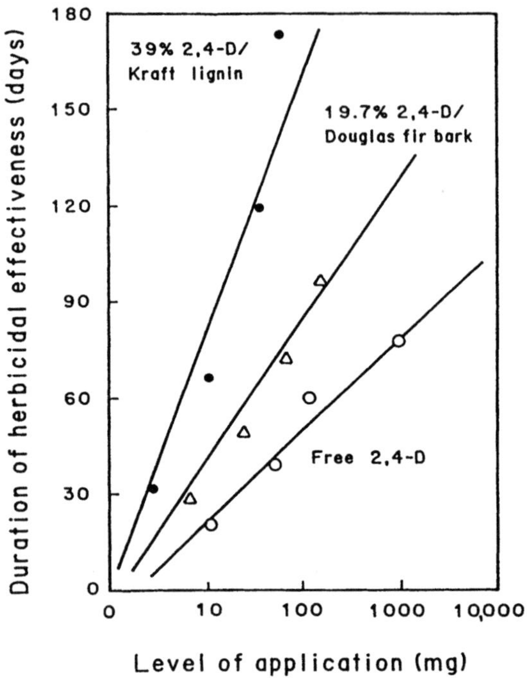

Fig. 3.8. Relationship between duration of herbicidal effectiveness and level of application of chemical combinations of 2,4-dichlorophenoxyacetic acid (2,4-D) and various polymers (Neogi and Allan, 1974)

Franzmann and Ringsdorf (1976) have bound chlorambucil [4-[4-/bis(2-chlorethyl)-amino/phenyl]butyric acid] to vinylpyrrolidone and vinylamine copolymers via an amide bond as follows:

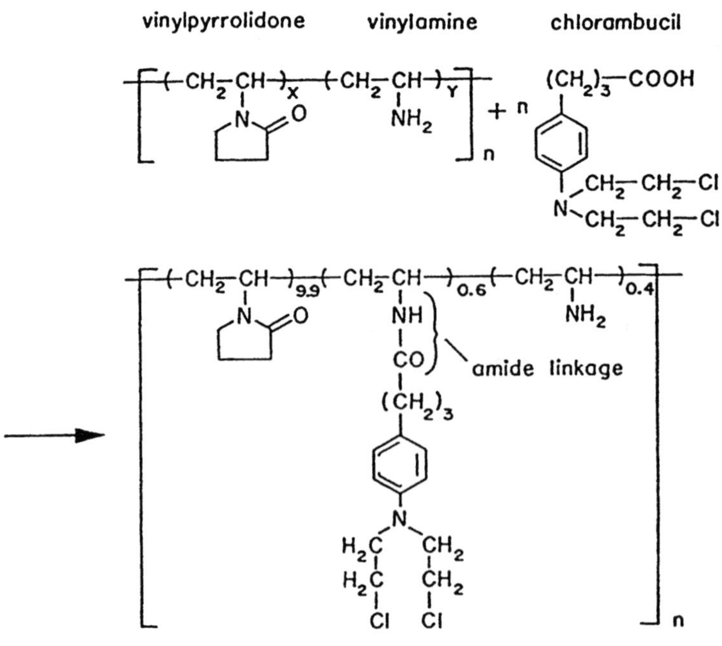

If the active agent and the preformed polymeric carrier do not contain suitable reactive groups permitting their direct reaction, one or both of them may need be derivatized to effect attachment. Systems of this type have been classified as type IV.B by Baker (1987) (Fig. 3.7). McCormick and Fooladi (1977) have created a polymer–herbicide system of this type by preparing isocyanate adducts of metribuzin and reacting it with the hydroxyl group of poly(vinyl alcohol), as illustrated below. Both linear and crosslinked poly(vinyl alcohol) polymers can be linked with

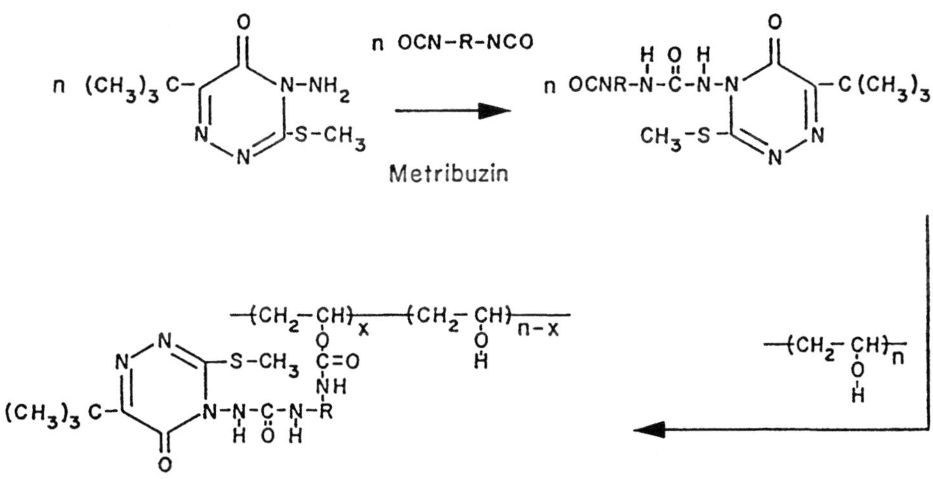

the herbicide. The linear polymers release the herbicide much faster than the crosslinked polymers, because of the longer duration required for water to diffuse into the latter to cause hydrolysis. The phytotoxicity to weeds of these compounds also reflects this behavior, as indicated by Fig. 3.9. The crosslinked-polymer products possess phytotoxicity similar to those of the free herbicide with rapid decay of activity while the linear-polymer products are effective over a larger period of time. However, it must be noted that the crosslinked products may exhibit phytotoxicity at a later stage as the polymer degrades (McCormick and Fooladi, 1977). Other systems of this type have been reported by Petersen et al. (1980) who have prepared the chloroformate derivative of Norethindrone and then reacted it with poly-N-(3-hydroxypropyl)-L-glutamine.

Type V.A systems (Fig. 3.7; Baker, 1987), involve conversion of the active agent into a polymerizable derivative, which can then be co- or homo-polymerized. This approach has been widely used in preparing polymer–drug (see, e.g., Duncan and Kopecek, 1984) and polymer–pesticide (see, e.g., Harris, 1980) combinations. Harris et al. (1976, 1977) have reported the copolymerization of vinyl derivatives of the herbicide 2,4-dichlorophenoxyacetic acid (2,4-D) with vinyl monomers containing hydrophilic residues. The 2-acryloyloxyethyl 2,4-D and methacrylic acid copolymers

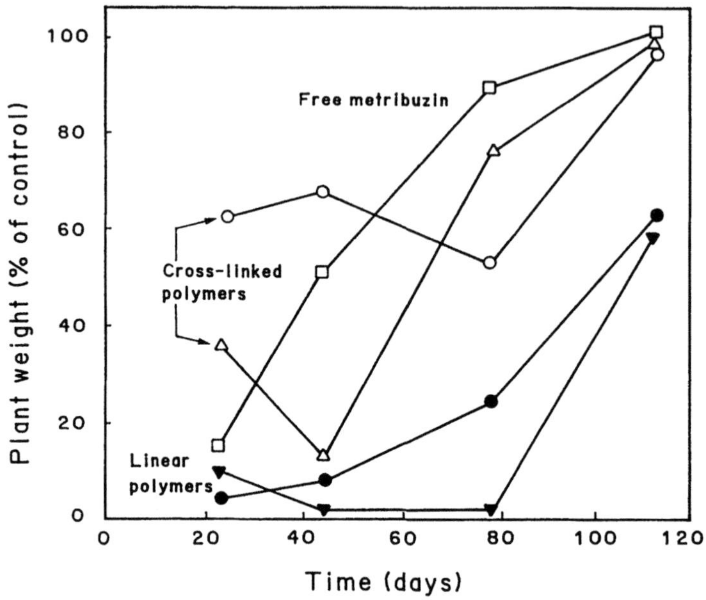

Fig. 3.9. Phytotoxicity of various herbicidal formulations of metribuzin and poly(vinyl alcohol) (McCormick and Fooladi, 1977)

slowly undergo hydrolysis under slightly alkaline conditions (pH = 8) at 30 °C. Increasing the percentage of the hydrophilic comonomer (methacrylic acid) in these systems has resulted in an increase in the rate of herbicide release (Fig. 3.10a). Copolymerization with trimethylamine methacrylamide has resulted in a relatively constant release rate over 200 days (Fig. 3.10b). It is interesting to note that homopolymers of 2-acryloyloxyethyl 2,4-D and of other acrylic acid 2,4-D esters do not undergo hydrolysis under the same conditions (Harris, 1980). Furthermore, increasing the length of the pendant side chain so as to increase the distance between the ester linkage and the hydrophobic backbone does not enhance the hydrolysis of this linkage in the homopolymer, even though it is less hindered sterically. However, incorporation of the hydrophilic comonomer has resulted in successful release, as indicated by Figs. 3.10a, b.

Other examples of type V.A systems are the acrylate derivatives of fungicides; they include pentachlorophenol and 8-hydroxyqinoline, homopolymerized, copolymerized with ethyl- or vinylacetate, and terpolymerized with methyl methacrylate and *n*-butyl acrylate (Pittman et al., 1982). Copolymers and terpolymer latexes of pentachlorophenol and 3,4′,5-tribromosalicylanilide acrylate were tested against *Auerobasidium pullulans* in agar dish accelerated growth studies; they have been found to prevent growth. Hoffman et al. (1979) prepared a polymerizable derivative of streptomycine sulfate (Str—CH=O) by reacting the aldehyde group of the drug with methacrylic acid hydrazide

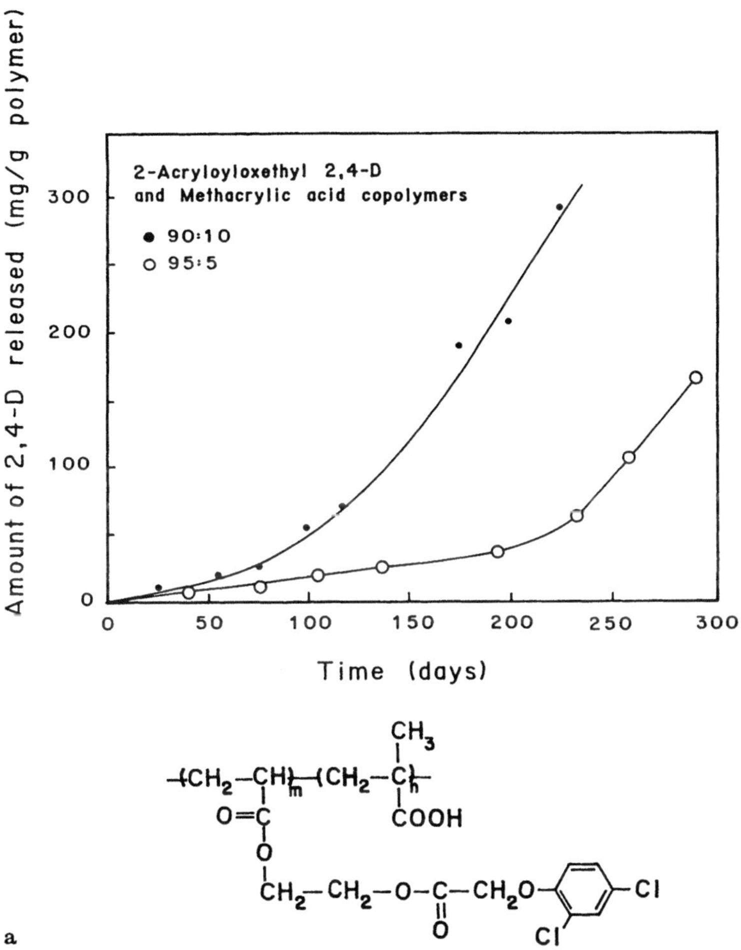

a

Fig. 3.10a. Release of 2,4-dichlorophenoxyacetic acid (2,4-D) from various copolymers of 2-acryloyloxyethyl 2,4-D and methacrylic acid by hydrolysis (Harris et al., 1977)

$$CH_2{=}\overset{\overset{\displaystyle CH_3}{|}}{C}{-}CO{-}NH{-}N{=}CH{-}Str$$

Water soluble copolymers were prepared by copolymerizing this derivative with methacrylamide or 2-(methylsulfinyl)ethyl methacrylate. Torchilin et al. (1977), in their study on immobilization of α-chymotrypsin prepared acylated derivatives of the enzyme and then copolymerized it with N-vinylpyrrolidone and N,N′-methylenebisacrylamide. Release of α-chymotrypsin from the resulting hydrogels has been well correlated with the rate of carrier solubilization. This is in contrast to the system in which the enzyme has been mechanically entrapped in the polymer (see description of type I systems).

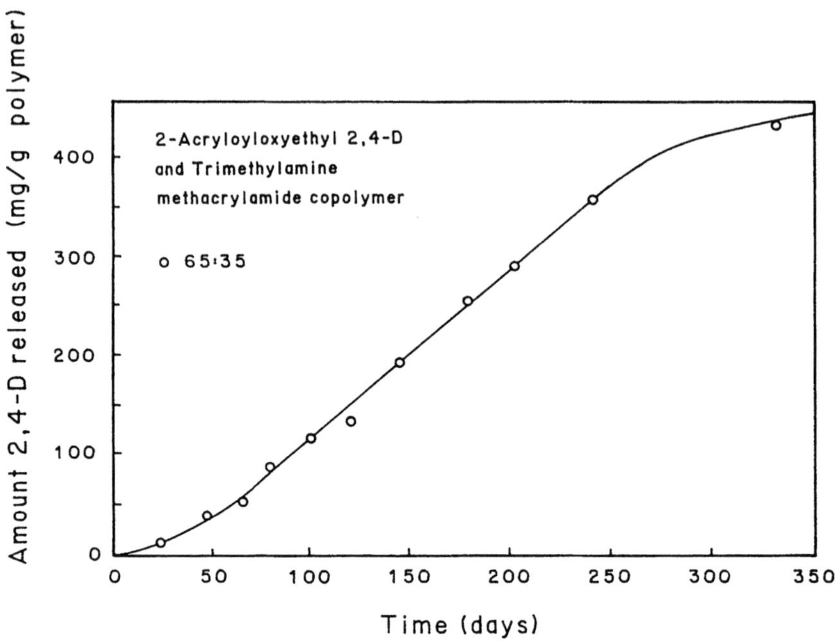

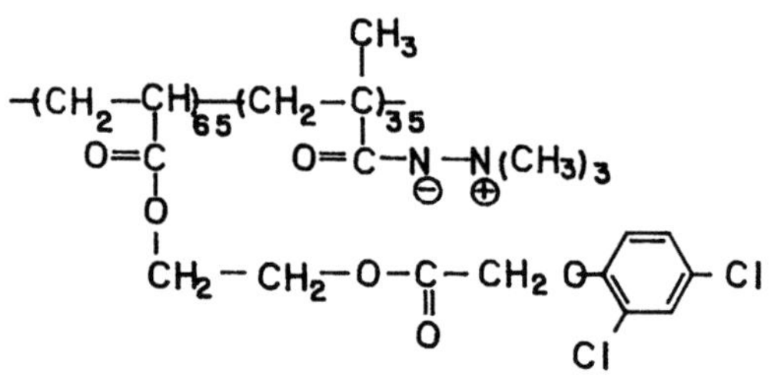

b

Fig. 3.10b. Release of 2,4-dichlorophenoxyacetic acid (2,4-D) from a copolymer of 2-acryloyloxyethyl 2,4-D and trimethylamine methacrylamide (Harris et al., 1977)

Each of the systems described thus far contains a backbone polymer that is biologically inactive. However, the active agent containing more than one functional group can be converted into a polymer entirely composed of the active agent (type V.B in Fig. 3.7). A system of this type would release the active agent through degradation of the main polymer chain instead of pendant group scission. The herbicide picloram [4-Amino-3,5,6-trichloropicolinic acid], can be homopolymerized to yield a highly effective product. Other such herbicides include endothall and chloramben illustrated below (Allan et al., 1980; Baker, 1987).

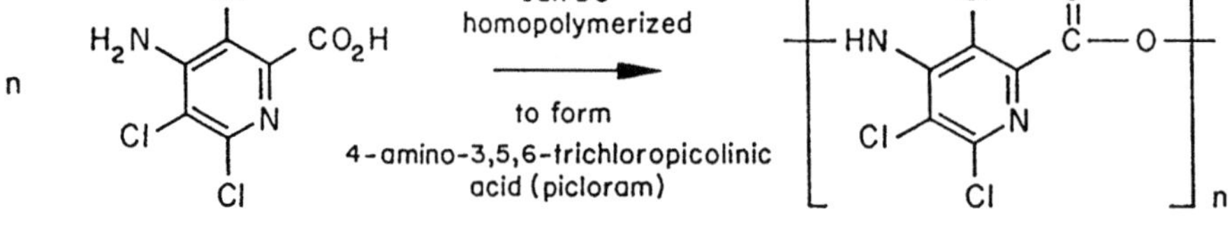

In most of the systems described above, the active agent, when bound directly to the polymer chain, exhibits either reduced or zero biological activity. It becomes active only after it has been split off. To facilitate cleavage, the active agent bound by any of the linkages listed earlier, must be separated from the polymeric backbone by a spacer (except in type V.B systems) allowing the cleavage effector to access the bond. Furthermore, while all linkages may be hydrolyzed, the rate of hydrolysis may vary from fast to insignificant. For instance, uncatalyzed hydrolysis of amide bonds is relatively slow; however, it can be increased by the introduction of specific groups adjacent to the amide bond (Duncan and Kopecek, 1984).

For polymer–drug combinations, the possibility of enzyme catalyzed hydrolysis makes available a relatively wide range of synthetic capabilities. Peptidic, glycosidic or phosphate bonds may also be used. Spacer groups can be created, which are cleaved specifically by certain enzymes in the target compartment. The relationship between the structure of the spacer sequence and its cleavability has been examined in detail for copolymers of N-(2-hydroxypropyl)methacrylamide containing *p*-nitroaniline (NAp) attached by various oligopeptide sequences (Duncan et al., 1983; Rejmanova et al., 1983). Release rates from some sequences cleaved by bovine spleen cathepsin B indicate that a wide range of release rates is obtainable (Fig. 3.11). Besides allowing the labile bond to be accessed by the degradation effector, the spacer sequence can be exploited to influence the hydrophilicity of the system.

A drug in combination with a polymeric carrier can also be targeted to a specific site within an organism, thus increasing the efficiency of delivery. This targeting can be achieved either through the inherent physico-chemical properties of the polymer

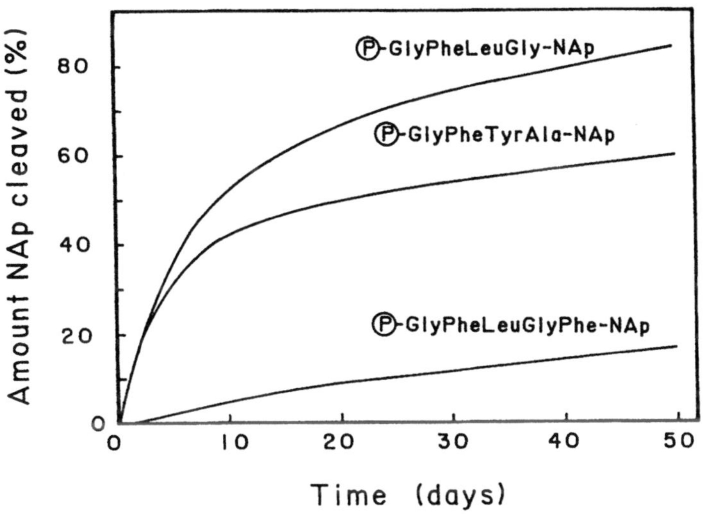

Fig. 3.11. Amount of *p*-nitroaniline (NAp) cleaved by bovine spleen cathespin B from copolymers of N-(2-hydroxypropyl)methacrylamide containing NAp attached by various oligopeptide sequences (Rejmanova et al., 1983)

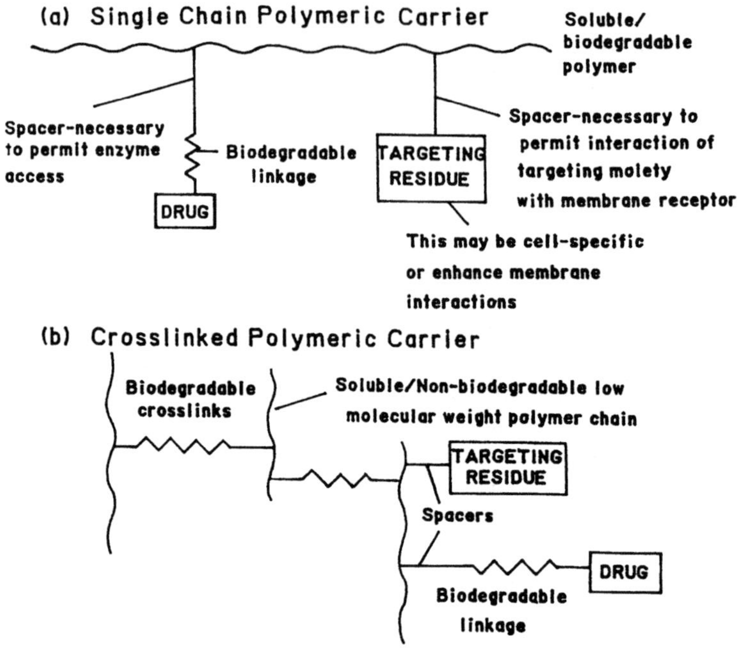

Fig. 3.12. Essential features of a targeted single-chain and cross-linked polymer-drug carrier (Duncan and Kopecek, 1984)

or through incorporation of specific targeting residues. Figure 3.12 shows a schematic of such a single chain or crosslinked polymeric carrier system. Targeting residues are groups that bind the specific receptors or antigens on the surface of the targeted cells. Carbohydrate moieties and antibodies are examples of such targeting residues (Duncan and Kopecek, 1984).

Various aspects of polymeric drug carriers have also been discussed by Ringsdorf (1975, 1978), Batz (1977), Kopecek (1984), while polymeric pesticide carriers have been extensively examined by Allan et al. (1980) and Harris (1980).

3.3 Models for Physically Immobilized Erosion-Activated Systems

Erodible devices can be manufactured by incorporating the active agent in a polymer which may undergo degradation when introduced into the target environment. The active agent can be released from such an erodible device either solely by erosion (surface detachment) or by a combination of diffusion and erosion. Expressions are available for the fractional dissolution rate of a solid tablet in which the active agent is held together by inerts and binders (Cooney, 1972).

Assuming a surface-detachment controlled process (slow dissolution rate, low solubility, and no diffusion resistance in medium), the expression

$$\frac{dM_t}{dt} = k_0 A (C_{bs} - C_b) \tag{3.1}$$

can be utilized to obtain equations for fractional dissolution rate from various geometries into infinite well-agitated sinks. Here, k_0 is the surface detachment rate constant, A is the surface area exposed to the medium, C_{bs} is the saturation concentration of active agent in the medium and C_b is the concentration of active agent in the medium. For a solid sphere of initial radius R_0 dissolving under surface-detachment control

$$\left(\frac{dM_t}{dt}\right)_r = k_0 4\pi (C_{bs} - C_b) \left[R_0 - \frac{k_0 (C_{bs} - C_b)\, t}{\varrho_a}\right]^2 \tag{3.2}$$

where ϱ_a is the amount of active agent per unit volume of the solid. Other shapes considered by Cooney were cylinders with various l/D ratios, cylinders with concentric bores, and spheres with internal cylindrical bores.

From a similar analysis it has been shown that, for monolithic erodible devices, a general expression can be written for a sphere of radius δ, a cylinder of radius δ and a slab of thickness 2δ (Hopfenberg, 1976)

$$\frac{M_t}{M_\infty} = 1 - \left(1 - \frac{k_0 (C_{bs} - C_b)\, t}{\varrho_a \delta}\right)^n \tag{3.3}$$

where n is 3 for a sphere, 2 for a cylinder, and 1 for a slab. The expressions above are valid for both the case of dissolved active agent and that of dispersed active agent since the release is assumed to be surface-detachment controlled.

Lee (1980, 1981) has used the heat balance integral method to give an approximate analytic solution for diffusional release from a surface erodible matrix with $C_{a0} > C_{as}$ in an infinite well-stirred medium. When the eroding front moves at a constant velocity, he obtained for a planar matrix of thickness δ, the relations

$$\frac{M_t}{M_\infty} = \left[1 - \frac{1}{2}\frac{C_{a0}}{C_{as}}\right]\beta + \frac{B(\delta/2)}{D_m}\,\frac{D_m t}{(\delta/2)^2} \tag{3.4}$$

and

$$\frac{D_m t}{(\delta/2)^2} = \frac{1}{6h}\left[3\left(\frac{C_{as}}{C_{a0}}\right) - 2\right]\left[\beta - \frac{1}{2h}\ln(1 + 2\beta h)\right] \tag{3.5a}$$

where

$$h = \frac{1}{2}\left[1 - \frac{C_{a0}}{C_{as}}\right]\frac{B(\delta/2)}{D_m}\,. \tag{3.5b}$$

B is the surface-erosion rate constant having the dimensions of velocity, while $\beta(\delta/2)$ is the separation between the extraction and erosion fronts. The term $B(\delta/2)/D_m$ is the ratio of erosion rate to matrix permeability and thus measures the relative contributions of the erosion and diffusion processes. Analysis of the effect of this term on the release shows that release generally starts with a typical first-order kinetics and shifts towards zero-order (limiting slope $B(\delta/2)/D_m$). The larger the value of this term, the sooner the shift towards zero-order occurs (Lee, 1981).

In order to increase the number of controllable design parameters at the disposal of a device designer, Thombre and Himmelstein (1984) have proposed the lamination of an erodible monolithic device ($C_{a0} \gg C_{as}$) with a membrane. Applying the pseudo steady-state assumption to both the extraction region and the membrane, they obtain the release rate as

$$\frac{d(M_t/M_\infty)}{d\tau} = \left(1 - \frac{C_{as}}{C_{a0}}\right)\left(-\frac{dR}{d\tau}\right) \tag{3.6}$$

where $dR/d\tau$ is the rate of movement of the diffusion front. It is given by the relation

$$-\frac{dR}{d\tau} = \frac{1}{\left(\frac{C_{a0}}{C_{as}} - 1\right)\left[\frac{1}{\alpha} + S(\tau) - R(\tau)\right]} \tag{3.7}$$

where

$$S(\tau) = 1 - \left[\frac{B(\delta/2)}{D_{m1}}\right]\tau\,, \tag{3.8a}$$

$$\tau = \frac{D_{m1}t}{(\delta/2)^2}\,, \tag{3.8b}$$

and

$$\alpha = \frac{(D_{m2}/h_m)\,(\delta/2)}{D_{m1}K}\,. \tag{3.8c}$$

In the above equations, $S(\tau)$ represents the position of the erosion front, while
D_{m1} = diffusivity of active agent in the eroding matrix,
D_{m2} = diffusivity of active agent in the membrane,
h_m = thickness of the membrane, and
K = partition coefficient between matrix and membrane $\equiv$ $(C_{\text{matrix}}/C_{\text{membrane}})$.

For $h_m = 0$, Eq. (3.6) simply gives the release rate, based on a pseudo steady-state analysis, from an erodible matrix (no lamination).

Analysis of their results by Thombre and Himmelstein (1984) showed that while the duration of zero-order release is not significantly extended by the lamination process, the life of the device is increased. The thickness of the membrane controls the time lag before the constant release rate period is realized.

Release from a monolithic device ($C_{a0} < C_{as}$) undergoing (bulk) chemical degradation due to action of the environmental fluid has been modeled by Thombre (1985). The system consists of a poly(ortho ester) matrix undergoing catalytically induced hydrolysis by an acid anhydride incorporated in the matrix; the reactions occur when the matrix is exposed to water. The active agent is unaffected by the reactions and diffuses out of the matrix. The diffusivities of the various species (water, acid anhydride, hydrogen ions, and active agent) are taken to be functions of the local extent of hydrolysis of the polymer. An infinite well-agitated sink is assumed. The system of governing differential equations has been solved by means of an implicit finite-difference approximation technique. The results indicate three phases in the delivery scheme: a short lag phase, an approximately zero-order rate phase over which a major part of the active agent is released, and a declining rate period at the tail-end of the release (Thombre, 1985).

In erodible reservoir devices, if diffusion does not contribute to release or is relatively slow, the active agent is released only after the membrane has been completely eroded. All the active agent is then released at one instant. Various release pattern can thus be obtained by varying the wall thickness in a population of microcapsules (Baker and Lonsdale, 1976). If, on the other hand, diffusion is faster than surface erosion, then the release rate at any time from a rectangular membrane is given by (Baker and Lonsdale, 1976)

$$\left(\frac{dM_t}{dt}\right)_r = \left(\frac{dM_t}{dt}\right)_0 \left(\frac{\delta}{\delta - Bt}\right) \quad \text{for } t < \delta/B \tag{3.9}$$

Table 3.1. Erodible reservoir devices

Controlling resistance	System description	Model description and solution type	Reference
Membrane mass transfer resistance	General; see Table 2.1 Any unerodible system whose initial release rate is known.	Rectangular reservoir: erosion rate known, PSSA, analytic solution based on initial release rate from corresponding unerodible device (Table 2.1)	Baker and Lonsdale (1976)
Reservoir, membrane, and boundary layer at sink mass transfer resistance	Finite source, infinite sink	Spherical reservoir: unsteady state, Fickian diffusion in reservoir and membrane, first-order erosion, temperature effects, numerical	Tojo et al. (1983)

where the subscript 0 denotes initial release rate which can be found from the expression for the corresponding non-erodible device case at zero time. δ is the thickness of membrane and B is the erosion rate with dimensions of velocity.

Tojo et al. (1983), in their mathematical simulation of a membrane-moderated controlled-release device, also considered the effect of a bioerodible membrane for a spherical reservoir device. A first-order degradation mechanism with an Arrhenius relationship for the temperature dependency of the rate constant was assumed. Numerically simulated results compared favorably with published experimental results. This and the models reviewed above are summarized in Tables 3.1 and 3.2.

Bioerodible polymers have a huge potential as carriers for controlled release. An erodible device when placed in the environment would eventually degrade leaving no residues and thus creating no pollution or environmental contamination. An erodible device, geometrically formulated such that the exposed area remains a constant at all times, can provide zero-order release rates if diffusion of the active-agent

Table 3.2. Erodible monolithic devices

Controlling resistance	System description	Model description and solution type	Reference
Surface detachment process	Finite source, infinite well-agitated sink, all concentrations of active agent	Rectangular, Cylindrical and Spherical geometries steady state, analytic	Hopfenberg (1976)
		Cylinders, Spheres with internal bores, other general shapes steady state, analytic	Cooney (1972)
Diffusive mass transfer resistance in surface-eroding polymer	Finite source, infinite well-agitated sink, active agent dispersed in polymer ($C_{a0} \geqq C_{as}$)	Rectangular geometry constant erosion front velocity, approximate analytic	Lee (1980a, 1981)
Diffusive mass transfer resistance in laminate and surface-eroding polymer	Finite source, infinite well-agitated sink, active agent dispersed in polymer ($C_{a0} \gg C_{as}$)	Rectangular geometry eroding polymer covered with a membrane, constant erosion front velocity, PSSA for both polymer and laminate, analytic	Thombre and Himmelstein (1984)
Diffusive mass transfer resistance in bulk eroding polymer	Finite source, infinite well-agitated sink, active agent dissolved in polymer ($C_{a0} \leqq C_{as}$)	Rectangular geometry erosion due to environmental fluid permeating polymer, diffusion with chemical reaction, matrix geometry, ersosion effect only on diffusion, numerical	Thombre (1985)

is negligible (see, e.g. Cooney, 1972). A simpler geometry (sphere or cylinder) can also produce zero-order release rate by having an appropriate concentration distribution to account for the change in area (no diffusional release case).

In the light of potential uses of erodible devices, more efforts to model these devices are needed. By relating the properties of the polymer to their performance under various environmental conditions, better and more efficient devices can be produced.

References

Allan, G. G., Beer, J. W., Cousin, M. J., Mikels, R. A.: The Biodegradative Controlled Release of Pesticides from Polymeric Substrates, pp. 7–62. In: Controlled Release Technologies: Methods, Theory, and Applications, Vol. II, A. F. Kydonieus (Ed.), CRC Press, Boca Raton, Florida (1980)

Baker, R. W.: Controlled Release of Biologically Active Agents, pp. 84–131, John Wiley and Sons, New York (1987)

Baker, R. W., Lonsdale, H. K.: Erodible Controlled Release Systems. In: Preprints of papers presented at the 171st meeting, ACS, Division of Organic Coatings and Plastics Chemistry *36* (1), 235–242 (1976)

Batz, H.-G.: Polymeric Drugs, Adv. in Pol. Sci. *23*, 25–53 (1977)

Cooney, D. O.: Effect of Geometry on the Dissolution of Pharmaceutical Tablets and Other Solids: Surface Detachment Kinetics Controlling, AIChE J. *18*, 446–449 (1972)

Duncan, R., Cable, H. C., Lloyd, J. B., Rejmanova, P., Kopecek, J.: Polymers Containing Enzymatically Degradable Bonds, 7. Design of Oligopeptide Side-Chains in Poly [N-(2-hydroxypropyl)methacrylamide] Copolymers to Promote Efficient Degradation by Lysosomal Enzymes, Makromol. Chem. *184*, 1997–2008 (1983)

Duncan, R., Kopecek, J.: Soluble Synthetic Polymers as Potential Drug Carriers, Adv. in Pol. Sc. *57*, 51–101 (1984)

Franzmann, G., Ringsdorf, H.: Pharmakologisch Aktive Polymere, 12. Depotformen von Chlorambucil durch Kovalente Bindung an Polymere, Makromol. Chem. *177*, 2547–2552 (1977)

Harris, F. W.: Polymers Containing Pendent Pesticide Substituents, pp. 63–82. In: Controlled Release Technologies: Methods, Theory, and Applications, Vol. II, A. F. Kydonieus (Ed.), CRC Press, Boca Raton, Florida (1980)

Harris, F. W., Aulabaugh, A. E., Case, R. D., Dykes, M. K., Feld, W. A.: Polymers Containing Pendent Herbicide Substituents: Preliminary Hydrolysis Studies. In: Controlled Release Polymeric Formulations, D. R. Paul and F. W. Harris (Eds.), ACS Symposium Series 33, 222–230 (1976)

Harris, F. W., Dykes, M. R., Baker, J. A., Aulabaugh, A. E.: Polymers Containing Pendant Herbicide Substitutes: Hydrolysis Studies II. In: Controlled Release Pesticides, H. B. Scher (Ed.), ACS Symposium Series 53, 102–111 (1977)

Heller, J.: Controlled Release of Biologically Active Compounds from Bioerodible Polymers, Biomaterials *1*, 51–57 (1980)

Heller, J., Baker, R. W., Gale, R. M., Rodin, J. O.: Controlled Drug Release by Polymer Dissolution. I. Partial Esters of Maleic Anhydride Copolymers — Properties and Theory, J. Appl. Polym. Sci. *22*, 1991–2009 (1978)

Heller, J., Baker, R. W.: Theory and Practice of Controlled Drug Delivery from Bioerodible Polymers, pp. 1–17. In: Controlled Release of Bioactive Materials, R. W. Baker (Ed.), Academic Press, New York (1980)

Heller, J., Trescony, P. V.: Controlled Drug Release by Polymer Dissolution. II: Enzyme-Mediated Delivery Device, J. Pharm. Sci. *68*, 919–921 (1979)

Hoffman, V., Ringsdorf, H., Seganova, A.: Pharmakologisch-Aktiv Polymere, 19. Polymere mit Kovalent Gebundenem Streptomycinsulfat, Makromol. Chem. *180*, 837–841 (1979)

Hopfenberg, H. B.: Controlled Release from Erodible Slabs, Cylinders, and Spheres. In: Preprints of papers presented at the 171st meeting, ACS, Division of Organic Coatings and Plastics Chemistry *36* (1) 229–234 (1976)

Kopecek, J.: Controlled Biodegradability of Polymers — A Key to Drug Delivery Systems, Biomaterials *5*, 19–25 (1984)

Langer, R.: Polymeric Delivery Systems for Controlled Drug Release, Chem. Eng. Commun. *6*, 1–48 (1980)

Langer, R., Peppas, N.: Chemical and Physical Structure of Polymers as Carriers for Controlled Release of Bioactive Agents: A Review:, J. Macromol. Sci. — Rev. Macromol. Chem. Phys. *C23*, 61–126 (1983)

Lee, P. I.: Diffusional Release of a Solute from a Polymeric Matrix — Approximate Analytical Solutions, J. Membrane Sci. *7*, 255–275 (1980)

Lee, P. I.: Controlled Drug Release from Polymeric Matrices Involving Moving Boundaries, pp. 39–48. In: Controlled Release of Pesticides and Pharmaceuticals, D. H. Lewis (Ed.), Plenum Publishing Corp., New York (1981)

McCormick, C. L., Fooladi, M.: Synthesis, Characterization, and Release Mechanisms of Polymers Containing Pendant Herbicides. In: Controlled Release Pesticides, H. B. Scher (Ed.), ACS Symposium Series 53, 112–125 (1977)

Neogi, A. N., Allan, G. G.: Controlled-Release Pesticides: Concepts and Realization, pp. 195–224. In: Controlled Release of Biologically Active Agents, A. C. Tanquary and R. E. Lacey (Eds.), Plenum Press, New York (1974)

Petersen, R. V., Andersen, C. G., Fang, S.-M., Gregonis, D. E., Kim, S. W., Feijen, J., Anderson, J. M., Mitra, S.: Controlled Release of Progestins from Poly (α-Amino Acid) Carriers, pp. 45–60. In: Controlled Release of Bioactive Materials, R. Baker (Ed.), Academic Press, New York (1980)

Pitt, C. G., Schindler, A.: The Design of Controlled Drug Delivery Systems Based on Biodegradable Polymers, pp. 17–30. In: Biodegradables and Delivery Systems for Contraception, E. S. I. Hafez and W. A. A. van Os (Eds.), MTP Press, Lancaster, England (1980)

Pitt, C. G., Schindler, A.: Biodegradation of Polymers, pp. 53–80. In: Controlled Drug Delivery, Vol. I, Basic Concepts, S. D. Bruck (Ed.), CRC Press, Boca Raton, Florida (1983)

Pittman, Jr., C. U., Lawyer, K. R., Ramachandran, K. S.: Polymer-Bound Fungicides for Paints. In: Biological Activities of Polymers, C. E. Carraher, Jr. and C. G. Gebelein (Eds.), ACS Symposium Series *186*, 35–53 (1982)

Rejmanova, P., Kopecek, J., Pohl, J., Baudys, M., Kostka, V.: Polymers Containing Enzymatically Degradable Bonds. 8. Degradation of Oligopeptide Sequences in N-(2-hydroxypropyl)methacrylamide Copolymers by Bovine Spleen Cathepsin B, Makromol. Chem. *184*, 2009–2020 (1983)

Ringsdorf, H.: Structure and Properties of Pharmacologically Active Polymers, J. Polym. Sci. Polym. Symp. *51*, 135–153 (1975)

Ringsdorf, H.: Synthetic Polymeric Drugs, pp. 197–223. In: Polymeric Delivery Systems, R. J. Kostelnik (Ed.), Gordon and Breach Science Publishers, New York (1978)

Schindler, A., Jeffcoat, R., Kimmel, G. L., Pitt, C. A., Wall, M. E., Zweidinger, R.: Biodegradable Polymers for Sustained Drug Delivery, pp. 251–286. In: Contemporary Topics in Polymer Science, Vol. 2, E. M. Pearce and J. R. Schaefgen (Eds.), Plenum Press, New York (1977)

Thombre, A. G.: Mechanism and Kinetics of Release from Two Controlled Drug Delivery Devices, pp. 9–106, PhD Dissertation, University of Kansas, Lawrence, Kansas (1985)

Thombre, A. G., Himmelstein, K. J.: Modelling of Drug Release Kinetics from a Laminated Device having an Erodible Drug Reservoir, Biomaterials *5*, 250–254 (1984)

Tojo, K., Miyanami, K., Fan, L. T.: Mathematical Simulation of Membrane-Moderated Controlled Release, Powder Technology *35*, 89–96 (1983)

Torchilin, V. P., Tischenko, E. G., Smirnov, V. N., Chazov, E. I.: Immobilization of Enzymes on Slowlv Soluble Carriers, J. Biomed. Mater. Res. *11*, 223–235 (1977)

Wise, D. L., Schwope, A. D., Harrigan, S. E., McCarthy, D. A., Howes, J. F.: Sustained Delivery of a Narcotic Antagonist from Lactic/Glycolic Acid Copolymer Implants, pp. 75–86. In: Polymeric Delivery Systems, R. J. Kostelnik (Ed.), Gordon and Breach Science Publishers, New York (1978)

Yaacobi, Y., Lotan, N., Sideman, S.: A Hybrid System for Zero Order Drug Delivery Regimen, Polym. Sci. Technol. *34*, 379–386 (1986)
Yolles, S., Eldridge, J., Leafe, T., Woodland, J. H. R., Blake, D. R., Meyer, F.: Long-Acting Delivery Systems for Narcotic Antagonists, pp. 177–193. In: Controlled Release of Biologically Active Agents, A. C. Tanguary and R. E. Lacey (Eds.), Plenum Press, New York (1974)

4. Swelling-Controlled Release

Consider a monolithic polymeric device containing an active agent introduced into the elution medium. If the elution medium is not thermodynamically compatible with the polymer or if the polymer is well above its glass transition temperature, the morphological structure is time independent and the device belongs to the diffusion-controlled category. However, if the elution medium or a constituent component is thermodynamically compatible with a polymer in the glassy state, its glass transition temperature may be lowered below the system temperature. Subsequently, the glassy polymer begins to undergo a glass-to-gel transition. The polymer chains in the gel state, being more mobile than those in the glassy state, allow the active agent to diffuse (more rapidly) out of the matrix. Such a device in which the change of polymer morphology, caused by interaction with the elution medium, controls or influences the release is classified as being swelling controlled. A distinction has been made between two types of systems formed by such swellable polymers (Peppas, 1984). A swellable controlled-release system comprising the first type, consists of a hydrophilic polymer that undergoes swelling more or less continuously throughout the matrix. This glass-to-gel transition loosens the matrix and the active agent is able to diffuse out, with the release rate determined by the diffusion process. A swelling-controlled system belonging to the second type, comprises a polymeric system exhibiting spatially discontinuous swelling. An interface separates the swollen gel region from the inner glassy core. This interface moves into the matrix as gelatinization proceeds. Release occurs by diffusion from the gel region; however, diffusivity is governed by the time-dependent macromolecular relaxation phenomena in this region and the release rate is, therefore, determined by this transition process (Peppas, 1984). This distinction between the two types of systems is often not clear-cut since a swellable system can also show time-dependent relaxation behavior, depending on its kinetics. A general representation of a swelling-controlled device along with the major steps in the release process is given in Fig. 4.1.

If chain relaxation and swelling are not restricted or constrained, the polymer matrix will eventually dissolve after completion of the gelatinization process. This can be prevented by cross-linking the polymer or using semi-crystalline polymers; the cross-links and crystallites act as permanent bonds holding the matrix together. However, chemical or enzymatic action by components in the elution medium may eventually lead to degradation of these bonds. In these cases, the device can be considered as being erodible swelling-controlled (Korsmeyer and Peppas, 1983).

To understand how the phenomenon of change in the polymer structure influences the release of active agent, the polymer relaxation process needs to be examined. In

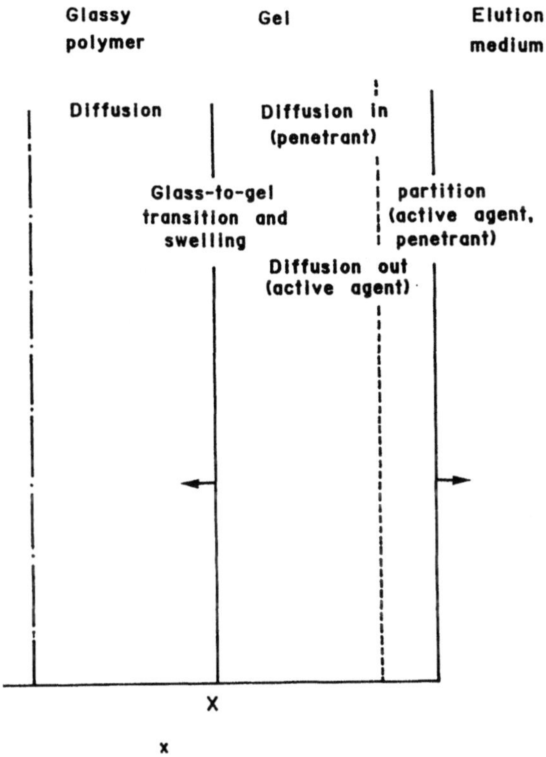

Fig. 4.1. Schematic representation of the functioning of a swelling-controlled release device

the subsequent discussion, the component in the external medium causing the change in polymer morphology will be referred to as the penetrant and identified by the subscript "*c*". The active agent will be identified by the subscript "*a*". The term "relaxation" will be used to indicate the transition of polymer chains from a glassy state to a gel state. If necessary, a distinction will be made between "relaxation without volume change" and "relaxation with volume change". The term "swelling" will be considered equivalent to "relaxation with volume change".

4.1 Penetrant Transport and the Polymer Relaxation Phenomena

The transport of penetrants into polymers has been extensively studied; experimental data on various systems abound. The nomenclature used for describing some of the observed phenomena can often be confusing because of the large variety of transport features possible. Some broad categories have been defined, however. The subsequent discussion will focus on a delineation of these categories without dealing with specific systems; it will be followed by a brief discussion of some models developed to explain these features.

4.1.1 Transport of Penetrants in Polymers

Polymers above their glass transition temperature respond rapidly to changes in their condition. A rubbery polymer film, when exposed to a constant activity reservoir of a smaller molecule capable of diffusing into the film (the penetrant), rapidly achieves a constant concentration at the exposed film surface. This sorption equilibrium is achieved in times much shorter than the characteristic times involved in the diffusion of penetrant (Frisch, 1980). This phenomenon of transport in rubber polymers is Fickian, and can be called either ideal Fickian or non-ideal Fickian on the basis of the constancy of the diffusion coefficient (McGregor, 1974).

The phenomenon of transport in glass polymers is much more complicated than that in rubbery polymers, due to significant contribution to the process from the characteristically larger times involved in chain mobility. Non-Fickian behavior, characterized by (among other factors) significant deviations from the linear relationship between the amount absorbed and square-root of time, is often observed below the glass transition temperature and also above the glass transition in certain cross-linked and crystalline polymers (Park, 1968). (For penetrant molecules that are small enough to move through the polymer by filling pre-existing microcavities and, thus, do not require flexibility in the macromolecular chains for their diffusion, Fickian diffusion is observed even below the glass transition temperature).

Certain glassy polymers, undergoing relaxation induced by absorption of a penetrant, commonly show a morphological discontinuity. An interface, often sharp, separates an inner glassy core from an outer relaxed gel-like region (Fig. 4.1). These two regions frequently have been observed in experiments (see, e.g., Kwei and Zupko, 1969; Baird et al., 1971; Thomas and Windle, 1977, 1978a, 1978b; Hopfenberg and Hsu, 1978; Franson and Peppas, 1983); nevertheless, the sharpness of the front varies. The interface between the two regions moves inwards into the glassy polymer as additional polymer undergoes relaxation. This change in the polymer's state occurs after a threshold value of the penetrant activity is reached, and the swelling is essentially localized at the moving interface (Astarita and Sarti, 1978; Thomas and Windle, 1978b; Gostoli and Sarti, 1982).

A useful classification scheme for the diffusion and sorption associated with polymers has been proposed (Alfrey, 1965; Alfrey et al., 1966; Frisch, 1980). Three basic classes of behavior have been established: case I, case II, and anomalous transport.

Case I, or Fickian diffusion behavior, is observed when the rate of transport or diffusion of the penetrant is much less than that of the relaxation process. A sorption equilibrium at the exposed surface of the polymer is rapidly established leading to time independent boundary conditions exhibiting no dependence on swelling kinetics (Frisch, 1980). The kinetics of this phenomenon can be characterized by a single parameter — the (concentration-dependent) diffusivity; if the phenomenon is accompanied by volume expansion, an additional parameter, namely the equilibrium swelling factor, must be considered (Alfrey et al., 1966). An ideal Fickian sorption phenomenon leads to a smooth and continuous concentration profile in the film; however, sharp advancing boundaries separating a glassy core from the gel region have been observed for Fickian diffusion with a strongly concentration-dependent diffusivity (Crank, 1951; Hopfenberg, 1970).

Case II transport is the other extreme type of behavior; diffusion of the penetrant through the gel region is very rapid compared with the relaxation process at the gel–glassy polymer interface. The interface marks the innermost limit of penetration or dispersal of the penetrant at any moment (Frisch, 1980). The glassy region contains almost no penetrant while the diffusivity of penetrant in the gel region is high. Therefore, the penetrant concentration profile in the system has a distinct shape; it is flat and equal to zero in the glassy core, and flat and equal to the level in equilibrium with the external medium in the gel region (Frisch et al., 1969; Thomas and Windle, 1978b; Yilmaz et al., 1983; Windle, 1985). The interface in the case II transport system advances at a constant characteristic velocity, while the gel behind the interface is essentially in an equilibrium state (Alfrey et al., 1966; Hopfenberg, 1970). An example of case II transport is illustrated in Fig. 4.2 for *n*-hexane sorption in polystyrene (Hopfenberg et al., 1975).

At the end of case II transport, a rapid increase in the penetrant sorption rate may sometimes be observed. In this situation, case II transport is said to have

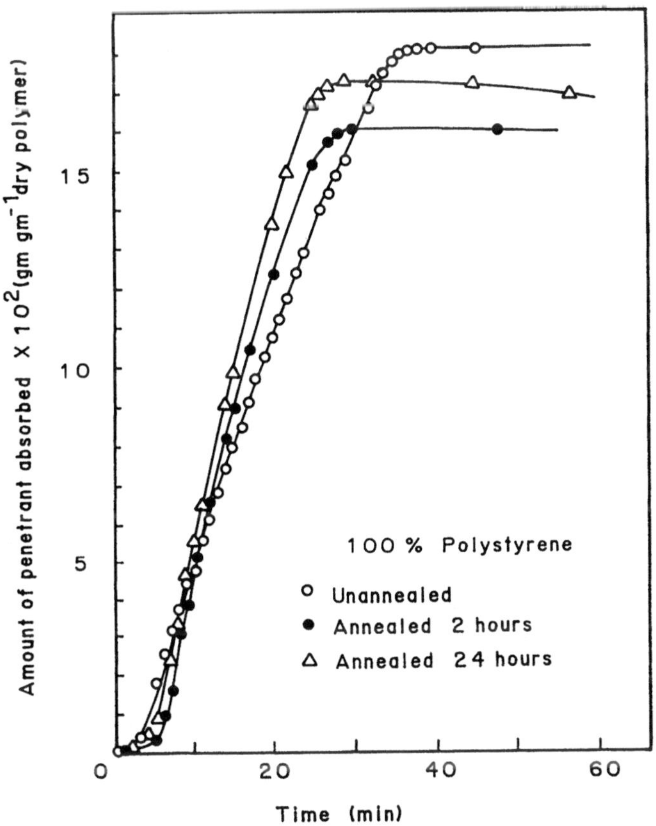

Fig. 4.2. Sorption of *n*-hexane vapors into annealed and unannealed films of 100% polystyrene at 40 °C and 0.98 activity (ratio of *n*-hexane partial pressure to equilibrium partial pressure), showing case II behavior; the films were annealed at 20 °C above their glass transition temperature (Hopfenberg et al., 1975)

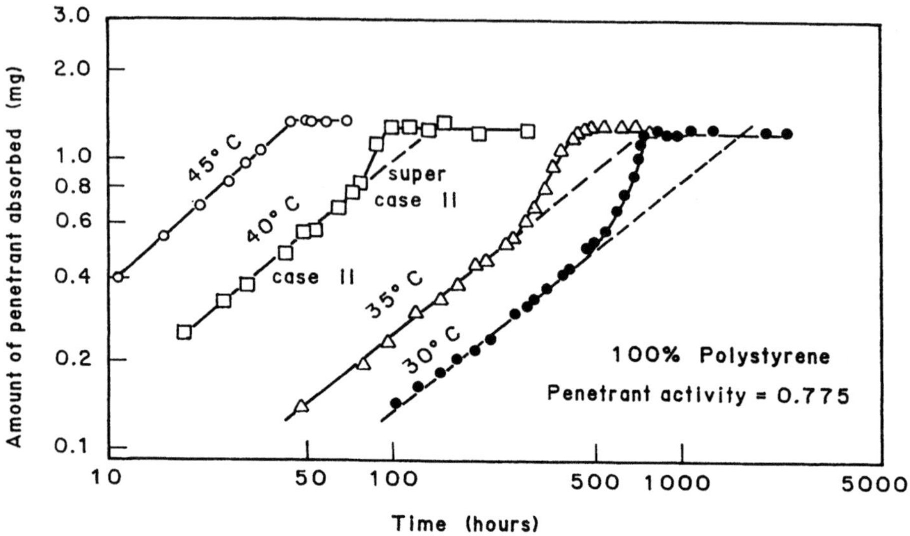

Fig. 4.3. Logarithmic plot of the kinetics of *n*-hexane sorption into polystyrene films showing evolution of case II into super case II transport (Jacques et al., 1974)

evolved into super case II transport; it is attributed to the expansion forces exerted by the swollen gel on the glassy core (Peterlin, 1979). Jacques et al. (1974), however, attribute this evolution to slow case II interface velocity, which allows a Fickian tail to develop ahead of the interface. Acceleration of the sorption process is initiated when the Fickian tails advancing from both sides of the film meet at the film midplane. In Fig. 4.3, the evolution of case II into super case II transport can be seen from the logarithmic plot of the kinetics of *n*-hexane sorption in polystyrene (Jacques et al., 1974).

The last category of transport behavior, termed anomalous transport, occurs when diffusion and relaxation rates in the polymer are comparable. Rearrangement of polymer chains, occurring slowly and simultaneously along with the diffusion process, gives rise to the time-dependent anomalous effects (Frisch, 1980). These effects can include:

i. Pseudo-Fickian behavior in which the sorption curve shows an anomalously small initial linear region when plotted against the square-root of time, or resembles Fickian curves but does not scale according to film thickness.
ii. Sigmoid behavior of the sorption curve when fractional sorption is plotted against the square-root of time. A simple inflection point is generally observed. Furthermore, the rate of desorption initially exceeds the rate of sorption; but it subsequently becomes slower and the curves intersect.
iii. Two-stage sorption behavior in which, after an initial period of rapid uptake, the sorption curve approaches a quasi-equilibrium followed by a slow approach to final equilibrium.
iv. Time-dependent exposed-surface concentration, related to the characteristic surface relaxation time.

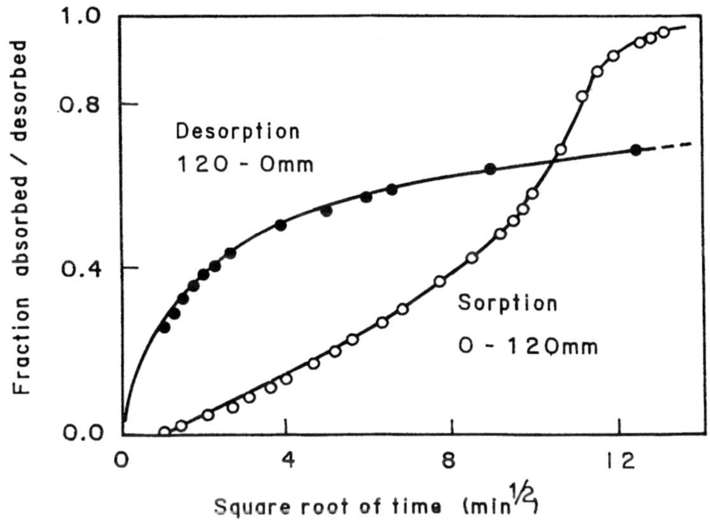

Fig. 4.4. Anomalous sorption and desorption behavior of acetone in cellulose acetate at 30 °C (Bagley and Long, 1955)

Sorption and desorption of acetone in cellulose acetate at 30 °C exhibit some of these features (Fig. 4.4) (Bagley and Long, 1955).

The three types of behavior of polymer–penetrant systems can be distinguished in a rectangular slab geometry by the shape of the fractional sorption-time curves, given by (see, e.g., Frisch, 1980)

$$\left(\frac{M_t}{M_\infty}\right)_c = k_c t^n \tag{4.1}$$

where
M_t = amount of penetrant absorbed in the polymer by time t
M_∞ = amount of penetrant absorbed at equilibrium, and
k_c, n = system parameters that depend on the mode of sorption.
For case I diffusion,
$n = 0.5$, and k_c is related to the diffusion coefficient, over the initial half of the sorption process.
For case II transport,
$n = 1.0$, and k_c is directly proportional to the constant velocity of the gel-glassy polymer interface.
Evolution of super case II transport from case II transport leads to
$n > 1.0$.
For anomalous transport,
$0.5 < n < 1.0$, and k_c reflects both the diffusional and relaxational processes.

Apart from the above mentioned classes of behavior, an extreme form of case II transport involving stress cracking of the polymer during the sorption process is sometimes observed (see, e.g., Bernier and Kambour, 1968; Michaels et al., 1968).

This phenomenon, termed "solvent crazing", occurs because the outer gel layer is constrained by the inner glassy core. The glassy core (in a cylindrical sample) prevents the gel from swelling (axially) and puts a compressive stress on the gel and a tensile stress on the core. As the gel–glassy polymer interface moves inward, the tensile stress increases until the tensile strength of the glassy core is reached, at which point the core cracks (Alfrey, 1965). Studies of the temperature dependence of this process have revealed that crazing possibly occurs through both primary bond breakage and chain separation (Michaels et al., 1968; Hopfenberg et al., 1969). Craze networks, which consist of interconnected microvoids, appear to nucleate at the case II front and represent localized plastic instability in the polymer; their appearance in certain systems, e.g., *n*-alkane polystyrene, may be associated with the propensity of certain polymers to develop highly localized regimes of plastic flow such as shear bonds (Windle, 1985).

For many polymer–penetrant pairs, all the above types of behavior can be observed by traversing a sufficiently wide range of temperature, penetrant activity, polymer molecular weight, and sample thickness (Hopfenberg et al., 1969; Astarita and Joshi, 1978; Sfirakis and Rogers, 1981). In addition, the thermomechanical history of the polymer also shows a significant effect on the sorption behavior (Hopfenberg et al., 1975). Hopfenberg and Frisch (1969) have attempted to identify regions exhibiting various behavioral features of polymer–penetrant transport on a plot of temperature against penetrant activity, as shown in Fig. 4.5; the following features are revealed:

i. Time-dependent anomalies and case II transport, including solvent crazing, are confined to regions of relatively high penetrant activities, with temperatures in the vicinity of and below the effective glass transition temperature of the system.

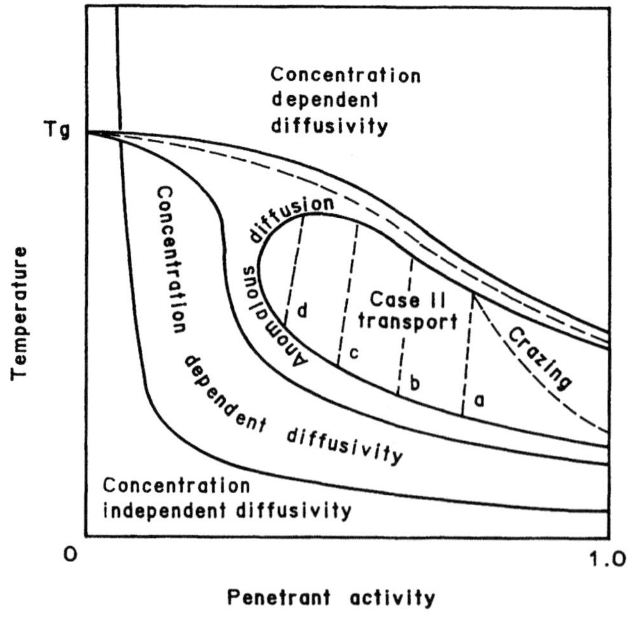

Fig. 4.5. Polymer-penetrant transport features in the various regions of temperature-penetrant activity plane; lines *a*, *b*, *c*, *d* are lines of constant activation energy (E) for case II transport with $E_a > E_b > E_c > E_d$ (Hopfenberg and Frisch, 1969)

The effective T_g is represented by the dashed line extending through the anomalous diffusion region.

ii. The region of case II transport is separated from the Fickian diffusion regions by a region where both relaxation and diffusion mechanisms are operative, thereby giving rise to diffusional anomalies.

iii. The activation energy characterizing case II transport (from an Arrhenius plot of interface velocity as a function of temperature) decreases as the penetrant activity is reduced.

iv. Concentration-independent diffusion coefficient is only apparent at very low temperatures and/or activities.

Since the features of the transport behavior are strongly influenced by the molecular weight, composition, and orientation of the polymer, additional temperature-activity planes would be required for each independent variable of the polymer (for a given penetrant) (Hopfenberg, 1970).

The various regions in the temperature-activity diagram (Fig. 4.5) can be distinguished by the ratio of two characteristic times (Vrentas et al., 1975). This ratio, *De*, of the characteristic relaxation time for the polymer-penetrant system, λ, and the characteristic diffusion time, Θ, is called the Deborah number

$$De = \frac{\lambda}{\Theta} . \tag{4.2}$$

This number represents the relative importance of the rate of diffusion to that of rearrangement of polymer chains in the diffusion process. The characteristic diffusion time is given by

$$\Theta = \frac{\delta^2}{D_c} , \tag{4.3}$$

where δ is the characteristic dimension of the polymer sample, and D_c is a mutual diffusion coefficient defined as

$$D_c = x_2 D_1^* + x_1 D_2^* . \tag{4.4}$$

x_1 and x_2 represent the mole fractions of the two components, penetrant (1) and polymer (2), respectively; D_1^* and D_2^*, the corresponding self-diffusion coefficients. The characteristic relaxation time is given by

$$\lambda = \frac{\int_0^\infty sG(s)\, ds}{\int_0^\infty G(s)\, ds} \tag{4.5}$$

where $G(s)$ is the shear relaxation modulus. The denominator in the equation represents the viscosity at zero shear rate (Vrentas et al., 1975).

The general characteristics of mass transport in a polymer-penetrant system can

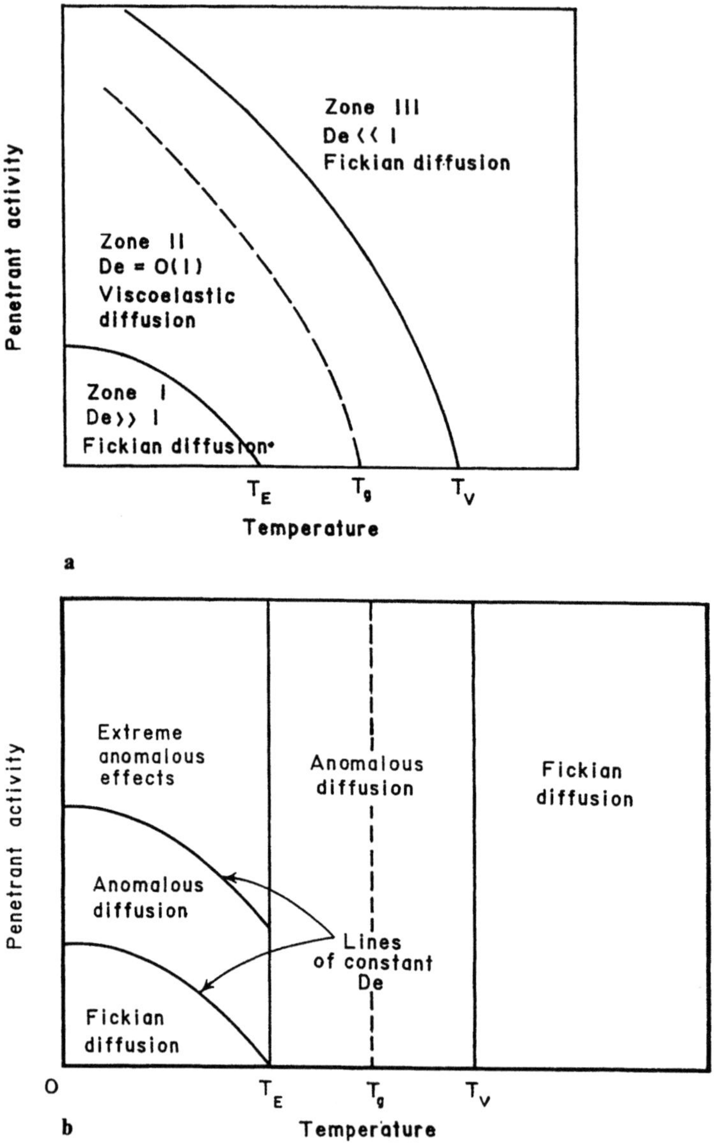

Fig. 4.6. **a** Classification of polymer-penetrant transport features based on temperature, penetrant activity, and Deborah number (Vrentas et al., 1975), **b** special from of temperature-penetrant activity diagram based on Deborah number showing the various types of polymer-penetrant behavior (Vrentas et al., 1975)

be represented in a temperature-activity diagram by lines of constant *De*, as shown in Fig. 4.6a. These lines can be drawn by calculating this dimensionless group as a function of temperature and composition for a given pressure, polymer molecular weight, and sample thickness. For $De \gg 1$ (of the order of 10) in zone I, the characteristic relaxation time is considerably larger than the characteristic diffusion time; effectively, no variation occurs in the polymer structure. The penetrant diffuses into an essentially elastic material by a vacancy mechanism; nevertheless, the process can be analyzed by the Fickian equation with a diffusion coefficient independent of concentration.

Conversely, in zone III where $De \ll 1$ (of the order of 0.1), the relaxation process is very rapid and appears to take place instantaneously. The penetrant molecules, therefore, diffuse in a purely viscous binary mixture (of polymer and penetrant) constituting the gel. This diffusion process is classically Fickian and the diffuion coefficient is a function of the concentration of penetrant/polymer.

For a Deborah number of the order of 1 (between 0.1 and 10) in zone II, the rearrangement of polymer chains takes place slowly. The penetrant molecules, therefore, diffuse in a viscoelastic binary mixture. The relaxation process influences the diffusive process, leading to anomalous transport (Vrentas et al., 1975).

The Deborah number is dependent, amongst other factors, on the concentration of penetrant (Vrentas et al., 1975). Since each element of the polymer sample can experience a range of concentrations during a sorption process, the Deborah number can differ from point to point and with time. The utility of a single value of this number for any system, therefore, may be limited. However, if the concentration change during the diffusion process is small, the value of *De*, calculated based on the average concentration of the polymer–penetrant mixture can indicate the general character of the transport process. This can be useful in screening polymers for designing specific delivery systems.

The changing value of the Deborah number also provides an indication of the presence of phenomenon such as case II transport. If this number changes from a large initial value to a small final value, the permeated part of the sample will behave like a viscous substance while the rest will behave like an elastic material. Thus, case II transport is observed when the diffusion in the viscous part of the sample (low *De*) is much faster than the polymer relaxation process (high *De*) at the interface between the viscous and elastic regions (Vrentas et al., 1975).

When utilizing the Deborah number, special forms of the temperature-activity diagrams need be drawn to include effects such as case II transport and solvent crazing. One such diagram is illustrated in Fig. 4.6b. Other diagrams can be drawn between sample thickness and temperature with lines of constant *De* delineating the regions of different types of transport (Vrentas et al., 1975). A good estimate of the size of polymer sample needed to observe a specific feature can thus be obtained. A procedure for constructing the Deborah number diagram has been demonstrated (Vrentas et al., 1977), and values of the Deborah number have been calculated for a poly(2-hydroxyethyl methacrylate-co-methyl methacrylate)–water system (Davidson and Peppas, 1986a, 1986b).

The phenomenon of transport in polymers and the accompanying structural changes have been extensively covered in the literature, e.g., Hopfenberg and Frisch (1969), an overview of behavioral features of a polymer–penetrant system as functions of polymer activity and temperature; Rebenfeld et al. (1976), a review of the polymer–penetrant interactions incorporating the diffusion phenomena, the controlling thermodynamic relationships, and the structural arrangements and mechanical behavior; Vieth et al. (1976), a review of dual-mode sorption theory; Frisch (1980), an overview of transport behavior in polymers and a review and classification of various theoretical models; Astarita and Nicolais (1983), a summary and review of the physics and mathematics of mass (and heat) transfer induced structural changes in polymeric materials and the kinetics thereof; Nagy (1985), a review of phenomenological and molecular aspects of swelling and mechanical behavior of swollen polymer gels; Windle (1985), a review of the various aspects of case II transport and of the theoretical models to describe this phenomenon.

4.1.2 Models for Transport of Penetrants in Glassy Polymers

A number of models, employing a variety of physicochemical principles, have been suggested to explain some or all the features of the transport process described in the preceding section. Among the earliest attempts is that of Crank (1953), involving the utilization of history-dependent diffusion coefficients. He has suggested that the relaxation of polymer chains occurs in two parts: an instantaneous jump reflecting a change in the concentration, and a slow drift towards an equilibrium value occurring even when the concentration remains invariant. On the basis of this assumption and adopting a first-order approach to equilibrium, the total change in the diffusion coefficient, D_c, is given by (Crank, 1953)

$$\left(\frac{\partial D_c}{\partial t}\right)_x = \left(\frac{\partial D_{ci}}{\partial C_c}\right)\left(\frac{\partial C_c}{\partial t}\right)_x + \alpha(D_{ce} - D_c) \tag{4.6}$$

where D_{ci} is the part of D_c that can change instantaneously and is a function of the concentration only. D_{ce} is the equilibrium diffusion coefficient, and α is a concentration-dependent relaxation parameter controlling the approach to equilibrium. D_{ci}, D_{ce}, and α are related to the penetrant concentration, C_c, by the expressions

$$D_{ci} = D_{c0} \exp(aC_c) \tag{4.7a}$$

$$D_{ce} = D_{c0} \exp(bC_c) \tag{4.7b}$$

and

$$\alpha = \alpha_0 \exp(\sigma C_c) \tag{4.7c}$$

where D_{c0}, α_0, a, b and σ are constants. Solutions of the Fickian diffusion equation with the diffusion coefficient given by Eqs. (4.6) and (4.7) could adequately predict sorption-desorption curves exhibiting sigmoid and pseudo-Fickian behavior. Experimental results for the methylene chloride-polystyrene system at 25 °C have been adequately described by this approach (Crank, 1953).

Another approach was employed by Wang et al. (1969) who modified the Fickian diffusion equation by including a convective contribution due to "partial stress" of the penetrant. From a consideration of the thermodynamics of irreversible processes, the following expression has been obtained for the flux of penetrant in one-dimension;

$$J = -BC_c\left[\left(\frac{\partial \mu}{\partial x}\right)_T - \frac{1}{C_c}\left(\frac{\partial s_{xx}}{\partial x}\right)\right] \tag{4.8}$$

where B is a mobility coefficient, μ the chemical potential, and s_{xx} is one component of the partial stress tensor of the penetrant. An assumption that s_{xx} is proportional to the total uptake of penetrant such that

$$s_{xx}(x) = S \int_0^x C_c(x', t)\, dx' \tag{4.9}$$

leads to a generalized diffusion equation

$$\frac{\partial C_c}{\partial t} = \frac{\partial}{\partial x}\left[D_c(C_c, x, t)\frac{\partial C_c}{\partial x} - B(C_c, x, t)\, SC_c\right] \tag{4.10}$$

where S is a constant. For a constant B, this equation reduces to

$$\frac{\partial C_c}{\partial t} = \frac{\partial}{\partial x}\left(D_c \frac{\partial C_c}{\partial x} - vC_c\right) \tag{4.11}$$

where v $(= BS)$ is interpreted as the convective velocity of the penetrant arising as a direct consequence of the internal stress (Frisch et al., 1969; Kwei and Zupko, 1969). Consequently, it represents a pseudo-convective contribution to the transport process. When the stress gradient dominates the process, i.e.,

$$\frac{\partial C_c}{\partial x} \ll \frac{\partial s_{xx}}{\partial x},$$

the differential equation describes the case II mechanism with v as the velocity of the gel–glassy polymer interface, also known as the case II velocity.

A linear form of Eq. (4.11), with constant D and v, has been solved along with the approximation that the interface moves according to the relation

$$X = 2k(D_c t)^{1/2} + vt \tag{4.12}$$

where k is a constant. The model successfully describes the time dependence of the penetration data for trichloroethylene and methyl ethyl ketone in an epoxy polymer (Kwei and Zupko, 1969), and the sorption-desorption of acetone in poly(vinyl chloride) (Kwei et al., 1972; Wang and Kwei, 1973).

In an attempt to account for the internal stresses generated from the differential swelling of the outer gel layer on the inner glassy core, Crank (1953) has assumed that only the diffusion coefficient is affected and not the surface concentration or internal solubility. Diffusion coefficients in the two regions are related to the changes in areas of the two regions and simplified to an effective diffusion coefficient that is a product of a function of the concentration and a function of time. Solution of the Fickian equation with this diffusion coefficient reproduces an experimental (sigmoid) sorption curve for methylene chloride in polystyrene, and the change in area with time. However, it fails to explain the absence of $(M_t/M_\infty)_c$ versus $(t^{1/2}\delta^{-1})$ scaling.

Peterlin (1979) has attempted to explain the case II behavior by postulating the existence of a constant interface velocity, v. A Fickian tail is assumed to extend into the glassy polymer ahead of the front, while the concentration gradient in the gel region is considered negligible because of the high diffusivity of the penetrant in the gel region. The Fickian diffusion equation, therefore, has been solved for the glassy region with the interface position being given by the relation,

$$X = 1 - vt . \tag{4.13}$$

The diffusion process is completed when this boundary reaches the center of the polymer. Peterlin (1979) has presented his simulation results for transport into spheres and films in terms of the concentration profiles and weight gain, both of which are functions of time.

A phenomenological model for the kinematics of polymer swelling and dissolution has been presented by Tu (1977) and Tu and Ouano (1977). They have regarded these phenomena as constituting a multiphase Stefan problem with a glassy polymer, a gel, and a liquid solution region wherein the dissolved polymer molecules are carried away. The penetrant at the gel–liquid interface is characterized by a dissociation concentration while its concentration at the gel–glassy polymer interface is given by the critical concentration level required for the transition to occur. The position of the gel–glassy polymer interface is determined implicitly within the problem. Analytic small time solutions have been obtained along with a complete numerical integration of this boundary value problem. For small times, the two interfaces move as $t^{1/2}$. At large times, the gel–glassy polymer interface is found to move at a constant rate, explicity determined in terms of the dissociation rate at the surface. The model has been found to be able to predict the experimentally measured boundary positions with time for a polystyrene–methyl ethyl ketone system (Tu and Ouano, 1977). No instantaneous case II transport is predicted by this model due to the existence of an initial period during which the gel–glassy polymer interface moves as $t^{1/2}$.

A general version of the history-dependence model of Crank (1953) has been presented by Cohen (1983, 1984) and Cohen and Stanley (1983). The model explicity incorporates the finite relaxation time of a glassy polymer in response to changing conditions. The mass conservation equation is written in terms of the dual-sorption model, leading to a generalized form in one-dimension as

$$\frac{\partial C_c}{\partial t} = \frac{\partial}{\partial x}\left(D_c \frac{\partial C_c}{\partial x}\right) + R(x, t, C_c, C_{ct}, C_{cx}) \tag{4.14}$$

where

$$D_c = \int_0^t K[(t, s, C_c(x, s)] \, f[C_c(x, s), C_{ct}(x, s)] \, ds . \tag{4.15}$$

$R(\cdot)$ represents a generalized kinetics for the dual-mode sorption process for the change from a dry polymer to a fully saturated one, i.e., for all the complex processes leading to

$$\text{site} + \text{molecule} \leftrightarrows \text{site-molecule complex} .$$

$K(\cdot)$ is any appropriate delay-kernel representing the history dependence implied by the finite relaxation times; a rapid response can be represented by taking the kernel as the Dirac function. The model can be reduced to describe several phenomena including dual-mode, sigmoid, and strain-dependent sorption. The rationale for the choice of kernel, K, and function, f, in various situations has been discussed (Cohen, 1983).

Apart from those discussed above, models attempting to describe certain specific features of non-Fickian transport in polymers include, among others:

- the time-dependent surface concentration model of Long and Richman (1960);
- the irreversible thermodynamical models of Frisch (1962) and Neogi (1983);
- the dual-mode sorption model of Vieth and Sladek (1965), Vieth et al. (1976) and Ahn et al. (1985);
- the infinitesimal elasticity theory of Alfrey et al. (1966);
- the diffusion-relaxation model of Berens and Hopfenberg (1978);
- the "degrees-of-swelling" model of Joshi and Astarita (1979);
- the moving boundary models of Peppas and Sinclair (1983), Tosun and Yilmaz (1982) and Yilmaz et al. (1983); and
- the moving boundary with n-th order relaxation kinetics model of Astarita and Sarti (1978) and Astarita and Joshi (1978).

Models have also been proposed to explain a wide range of polymer–penetrant behavior. Some of these models are discussed below.

The stress-relaxation and history-dependence models of Crank (1953) have been combined by Petropoulos and Roussis (1978) to yield a model capable of predicting sorption behavior ranging from the Fickian through case II transport to super case II transport. The polymer is considered as a linear viscoelastic medium, capable of expanding in the radial direction but constrained in the transverse direction, thereby creating a transverse differential swelling stress, f. This stress influences the diffusion coefficient given by

$$D_c(C_c, f) = D_c(0, 0) \exp (k_2 C_c + k_3 f) \tag{4.16}$$

with k_2 and k_3 being constants. The local transverse stress, f, at any position is a function of the diffusion and relaxation processes and obtained from analysis of an equivalent mechanical model consisting of a Maxwell element in parallel with a spring. The mass balance equation (Fickian diffusion equation in polymer-fixed coordinates) is solved numerically, along with the force balance equation, to determine sorption-time curves as well as the distribution of penetrant and transverse swelling stress in the polymer. The sorption curves display a marked deviation from the Fickian behavior as the effect of stress is increased. They acquire a quasi-linear (case II) middle portion, extending upwards concavely (super case II). The super case II transport arises from a strong acceleration of the diffusion process during the later stages of sorption; this, in turn, results from the more uniform distribution of penetrant at this stage, and a consequent reduction of the compressive stresses (Petropoulos and Roussis, 1978). The ideal case II behavior, starting from time zero is not predicted, however. The concentration-dependence of the diffusion coefficient enables a prediction of the formation of the front; in contrast, its rate of propagation is determined by the stress-relaxation term.

In an extension of their group's work, Gostoli and Sarti (1982) have modeled the transport phenomenon as a diffusion problem with a moving interface. The rate of movement of the interface is related to the stress exerted on this interface by the penetrant, which is considered as the sum of the osmotic tension and the differential swelling stress. Consequently, the front velocity is given by

$$\left[\frac{dX}{dt}\right] = K(a\pi + b\sigma_g - \sigma_c) \tag{4.17}$$

where π is the osmotic tension, σ_g the differential swelling stress, and σ_c is a threshold uniaxial tensile stress. K is a kinetic constant while the constants a and b account for the fact that π is isotropic while σ_g is biaxial. The Fickian diffusion equation has been solved numerically for the gel region with Eq. (4.17) and a boundary condition representing flux balance at the interface. The diffusive tail in the glassy core is neglected. A range of behavior including Fickian diffusion, case II, and super case II transport is predicted. The case II transport arises when the process is governed by osmotic stress only. When the differential swelling stress becomes relevant, an acceleration effect is observed prior to saturation. This effect leads to super case II transport for high values of the Deborah number of the gel layer.

The model of Gostoli and Sarti (1982) assumes, *a priori*, the existence of the gel-glassy polymer front and then attempts to generate a suitable constitutive equation for the front velocity whereas Thomas and Windle (1982) have formulated a model to predict the formation of this front. Thomas and Windle (1982) consider the kinetics of swelling as being controlled by the mechanical viscous resistance of the polymer against an increase in its volume and a change in its shape in response to the osmotic pressure. Transport is considered to occur by diffusion down an activity gradient, and is coupled to the time-dependent mechanical deformation of the glassy polymer in response to the swelling stress. Thus, the two basic parameters in the theory are the diffusion coefficient within the glassy polymer and its flow viscosity; the viscosity is assumed to decrease exponentially with penetrant concentration. The model involves the numerical solution of a Fickian equation and a parallel equation for the concentration variable in terms of the mechanical response to swelling. A complete range of anomalous behavior between Fickian and case II extremes is described. The existence and nature of a moving front are determined by the magnitude of deformation resistance. The higher the resistance (or the steeper the dependence of viscosity on penetrant concentration), the sharper the front and the smaller the Fickian tail ahead of the front. The model is versatile, but cumbersome to compute.

It is evident from the preceding review that various approaches, from the simple to the complicated, have been adopted to model the phenomenon of transport of a penetrant in a glassy polymer. Petropoulos and Roussis (1974) have reviewed some of the early modeling efforts; Windle (1985) has examined the various aspects of osmotic pressure theories, while Frisch (1980) has compared a number of these models and suggested possible directions for further theoretical and experimental efforts.

4.2 Active Agent Release from Swellable Polymers

Release of an active agent from a polymer undergoing macromolecular relaxation induced by the sorption of a thermodynamically active penetrant, is influenced by the relaxation process as well as the countercurrent transport of the penetrant. Swelling-controlled devices can be produced by dissolving or dispersing the active agent in a polymer solution. The solvent can then be evaporated to produce a glassy polymeric matrix (Hopfenberg and Hsu, 1978). The active agent can also be loaded (dissolved) into a (copolymerized) polymer sample in the gel state by soaking the gel in a solution containing the active agent and allowing the two to equilibrate. After loading, the penetrant in the polymeric gel can be evaporated, thereby leaving a penetrant-free glassy polymeric matrix loaded with the active agent (Lee, 1984a). Other devices may be formed by compressing particles of the solvent-free polymer and the active agent, thus producing a porous matrix (Lapidus and Lordi, 1968; Bamba et al., 1979; Korsmeyer et al., 1983). The active agent is almost immobile in the glassy polymer; however, it is able to diffuse out as the glassy polymer is converted into the gel. In porous matrices formed by compression, diffusion through the gel is accompanied by diffusion through the penetrant-filled pores.

Among the first investigations of active agent release from swellable polymers was that reported by Lapidus and Lordi (1968). They compressed four candidate drugs with hydroxypropyl methyl cellulose ether powder to produce hydrophilic matrices that undergo swelling upon hydration. This matrix is able to resist erosion by attrition of the gel layer. The release rate has been found to be a function of the square-root of time, indicating that, effectively, it was diffusion controlled. One of the first examples of the true swelling-controlled release system has been formulated from poly(2-hydroxyethyl methacrylate) hydrogels crosslinked with tetraethyleneglycol dimethacrylate and containing tripelennamine-hydrochloride (Good, 1978). A signifi-

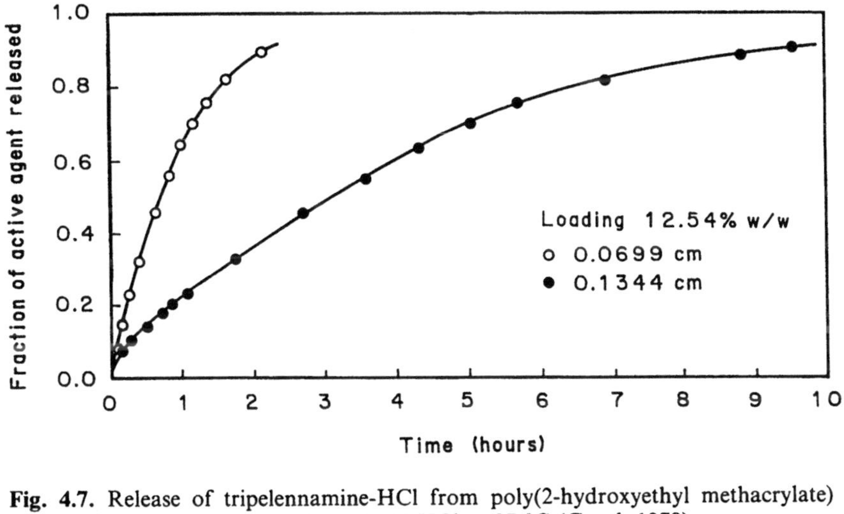

Fig. 4.7. Release of tripelennamine-HCl from poly(2-hydroxyethyl methacrylate) sheets of various thicknesses immersed in 0.007 N HCl at 37 °C (Good, 1978)

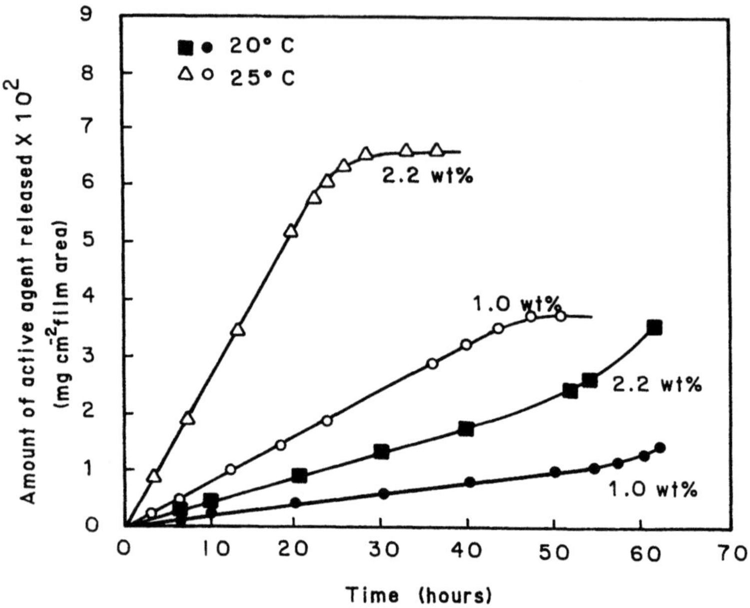

Fig. 4.8. Release of Sudan Red IV dye from polystyrene films immersed in *n*-hexane (Hopfenberg and Hsu, 1978)

cant portion of the release data is linear with time, indicating that release in this period is not governed by diffusion but by the rate of gelatinization (Fig. 4.7).

The capability of swelling-controlled release devices to produce zero-order release rates has been well demonstrated for release of Sudan Red IV dye from polystyrene (Hopfenberg and Hsu, 1978). The matrices are prepared by dissolution of dye and polymer in a solution, homogenization, and evaporation of the solvent, trichloroethylene. Constant release rates of the dye are obtained from various film thicknesses and loading (Fig. 4.8) when these devices are immersed in *n*-hexane. The diffusivity of dye in the gel layer is very high compared to that in the glassy polymer, and the rate of diffusion in the gel faster than the rate of movement of the gel–glassy polymer interface. Transport of *n*-hexane in polystyrene shows simple case II behavior unaffected by the presence of the dye. The latter part of some release curves in Fig. 4.8 reveals acceleration of the release. This is a consequence of the evolution of case II transport (of *n*-hexane) into super case II transport, thereby increasing the rate of gelatinization and consequently the rate of release of the dye.

Release of the active agent from swellable polymers as a function of time can also be expressed in the form analogous to Eq. (4.1)

$$\left(\frac{M_t}{M_\infty}\right)_a = k_a t^n \qquad (4.18)$$

where

M_t = amount of active agent released by time t

M_∞ = amount of active agent released at infinite time, theoretically corresponding to the total amount of active agent incorporated into the matrix, and
k_a, n = system parameters that depend on the nature of the polymer/penetrant/active agent interaction.

The parameter n can take a range of values, which indicate the type of active agent transport occurring in the matrix. The values of n and the corresponding release mechanisms are (Peppas, 1984)

$n = 0.5$; Fickian diffusion
$n = 1.0$; Pseudo-case II transport
$n > 1.0$; Pseudo-super-case II transport
$0.5 < n < 1.0$; Anomalous transport

Thus, the results of Lapidus and Lordi (1968) correspond to Fickian release, those of Good (1978) to release by an anomalous transport mechanism, and those of Hopfenberg and Hsu (1978) to the pseudo-case II mechanism. The latter parts of some release curves in Fig. 4.8 correspond to the pseudo-super-case II release mechanism.

When the active agent is effectively mobile only in the gel region, its rate of release is governed by the rate of gel formation and its diffusivity in the gel. The relative rate of gel formation and active agent mobility are characterized by a Peclet-type dimensionless group, called the swelling interface number, defined as (Hopfenberg et al., 1981; Korsmeyer and Peppas, 1981, 1983)

$$Sw = \frac{v' \delta_g(t)}{D_{ga}} \tag{4.19}$$

where v' is the velocity of the gel–glassy polymer interface, D_{ga} the diffusivity of active agent in the gel, and $\delta_g(t)$ the time-dependent thickness of the gel region. When the rate of active agent transport through the gel region is higher than that of movement of the gel-glassy polymer interface, we have $Sw \ll 1$; this corresponds to a pseudo-case II active agent transport mechanism. The release rate is then independent of time, if v' is a constant (case II). $Sw \gg 1$ represents the situation where the interface advances at a rate higher than the active agent transport rate in the gel. Diffusion, Fickian in nature, then occurs through a "quasi-equilibrium" swollen gel. For $Sw \approx 1$, anomalous transport of the active agent is observed (Korsmeyer and Peppas, 1983). The swelling interface number must be utilized, however, in combination with the Deborah number [Eq. (4.2)] to assess the nature of the release mechanism from a particular polymer/penetrant/active agent combination (Davidson and Peppas, 1986a, 1986b).

Among the polymers being examined for utilization in swelling-controlled release matrices are hydrogels, especially those based on the methacrylate group. Reviews of some of these hydrogels and their properties relevant to pharmaceutical application are available (Graham and McNeill, 1984; Roorda et al. 1986a; Baker, 1987). A series of studies has been made by Peppas and his coworkers on a variety of hydrogel and other polymeric systems to assess their performance for controlled active agent release (see, e.g., Franson and Peppas, 1983; Peppas and Franson, 1983; Korsmeyer and Peppas, 1984; Davidson and Peppas, 1986a, 1986b; Segot-Chicq and Peppas, 1986; Urdahl and Peppas, 1987). Korsmeyer and Peppas (1984) reported

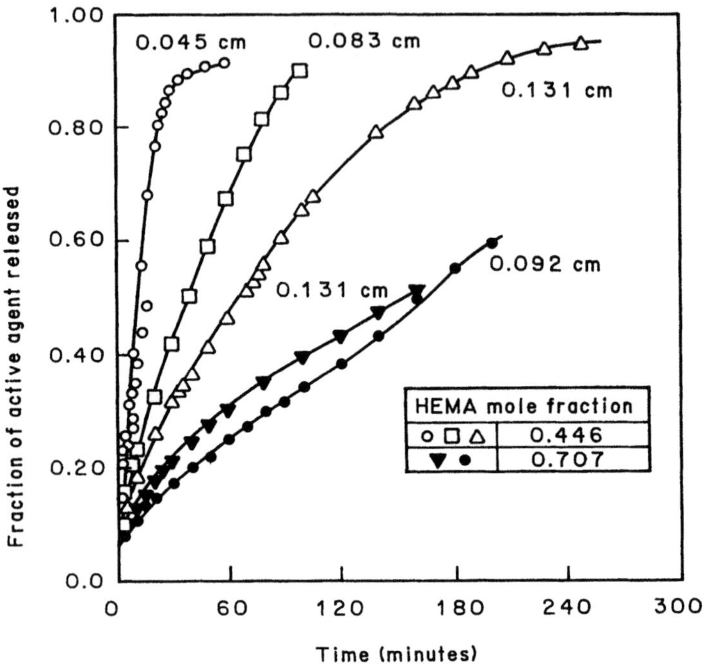

Fig. 4.9. Release of theophylline from poly(2-hydroxyethyl methacrylate-co-N, vinyl pyrrolidone) copolymers of varying thicknesses immersed in water at 37 °C (Korsmeyer and Peppas, 1984)

that matrices of poly(2-hydroxyethyl methacrylate-co-N, vinyl pyrrolidone) [P(HEMA-co-NVP)] exhibit a quasi-zero-order release of theophylline, with the rate dependent on the copolymer composition (Fig. 4.9). In an interesting study, Lee (1984a, 1984b) utilized non-uniform initial active agent distributions to obtain zero-order release from crosslinked poly(HEMA) matrices. The theoretical profiles illustrating the effect of initial active agent concentration distribution on the characteristics of release from spherical matrices are demonstrated in Fig. 4.10.

A novel approach to the utilization of swellable polymers to achieve zero-order release has been reported (Roorda et al., 1986b). The device does not belong to the category of conventional swelling-controlled release systems, but resorts to the solvent crazing and rupture phenomena in hydrogels to obtain zero-order release. The investigators immobilized oxprenolol-HCl in cross-linked poly(HEMA) matrices. On immersion in water, the glass-to-gel transition is initiated and the device begins to rupture. Rupturing increases the surface area available for release, thus counteracting the effect of increasing distance of diffusion and leading to near zero-order release over certain portions of the release period. The rupture formation is a function of the contents of the crosslink and active agent, providing a method to manipulate the extent of rupturing and, thus, the release rate. While Roorda et al. (1986b) reported good reproducibility, the sensitivity and the extreme nature of the phenomenon utilized in this approach could limit its applicability.

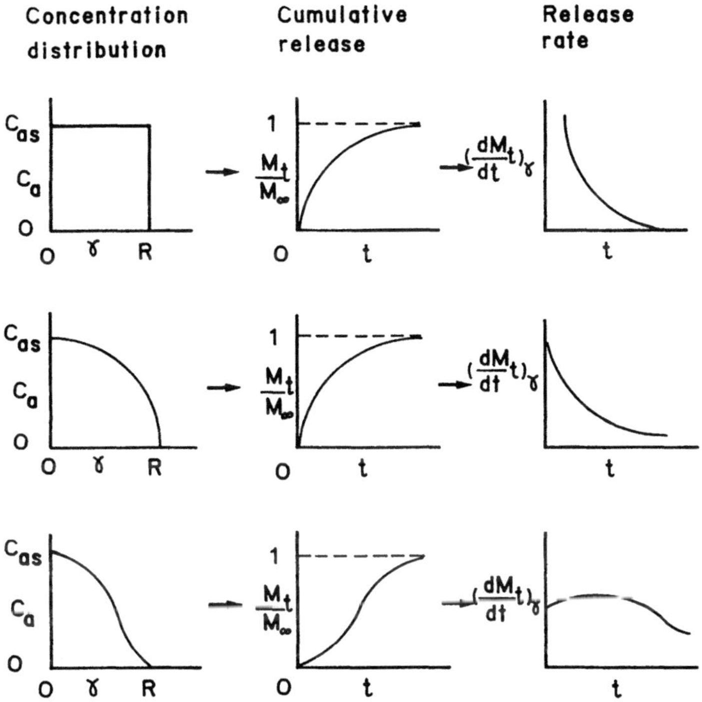

Fig. 4.10. Theoretical profiles illustrating the effect of initial active agent distribution on the character of release from spherical matrices (Lee, 1984b)

An important aspect of the active agent's release is its diffusivity in the polymeric gel. The free-volume theory was first applied to its analysis by Yasuda et al. (1964, 1969) and Yasuda and Lamaze (1971). The active agent (solute) in a gel diffuses through the continuous (liquid) phase constituting the gel, with the polymer chains serving as a sieve or screen. The ratio of diffusivity of active agent in the swollen (hydrated) polymer membrane, D_{ga}, to that in the pure liquid penetrant (water) D_{la} is given as

$$\frac{D_{ga}}{D_{la}} = \psi(q_a) \exp\left[- B\left(\frac{q_a}{V_{fl}}\right)\left(\frac{1}{H} - 1\right)\right] \tag{4.20}$$

where V_{fl} is the free volume of pure liquid (water), H is the degree of swelling (hydration), and B is a porportionality factor between the cross-section, q_c, and the characteristic volume of the hole required for movement of the active agent molecule. $\psi(q_a)$ represents the "sieve mechanism" by which small molecules are permitted to diffuse and larger molecules are not, owing to the absence of holes of appropriate size in the networks. Thus, $\psi(q_a)$ represents the probability of finding a hole of size q_a or greater; various distributions can be adopted to determine this probability (Yasuda et al., 1969). An increase in the size of the active agent leads to a rapid decrease in the diffusivity.

The free-volume theory of Yasuda and his coworkers (1969, 1971) has been extended to include the effect of crosslinks and number-average molecular weight between crosslinks in determining the size of the mesh or sieve opening (Peppas and Reinhart, 1983; Reinhart and Peppas, 1984); the resultant expression is

$$\frac{D_{ga}}{D_{la}} = k_1 \frac{\bar{M}_x - \bar{M}_x^*}{\bar{M} - \bar{M}_x^*} \exp\left(\frac{-r_a^2 k_2}{Q-1}\right) \tag{4.21}$$

where

$\bar{M}_x$ = number-average molecular weight between crosslinks,
$\bar{M}$ = number-average molecular weight for the original polymer before crosslinking,
$\bar{M}_x^*$ = value of $\bar{M}_x$ below which the active agent does not diffuse,
r_a = characteristic radius of the active agent,
Q = swelling ratio, given as the reciprocal of the polymer volume fraction in the swollen gel, and
k_1, k_2 = characteristic constants of the system.

A slightly modified expression has been provided for "moderately" swollen networks (Peppas and Moynihan, 1985)

$$\frac{D_{ga}}{D_{la}} = k_1 \xi \exp\left[-\frac{k_2 r_a^2 \xi}{\bar{M}^2(Q-1)}\right] \tag{4.22}$$

where ξ is the mesh size representing the space between four tetrafunctional crosslinks or four entanglements in the membrane.

An experimental study to identify factors influencing the diffusivity of active agent in a swollen membrane has been reported (Paul et al., 1976). An overview of the theories of active agent transport in polymeric membranes, including the influence of polymer structure on it, has been provided by Peppas and Meadows (1983).

4.3 Models for Swelling-Controlled Release Systems

For convenience, the available models for swelling-controlled systems are classified here into one-region and two-region models. A one-region model visualizes the polymer as a single phase with two species being transported in it. A two-region model portrays the polymer as consisting of distinct gel and glassy regions, the diffusivities of the two species being different in each region. The principal features of various important models are summarized in Table 4.1. The majority of the models consider the transport of both the penetrant and active agent as being governed by the Fickian diffusion process. Thus, the general governing equations of these models are of the form

$$\frac{\partial C_i}{\partial t} = \nabla \cdot (D \, \nabla C_i) \tag{4.23}$$

4.3.1 One-Region Models

A polymer matrix immersed in a well-stirred infinite medium has been analyzed by resorting to the so-called refined heat balance method (Lee, 1981). By assuming that diffusion of the penetrant obeys the Fick's law in one dimension, and that a linear relationship exists between the change in dimensions of the matrix and the extent of penetrant uptake, the position of the moving polymer-medium interface is defined by

$$S(t) = \delta/2 + Lt^{1/2} \tag{4.24}$$

Table 4.1. Swelling-controlled devices

Controlling resistances	System description	Model description and solution type	Reference
Fickian diffusion of penetrant	Infinite well agitated sink, dissolved active agent ($C_{a0} \leqq C_{as}$)	Semi-infinite slab no volume change, single polymeric region diffusivities functions of concentration (free volume relationships), analytic	Frisch (1978)
	Infinite unagitated sink, dissolved active agent ($C_{a0} \leqq C_{as}$)	Semi-infinite slab no volume change, distinct gel-polymer regions, analytic; with volume change, analytic	Rudolph (1979) Rudolph (1980)
	Infinite well-agitated sink, dissolved active agent ($C_{a0} \leqq C_{as}$)	Rectangular geometry one face exposed, changing volume, distinct gel-polymer regions, analytic	Peppas et al. (1980)
	Infinite well-agitated sink, dissolved active agent ($C_{a0} \leqq C_{as}$)	Rectangular geometry one face exposed, changing volume, single polymeric region, numerical, analytic for early part of release	Wang et al. (1980)
	Infinite well-agitated sink, dispersed active agent ($C_{a0} > C_{as}$)	Rectangular geometry changing volume, distinct gel-polymer regions, approximate analytic	Lee (1981)
	Infinite well-agitated sink, dissolved active agent ($C_{a0} \leqq C_{as}$)	Rectangular geometry one face impermeable, changing volume, single region, diffusivities from free volume theory, numerical	Vrentas et al. (1985)
	Infinite well-agitated sink, dissolved active agent ($C_{a0} \leqq C_{as}$)	Rectangular geometry changing volume, single region, diffusivities functions of concentration	Korsmeyer et al. (1986a)

Table 4.1 (continued)

Controlling resistances	System description	Model description and solution type	Reference
		(free volume relationships), one-dimensional expansion followed by isotropic three-dimensional expansion, numerical	
Anomalous diffusion	Infinite well-agitated sink, dissolved active agent ($C_{a0} \leqq C_{as}$)	Rectangular geometry, changing volume, two regions, diffusivity step function of penetrant concentration at critical gelatinization concentration, instantaneous volumetric expansion: one-dimensional followed by isotropic three-dimensional, convective drift created by stress, reducible to Fickian and case II, numerical	Singh and Fan (1986)
Fickian diffusion of penetrant	Infinite well-agitated sink, dissolved active agent ($C_{a0} \leqq C_{as}$)	Rectangular geometry changing volume, two regions, diffusivities functions of penetrant concentration (free volume relationships), relaxation-controlled volumetric expansion: one-dimensional followed by isotropic three-dimensional, position of gel-glassy polymer interface based on mechanical equilibrium criterion, numerical	Lustig and Peppas (1987)

where $(\delta/2)$ is the half thickness of the matrix, and L, a proportionality constant. The equation is obtained from the observation that the amount of penetrant absorbed into the polymer is proportional to $t^{1/2}$ when transport is governed by Fickian diffusion. A change in dimensions of the polymer is assumed to have no effect on the properties of the polymer, and thus separate regions are not defined.

Transport of the active agent is also visualized as governed by the one-dimensional Fickian diffusion equation (Lee, 1981)

$$\frac{\partial C_a}{\partial t} = \frac{\partial}{\partial x}\left(D_a \frac{\partial C_a}{\partial x}\right) \tag{4.25}$$

with the boundary conditions similar to those for a monolithic dispersed active agent system ($C_{a0} > C_{as}$). Thus at the surface of the matrix, $S(t)$,

$$C_a[S(t), t] = 0 \tag{4.26a}$$

and at the diffusion front, $R(t)$,

$$C_a[R(t), t] = C_{as} \tag{4.26b}$$

and

$$D_a \left. \frac{\partial C_a}{\partial x} \right|_{R(t)} = (C_{a0} - C_{as}) \frac{dR(t)}{dt} \tag{4.26c}$$

where, D_a is the diffusivity of the active agent in the matrix, C_{a0} the initial active agent loading, and C_{as} the saturation solubility of the active agent in the polymer. Solution of these governing equations gives the fraction of active agent released to be a function of the square-root of time, i.e. (Lee, 1981),

$$\frac{M_t}{M_\infty} = \left[\beta\left(1 - \frac{1}{2}\frac{C_{as}}{C_{a0}}\right) - \left(\frac{L}{2\sqrt{D_a}}\right)\right]\sqrt{\tau} \tag{4.27}$$

where

$$\beta = \frac{-\left(1 - \frac{C_{a0}}{C_{as}}\right)\frac{L}{2\sqrt{D_a}} + \left[\left(1 - \frac{C_{a0}}{C_{as}}\right)^2\left(\frac{L}{2\sqrt{D_a}}\right)^2 + 2\frac{C_{a0}}{C_{as}} - \frac{4}{3}\right]^{1/2}}{\frac{C_{a0}}{C_{as}} - \frac{2}{3}} \tag{4.28a}$$

and

$$\tau = \frac{4D_a t}{\delta^2}\,. \tag{4.28b}$$

An analysis of the migration of two species (water and monomer) in a plastic container wall has been presented by Wang et al. (1980). Results of this analysis may be adapted for modeling controlled release by taking water as the penetrant and the monomer as the active agent. The model thus describes a planar matrix without end effects, with transport of the penetrant into the matrix from the surfaces at $x' = 0$ and $x' = L(\tau)$. The diffusion fluxes are described by constitutive equations for the classical Fickian diffusion. These are

$$\frac{\partial y_c}{\partial \tau} = \left(\frac{D_c}{D_a}\right)\frac{\partial^2 y_c}{\partial x'^2} - \left(\frac{V\delta}{D_a}\right)\frac{\partial y_c}{\partial x'} \tag{4.29}$$

and

$$\frac{\partial y_a}{\partial \tau} = \frac{\partial^2 y_a}{\partial x'^2} - \left(\frac{V\delta}{D_a}\right)\frac{\partial y_a}{\partial x'}, \tag{4.30}$$

where y_c and y_a are the mass fraction of penetrant and active agent in the matrix, respectively. V is the mass velocity defined by (Wang et al., 1980)

$$V = -\frac{D_a}{\delta}\left[\frac{\frac{\partial y_a}{\partial x'} + \left(\frac{D_c}{D_a}\right)\frac{\partial y_c}{\partial x'}}{1 - y_a - y_c}\right]_{x'=0} \tag{4.31}$$

and δ is the original thickness of the matrix. Absorption of the penetrant is assumed to swell the matrix in proportion to the amount absorbed. Thus the surface at $x' = L(\tau)$ moves at a rate given by

$$\frac{dL}{d\tau} = \left[\frac{\frac{\partial y_a}{\partial x'} + \left(\frac{D_c}{D_a}\right)\frac{\partial y_c}{\partial x'}}{1 - y_a - y_c}\right]_{x'=L} + \frac{V\delta}{D_a}. \tag{4.32}$$

The polymer properties, however, are assumed to be unchanged, and thus no internal fronts are considered. In the above equations

$$x' = \frac{x}{\delta} \tag{4.33a}$$

and

$$\tau = \frac{D_a t}{\delta^2}. \tag{4.33b}$$

The equilibrium at each surface of the matrix is characterized by a partition coefficient for each species. The sink at $x' = 0$ is considered to be finite and well stirred while that at $x' = L$ to be infinite and well stirred.

The two partial differential equations [(4.29) and (4.30)] have been solved by an implicit finite difference technique in conjunction with an iteration method since the equations are coupled. An analytic solution has also been presented for the early stages of transport by approximating the geometry of the system with a semi-infinite slab and considering the boundary condition at $x' = 0$ to be invariant (Wang et al., 1980).

Frisch (1978) has considered the simultaneous nonlinear Fickian diffusion of active agent and penetrant in a semi-infinite polymer slab. The polymer is assumed to be immersed in an infinite well-stirred medium and the volume change due to sorption of penetrant is assumed negligible. The two species migrate through purely diffusive mechanisms; thus the governing differential equations are

$$\frac{\partial C_c}{\partial t} = \frac{\partial}{\partial x}\left(D_c \frac{\partial C_c}{\partial x}\right) \tag{4.34}$$

and

$$\frac{\partial C_a}{\partial t} = \frac{\partial}{\partial x}\left(D_a \frac{\partial C_a}{\partial x}\right). \tag{4.35}$$

These equations are coupled through the concentration dependence of D_c and D_a. The expressions for D_c and D_a, arising from a "Fujita-like" free-volume treatment are

$$D_c \approx D_c^0 \exp\left[\beta(C_c - C_{cs})\right] \tag{4.36a}$$

and

$$D_a \approx \alpha D_c^0 \exp\left[a\beta(C_c - C_{cs})\right]. \tag{4.36b}$$

Here C_{cs} is the equilibrium solubility of penetrant in polymer, while D_c^0 and αD_c^0 are the values of D_c and D_a respectively, at $x = 0$ ($C_c = C_{cs}$). Analytic solutions for Eqs. (4.34) and (4.35) have been presented (Frisch, 1978) and the effect of various simplifications assessed. These include the absence of active agent concentration dependence in the diffusion coefficient, the absence of cross diffusion terms, and the assumption of well-stirred medium. Numerical simulation results of a model employing similar functional forms for diffusivities have also been given by Korsmeyer (1983).

An analysis of a devolatilization scheme to enhance the removal of a volatile impurity from polymers has been presented by Vrentas et al. (1985). The scheme involves the introduction of a second solvent (penetrant) into the polymer, thus loosening the polymer matrix and promoting more rapid transport of the contaminant molecules. By considering the contaminant as an active agent, the analysis can be extended to the modeling of controlled release in a polymer–penetrant system.

The system analyzed consists of a flat polymer matrix (no end effects) with one surface impermeable to transport. Migration of the two species is purely (Fickian) diffusive; thus the governing differential equations are as given above, Eqs. (4.34) and (4.35). Equilibrium at the matrix-medium interface is assumed to be instantaneous. The volume-average velocity is taken to be zero everywhere in the matrix, and the thickness of the matrix changes in proportion to the amounts absorbed and desorbed. Thus, the rate of change of thickness is given by the jump mass balance at the interface

$$\frac{dX}{dt} = \left[\frac{\hat{v}_a D_a \dfrac{\partial C_a}{\partial x} + \hat{v}_c D_c \dfrac{\partial C_c}{\partial x}}{1 - C_c(t)\,\hat{v}_c}\right]_{x=X(t)} \tag{4.37}$$

where $\hat{v}_i$ is the partial specific molar volume of component i and is taken to be equal to the specific molar volume of the pure component. Assuming that the concentrations of penetrant and active agent in the matrix are small, the diffusivities can be equated to the expressions for their self-diffusion coefficients obtained from the free-volume theory of transport (Vrentas et al., 1985).

The model has been simulated numerically and the effect of various parameters including the free-volume properties, studied. It is shown that addition of a faster diffusing penetrant (solvent) has a very pronounced effect on transport of the active agent (contaminant) through an increase in the free volume available in the system (Vrentas et al., 1985). These results can be useful in controlled release since by an analysis of the (free volume) transport properties of the active agent, polymer, and fluid in the target medium, suitable combinations of polymer-active agent may be identified.

Recently, Korsmeyer et al., (1986a, 1986b) have presented a model for active agent and penetrant diffusion in swellable polymers. The process has been modeled as two-component diffusion (non-ideal Fickian) in a continuous medium, with structural features of the polymer and non-ideal concentration effects incorporated in the diffusion coefficients. These diffusion coefficients are given by "Fujita-like" free-volume expressions, Eqs. (4.36a) and (4.36b), while the governing differential equations are given by Eqs. (4.34) and (4.35). The swelling process is initially restricted to one dimension by the glassy core; when the concentration of penetrant at the center of the matrix reaches a critical level, the matrix relaxes to an isotropically swollen state and further swelling is three-dimensional. These dimensional changes are accounted for by allowing each spatial element to expand according to the amount of penetrant contained in it, during the simulation (Korsmeyer et al., 1986a).

The model has been simulated numerically by resorting to implicit and explicit finite difference techniques. Results have been presented as plots of concentration profiles, dimensional changes, fraction penetrant uptake, and fraction active agent released as functions of time. The shapes of these plots are quite similar to those obtained from swelling experiments on P(HEMA-co-NVP), loaded with theophylline, in water. The model has been fitted to the experimental data through adjustment of the parameters. The active agent release rate and the extent of swelling cannot be simultaneously fitted by a single set of parameters for any one particular experiment. However, the set of parameters rendering the model to fit to the swelling behavior, can also describe adequately the active agent release rates if the data from all experiments (including those with different copolymer ratios) are pooled (Korsmeyer et al., 1986b).

4.3.2 Two-Region Models

Experimental observations have confirmed that two separate regions are often formed when a polymer absorbs a thermodynamically compatible penetrant and undergoes macromolecular relaxation. These are the relaxed gel-like region and the region of unrelaxed glassy polymer (see, e.g. Kwei and Zupko, 1969; Baird et al., 1971; Hopfenberg and Hsu, 1978; Thomas and Windle, 1978a, 1978b; Franson and Peppas, 1983). Diffusivities of the migrating species may be substantially different in the two regions due to the differences in their structure. Anomalous diffusion may also be observed due to stress effects that accompany the relaxation phenomena. The mathematical treatment of such a system, therefore, involves a moving boundary or boundaries. The moving boundary is the gel–glassy polymer interface if relaxation is not accompanied by volume expansion; the moving boundaries are the gel-medium and gel–glassy polymer interfaces if volume expansion is significant. A governing differential equation need be written for each species in each region under isothermal conditions.

A two-region model is available for a semi-infinite polymer slab in an infinite unagitated medium (Rudolph, 1979). The volume of the slab is assumed to be unaffected by polymer relaxation in this model. Diffusion of the active agent and penetrant are described by four partial differential equations. These are:

$$\frac{\partial C_{gc}}{\partial t} = D_{gc}\frac{\partial^2 C_{gc}}{\partial x^2}, \qquad 0 < x \leqq X(t) \tag{4.38}$$

for penetrant in the gel;

$$\frac{\partial C_{ga}}{\partial t} = \frac{\partial}{\partial x}\left(D_{ga}\frac{\partial C_{ga}}{\partial x}\right), \qquad 0 < x \leqq X(t) \tag{4.39}$$

for active agent in the gel;

$$\frac{\partial C_{pa}}{\partial t} = D_{pa}\frac{\partial^2 C_{pa}}{\partial x^2}, \qquad X(t) < x < \infty \tag{4.40}$$

for active agent in glassy polymer; and

$$\frac{\partial C_{la}}{\partial t} = D_{la}\frac{\partial^2 C_{la}}{\partial x^2}, \qquad -\infty < x < 0 \tag{4.41}$$

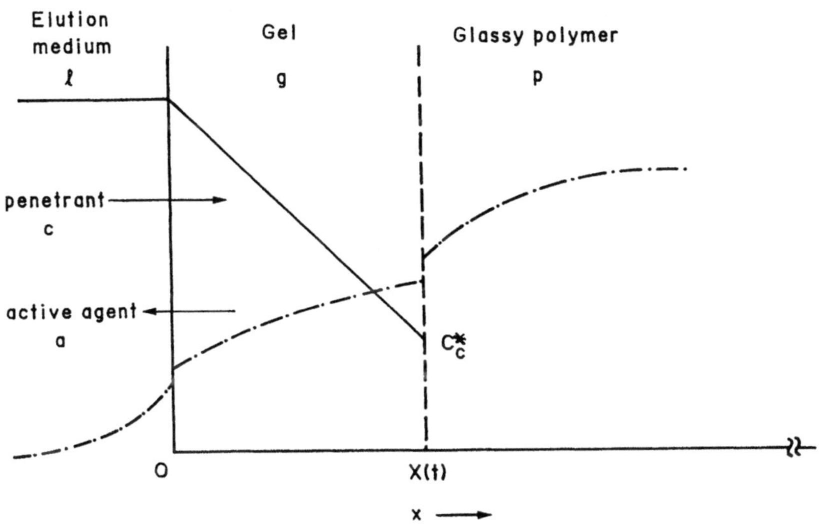

Fig. 4.11. Swelling-controlled system modeled by Rudolph (1979)

for active agent in the unagitated medium. In these expressions, $X(t)$ is the position of the gel-glassy polymer interface at any time t, the origin being placed on the gel-medium interface as shown in Fig. 4.11. Penetrant transport in the glassy polymer region is neglected. Diffusivity of active agent in gel, D_{ga}, is taken to be a function of the penetrant concentration, C_{gc}. Active agent and penetrant equilibria at the gel–glassy polymer and gel–medium interfaces are described by partition coefficients. The movement of the gel–glassy polymer interface at $X(t)$ is defined by a penetrant mass balance under the assumption that the penetrant does not diffuse in the glassy region. Thus

$$D_{gc} \left. \frac{\partial C_{gc}}{\partial x} \right|_{x = X(t)^-} = -C_c^* \left[\frac{dX(t)}{dt} \right] \tag{4.42 a}$$

where C_c^* is the critical concentration of the penetrant at the interface (see Fig. 4.11). The active agent mass balance at this interface is given by

$$D_{pa} \left. \frac{\partial C_{pa}}{\partial x} \right|_{x = X(t)^+} - D_{ga} \left. \frac{\partial C_{ga}}{\partial x} \right|_{x = X(t)^-} = [C_{ga} - C_{pa}]_{X(t)} \left[\frac{dX(t)}{dt} \right]. \tag{4.42 b}$$

Analytical solutions in terms of concentration of the two species and amount of the active agent released are available. Concentration profiles are also given for certain special cases. The concentrations of the two species are determined to be functions of $(x/t^{1/2})$, and the amount of active agent accumulated in the medium at any time is found to be proportional to $t^{1/2}$.

The above model was extended by Rudolph (1980) to include the effect of changing volume of the polymer. The governing differential equations for transport of active agent in this system are unchanged from the no-volume-change case above. However, the regions over which these are valid differ. Thus

$$\frac{\partial C_{ga}}{\partial t} = \frac{\partial}{\partial x} \left(D_{ga} \frac{\partial C_{ga}}{\partial x} \right), \qquad X_A(t) < x < X_B(t) \tag{4.43}$$

describes transport of active agent in the gel,

$$\frac{\partial C_{pa}}{\partial t} = D_{pa} \frac{\partial^2 C_{pa}}{\partial x^2}, \qquad X_B(t) < x < \infty \tag{4.44}$$

for active agent in the glassy polymer, and

$$\frac{\partial C_{la}}{\partial t} = D_{la} \frac{\partial^2 C_{la}}{\partial x^2}, \qquad -\infty < x < X_A(t) \tag{4.45}$$

for active agent in the unagitated medium. At time $t = 0$, $X_A(t)$ and $X_B(t)$ are coincident with the origin, i.e.,

$$X_A(0) = X_B(0) \equiv 0 \tag{4.46}$$

No equation is given for the transport of penetrant in gel. However, the concentration of penetrant in gel, C_{gc}, is taken to be proportional to $(x/t^{1/2})$ (see, Rudolph, 1979; reviewed above). The penetrant does not diffuse in the glassy polymer. Diffusivity of the active agent in gel, D_{ga}, is taken to be a function of the concentration of penetrant, C_{gc}. Active agent equilibria at the gel-medium interface, $X_A(t)$, and at the gel–glassy polymer interface, $X_B(t)$, are described by partition coefficients. The flux continuity expressions for active agent at the interfaces are given by

$$D_{la} \left. \frac{\partial C_{la}}{\partial x} \right|_{x=X_A} - D_{ga} \left. \frac{\partial C_{ga}}{\partial x} \right|_{x=X_A} = [C_{ga} - C_{la}]_{X_A} \left[\frac{dX_A}{dt} \right] \tag{4.47a}$$

and

$$D_{ga} \left. \frac{\partial C_{ga}}{\partial x} \right|_{x=X_B} - D_{pa} \left. \frac{\partial C_{pa}}{\partial x} \right|_{x=X_B} = [C_{pa} - C_{ga}]_{X_B} \left[\frac{dX_B}{dt} \right]. \tag{4.47b}$$

The above system of equations have been solved analytically by Rudolph (1980), by applying the similarity transform,

$$\eta = \frac{x}{2\sqrt{t}}.$$

Thus the rate of movement of both boundaries is proportional to $t^{1/2}$. The amount of active agent released is also proportional to $t^{1/2}$. Concentration profiles have been presented for certain special cases based on diffusivity values.

The case of Fickian diffusion in a rectangular matrix which undergoes significant volume expansion has been analyzed by Peppas et al. (1980). Only one face on the matrix is assumed to be exposed to the well-agitated infinite medium. A set of three partial differential equations describes the diffusive transport of the penetrant and active agent in the matrix. One is for the penetrant through the gel, one for the active agent in the gel, and one for the active agent in glassy polymer; these three equations are similar to Eqs. (4.38), (4.39), and (4.40) respectively. The position of the gel-glassy polymer interface, $X(t)$, is specified by (see Fig. 4.12)

$$\left. \frac{\partial C_{gc}}{\partial x} \right|_{x=X(t)} = 0 \tag{4.48}$$

The gel–medium interface is defined by a volume balance on the penetrant absorbed; the volume expansion of polymer is proportional to the volume of penetrant absorbed. By assuming that the diffusivity of the active agent in the glassy polymer is negligibly small compared to that in gel and thus reducing the problem to one of pseudo steady-state, analytic expressions have been obtained for the profile and release rate of the active agent. The release rate is again found to be proportional to $t^{1/2}$. The model has been verified experimentally for tablets of KCl dispersed in a cellulose derivative, placed in water.

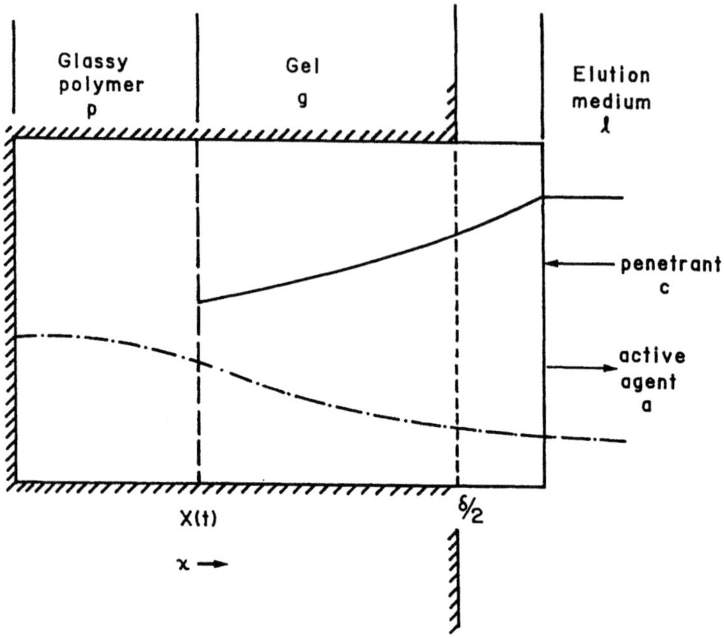

Fig. 4.12. Swelling-controlled system modeled by Peppas et al. (1980)

All the models considered so far describe the transport of penetrant and active agent by Fickian diffusion. Anomalous transport effects caused by the relaxation process have not been considered. One approach to account for these anomalous transport effects may be to resort to the generalized one-dimensional equation for transport of penetrant presented by Wang et al. (1969). The equation

$$\frac{\partial C_c}{\partial t} = \frac{\partial}{\partial x}\left[D_c(C_c, x, t)\frac{\partial C_c}{\partial x} - B(C_c, x, t)\, SC_c\right] \tag{4.10}$$

has been obtained by combining the mechanisms of (case I) Fickian and case II transport (see the discussion earlier in this chapter). For a constant B, Eq. (4.10) reduces to

$$\frac{\partial C_c}{\partial t} = \frac{\partial}{\partial x}\left(D_c \frac{\partial C_c}{\partial x} - vC_c\right) \tag{4.11}$$

where v ($= BS$) is interpreted as the convective velocity of solvent penetration arising as a direct consequence of the internal stress (Frisch et al., 1969; Kwei and Zupko, 1969; Kwei et al., 1972; Wang and Kwei, 1973). Thus it represents a pseudo-convective contribution to the transport process. When stress gradient dominates the process, the differential equation describes the case II mechanism with v as the velocity of the advancing gel–glassy polymer interface, also known as the case II velocity.

Equations (4.10) and (4.11) have been derived from consideration of the irreversible thermodynamics of transport processes based on the work of Bearman (1958), Bearman and Kirkwood (1958), and Frisch (1962). The concept, however, can be simply looked upon as one of a force field (internal stresses created by chain relaxation) inducing a linear drifting or convection of penetrant which is superimposed on its random motion (diffusion) (see, e.g., Feynman et al., 1963; Omar, 1978; Yariv, 1982). An analogous expression may also be obtained by stochastically modeling the diffusion process; the resultant Kolmogorov forward diffusion (Fokker-Planck) equation is

$$\frac{\partial}{\partial t} C_c(x, t) = \frac{\partial^2}{\partial x^2} [D_c(x, t)\, C_c(x, t)] - \frac{\partial}{\partial x} [V(x, t)\, C_c(x, t)] \tag{4.49}$$

(see, e.g., Bharucha-Reid, 1960; Seinfeld and Lapidus, 1974; Fan, 1979; van Kampen, 1981). Equation (4.11) is thus a special case of Eq. (4.49) when D_c is position invariant.

A general phenomenological model based on Eq. (4.11) has been proposed to describe the release of active agent from a swellable polymer (Singh and Fan, 1986). The system under consideration involves a rectangular monolithic matrix of thickness δ, with negligible end effects (flat plate) placed in a well-stirred infinite medium. The active agent "a" is initially dissolved in the polymer with a uniform concentration $C_{a0} \leqq C_{as}$, where C_{as} is the saturation concentration of active agent in the polymer. The system is depicted in Fig. 4.13. Absorption of the penetrant "c" from the external medium causes the polymer to undergo a glass to

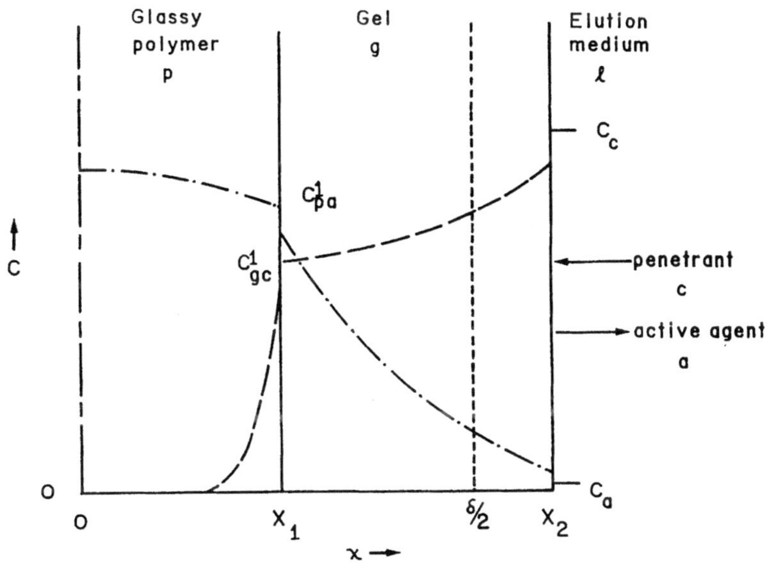

Fig. 4.13. Swelling-controlled system modeled by Singh and Fan (1986)

gel transition, creating two distinct regions in the matrix; the outer relaxed polymer or gel region and the inner core of the unrelaxed glassy polymer.

It has been assumed that a critical concentration of penetrant is required for the relaxation to occur (Peppas, 1984), and for the sake of generality the concentration of penetrant at the gel–glassy polymer interface is taken to be equal to or greater than this critical concentration. The concentration of penetrant on the glassy polymer side of the interface is related to that on the gel side through an equilibrium partition relationship. The analysis is simplified by restricting the relaxation phenomenon to occur only at the interface. The gel region is constrained by the glassy polymer core and the volume of the gel region at any instant is taken to be equal to its equilibrium volume under the constrained state, i.e., the volume change accompanying the glass-to-gel transition is assumed instantaneous. The glassy core also restricts the volumetric expansion to one dimension. The gel–glassy polymer interface moves inwards into the glassy polymer core as an increasing amount of penetrant is absorbed (Singh and Fan, 1986).

The three regions of the model described in the preceding paragraph are the medium (identified by l), the gel (g), and the unswollen polymer (p) as shown in Fig. 4.13. The coordinate system is fixed with the origin at the axis of symmetry of the matrix. The gel-glassy polymer ($g-p$) interface position is defined by $X_1(t)$, and the gel-medium ($g-l$) interface by $X_2(t)$. The concentration of species i in region j is denoted by $C_{ji}(x, t)$, and the diffusivity of i in region j by D_{ji}. The convection velocity in region j is designated as V_j. The differential equations describing the transport of penetrant ($i \equiv c$) and active agent ($i \equiv a$) in the different regions, as given by Singh and Fan (1986), are:

$$\frac{\partial C_{gi}}{\partial t} = -\frac{\partial}{\partial x}\left[-D_{gi}\frac{\partial C_{gi}}{\partial x} + V_g C_{gi}\right], \qquad X_1^+ \leqq x \leqq X_2^-, \qquad t > 0 \tag{4.50}$$

for transport in the gel region (g), and

$$\frac{\partial C_{pi}}{\partial t} = -\frac{\partial}{\partial x}\left[-D_{pi}\frac{\partial C_{pi}}{\partial x} + V_p C_{pi}\right], \qquad 0 \leqq x \leqq X_1^-, \qquad t \geqq 0 \tag{4.51}$$

for transport in the glassy polymer (p).

At time $t = 0$, the $g-p$ and $g-l$ interface are coincident with the surface of the matrix, i.e.,

$$X_1(0) = X_2(0) = \frac{\delta}{2}. \tag{4.52}$$

The other initial conditions are

$$C_{pc}(x, 0) = 0 \tag{4.53a}$$

and

$$C_{pa}(x, 0) = C_{a0}. \tag{4.53b}$$

The continuity of diffusing species concentrations on both sides of the $g - p$ and $g - l$ interface under the condition that they are in equilibrium, characterized by the equilibrium relationships, f_{1i} and f_{2i}, leads to the boundary conditions

$$C_{gi}(X_1^+, t) = C_{gi}^1 \equiv f_{1i}\{C_{pi}(X_1^-, t)\} \tag{4.54a}$$

and

$$C_{li}(X_2^+, t) = C_i \equiv f_{2i}\{C_{gi}(X_2^-, t)\} \tag{4.54b}$$

where C_i is the concentration of diffusing species in the medium and is taken to be constant. According to the earlier stated assumption regarding concentration of penetrant at the $g - p$ interface

$$C_{gc}(X_1^+, t) \geqq C_{gc}^*$$

where C_{gc}^* is the critical penetrant concentration for relaxation of polymer to occur. The boundary condition given by

$$\left[-D_{pi}\frac{\partial C_{pi}}{\partial x} + V_p C_{pi}\right]_{x=0} = 0 \tag{4.55}$$

implies that no transport occurs across the center of the matrix (symmetry criterion). Since equilibrium is assumed to prevail at the $g - p$ interface, as stated in connection with Eq. (4.54a)

$$C_{pi}(X_1^-, t) = C_{pi}^1 \equiv f_{1i}^{-1}(C_{gi}^1)\,. \tag{4.56}$$

Finally, the continuity of fluxes on both sides of the $g - p$ interface gives rise to

$$\left[-D_{pi}\frac{\partial C_{pi}}{\partial x} + V_p C_{pi}\right]_{X_1^-} + (C_{gi}^1 - C_{pi}^1)\, V_1(t) = \left[-D_{gi}\frac{\partial C_{gi}}{\partial x} + V_g C_{gi}\right]_{X_1^+} \tag{4.57}$$

where $V_1(t)$ is the velocity of the $g - p$ interface $X_1(t)$. Thus,

$$V_1(t) = \frac{dX_1(t)}{dt}\,. \tag{4.58}$$

The above set of equations is valid only while the glassy polymer region exists. If the time instant when the glassy polymer region vanishes is denoted by t^*,

$$X_1(t^*) = 0\,. \tag{4.59}$$

Then the equations are valid only for

$$t < t^*\,.$$

The above equations are applied to both the penetrant and the active agent, thus giving rise to four coupled partial differential equations. To solve these equations, however, the convective velocities V_g and V_p needed be related to the known parameters of the polymer–penetrant system (Singh and Fan, 1986). These velocities have been considered to comprise two terms: a stress-induced drift velocity v, created by the constraining effect of the glassy core in the region around the $g-p$ interface (Alfrey et al., 1966; Thomas and Windle, 1977), and a volume-change-induced drift (bulk) velocity.

In the generalized diffusion equation, equation (4.11), proposed by Wang et al. (1969) and modified by Kwei et al. (1972), v is termed the velocity of case II transport; its value can be shown to be equal to the constant velocity of the $g-p$ interface when penetrant transport is governed by the stress gradient only (case II transport) (Frisch et al., 1969). Thus, the stress induced drift velocity v has been considered to be characteristic of the polymer-penetrant system, defined under the constant volume condition. For the sake of generality, the value of this drift velocity has been denoted by v_p in the glassy polymer region and by v_g in the gel region. The definition of V_g and V_p may then be completed by defining the corresponding volume change-induced drift (bulk) velocities (Singh and Fan, 1986).

It has been assumed that the initial extent of swelling is the same as the equilibrium extent of swelling when the relaxation phenomenon is accompanied by a change in volume. This is equivalent to assuming that the volume expansion process at the $g-p$ interface occurs instantaneously. The actual rate of volume "generation" at the $g-p$ interface ($\equiv$ change in thickness) is then governed only by the rate of movement of this interface. The glassy polymer region remains stationary with respect to the origin since the volume expansion occurs on the gel side of the $g-p$ interface; the velocity of the thickness-change is apparent only in the gel region (see Fig. 4.13). This velocity is equal to the rate of outward movement of the $g-l$ interface at X_2; it is denoted by $V_2(t)$, i.e.,

$$V_2(t) = \frac{dX_2(t)}{dt} . \tag{4.60}$$

The two drift velocities, v and V_2, have now been defined, independent of each other. One is the stress-induced drift velocity under the condition that the relaxation process causes no change in volume, and the other is the volume-change-induced drift velocity based on the movement of the $g-l$ interface. In the general system, therefore, both these drift velocities must be accounted for when the relaxation phenomenon is accompanied by volume change. In the gel region, these two velocities are superimposed on each other; thus the effective convective velocity is

$$V_g = v_g + V_2(t) . \tag{4.61}$$

Since the volume-change-induced drift velocity, V_2, is not apparent in the glassy polymer region,

$$V_p = v_p . \tag{4.62}$$

The $g - l$ interface velocity, $V_2(t)$, has also been related to the velocity of the $g - p$ interface, $V_1(t)$, by an overall balance of the polymer mass in the gel region. In general, this balance can be written as

$$\left[\frac{\delta}{2} - X_1(t)\right] \varrho_p = [X_2(t) - X_1(t)]\, \varrho_g(t) \tag{4.63}$$

where $\varrho_g(t)$ is the time-dependent density of the gel region. However, when the volume change is instantaneous, as has been assumed in this model, $\varrho_g(t)$ is equivalent to the equilibrium density of gel constrained by glassy core, i.e.,

$$\varrho_g(t) = \varrho_{gs} \, .$$

Thus, Eq. (4.63) has been rewritten as

$$X_2(t) = \frac{\delta}{2}\alpha - (\alpha - 1)\, X_1(t) \tag{4.64}$$

where α is the expansion factor, defined as the ratio of density of glassy polymer to that of the gel, i.e.,

$$\alpha = \frac{\varrho_p}{\varrho_{gs}} \, . \tag{4.65}$$

Differentiating Eq. (4.64) with respect to time, the relation

$$V_2(t) = -(\alpha - 1)\, V_1(t) \tag{4.66}$$

is obtained. Substituting this relation into Eq. (4.61) gives

$$V_g = v_g - (\alpha - 1)\, V_1(t) \, . \tag{4.67}$$

At this point, two equations representing the overall instantaneous mass balances of the penetrant and active agent may be written as follows:

$$\begin{bmatrix} \text{flux of the species at} \\ \text{the } g - l \text{ interface} \end{bmatrix} = \begin{bmatrix} \text{rate of accumulation} \\ \text{of the species in the matrix} \end{bmatrix}$$

After appropriate substitutions, this has been rewritten as

$$\left[-D_{pi}\frac{\partial C_{pi}}{\partial x}\right]_{X_1^-} - \left[-D_{gi}\frac{\partial C_{gi}}{\partial x}\right]_{X_1^+} = V_1(C_{pi}^1 - \alpha C_{gi}^1) - v_p C_{pi}^1 + v_g C_{gi}^1 \, . \tag{4.68}$$

This expression is equivalent to Eq. (4.57) if V_p and V_g are substituted into it from Eqs. (4.62) and (4.67), respectively. Thus the instantaneous overall mass balance equation (4.68) can be employed to validate the jump boundary condition, Eq. (4.57), derived from the flux continuity (Singh, 1987).

The complete system of governing equations, therefore, comprises the following set for each of the diffusing species: the governing partial differential equations, Eqs. (4.50), (4.51); and the initial and boundary conditions, Eqs. (4.52) through (4.57) along with Eqs. (4.62), (4.64), and (4.67).

After

$$t = t^* ,$$

the entire polymer matrix is converted to gel. From this instant onward, the governing partial differential equations revert to simple Fickian diffusion forms. However, instead of the one-dimensional linear expansion observed before t^*, the volumetric expansion subsequent to the isotropic relaxation step at $t = t^*$, is three-dimensional. For a unit initial cross-sectional area, the length at t^* is

$$L^* = \alpha\delta \tag{4.69}$$

while the cross-sectional area remains unchanged due to the expansion being one-dimensional. When the gelatinized matrix relaxes to an isotropically swollen state at t^*, the length of the system will change to L' given by

$$L' = (L^*)^{1/3} , \tag{4.70}$$

the volume of the system thus having been conserved. Further volumetric expansion after t^* is accounted for by allowing the volume of the system to increase uniformly in proportion to the amount of penetrant absorbed. Thus the volume at time $t > t^*$ is given by (Singh and Fan, 1986)

$$\text{volume}\big|_{t>t^*} = \text{volume}\big|_{t=t^*} + \left\{\begin{bmatrix}\text{penetrant}\\ \text{accumulation}\end{bmatrix}_t - \begin{bmatrix}\text{penetrant}\\ \text{accumulation}\end{bmatrix}_{t^*}\right\}\begin{bmatrix}\text{specific volume}\\ \text{of penetrant in system}\end{bmatrix}$$

or

$$L^3(t) = L'^3(t^*) + 2\left[L^2(t)\int_0^{L(t)/2} C_{gc}(x, t)\,dx - 1\cdot\int_0^{L^*/2} C_{gc}(x, t^*)\,dx\right]v_c \tag{4.71}$$

where

$$v_c = \frac{(\alpha - 1)\,\delta}{2\int_0^{L^*/2} C_{gc}(x, t^*)\,dx} . \tag{4.72}$$

In finite difference terms, Eq. (4.71) may be rewritten as

$$L^3(t_{i+1}) = L^3(t_i) - \left[-2L^2 D_{gc}\left.\frac{\partial C_{gc}}{\partial x}\right|_L\right]_{t_i}\Delta t_i \cdot v_c . \tag{4.73}$$

The above equation has been used directly in the simulation procedure.

The governing equations of the model have been reduced to represent various special cases based on polymer–penetrant system behavior (Singh and Fan, 1986). These are:

i. Fickian diffusion without volume change: For this type of behavior, the flux contribution due to stress gradients has been neglected, i.e.,

$$v_g = v_p \equiv 0 .$$

Due to the unchanged volume, $\alpha = 1$. The governing differential equations revert to the simple Fickian diffusion form. Since the relaxation and swelling behavior is diffusion limited, the concentration of penetrant at the $g - p$ interface has been taken to be equal to the minimum required for the glass-to-gel transition to occur (= critical concentration).

Singh and Fan (1986) have compared their equations for this case to those presented by Crank (1975) and Rudolph (1979); their equations have been shown to reduce to the forms identical to those of Crank and Rudolph. A comparison with the equations by Peppas and Sinclair (1983) indicates that the penetrant continuity equations at the $g - p$ interface are not in agreement (Singh, 1987).

ii. Fickian diffusion with volume change: The simplifications for this type of behavior are similar to those of the previous case, except $\alpha \neq 1$. Singh and Fan (1986) compared their equations for penetrant transport for this case to those by Peppas and Sinclair (1983); they have found that the equations for penetrant transport and continuity are not in agreement (also see, Singh, 1987).

iii. Case II transport without volume change: Case II transport is a consequence of the fact that diffusion of the penetrant in the gel is much faster than the polymer chain relaxation process at the $g - p$ interface (Vrentas et al., 1975; Frisch, 1980). In the case of negligible diffusional resistance in the gel phase, this interface represents the innermost limit of dispersal of the penetrant (Frisch et al., 1969; Frisch, 1980). Experimentally observed concentration profiles also indicate little or no penetrant in the glassy region for case II transport (Thomas and Windle, 1978a, 1978b). Consequently, Singh and Fan (1986) have neglected penetrant transport in the unrelaxed glassy polymer region (no diffusion or convection). Since no change of volume occurs between the glass and gel phases, the value of α is one. The velocities V_g and V_p, therefore, are given by

$$V_g = v_g \tag{4.74a}$$

and

$$V_p = v_p \equiv 0 . \tag{4.74b}$$

The governing differential equations, Eqs. (4.50) and (4.51), become

$$\frac{\partial C_{gc}}{\partial t} = -\frac{\partial}{\partial x}(v_g C_{gc}) , \qquad X_1^+ \leqq x \leqq \frac{\delta^-}{2} ; \qquad t > 0 , \tag{4.75}$$

$$\frac{\partial C_{pc}}{\partial t} = 0 \qquad 0 \leqq x \leqq X_1^- ; \qquad t \leqq 0 , \tag{4.76}$$

$$\frac{\partial C_{ga}}{\partial t} = -\frac{\partial}{\partial x}\left(-D_{ga}\frac{\partial C_{ga}}{\partial x} + v_g C_{ga}\right), \qquad X_1^+ \leqq x \leqq \frac{\delta^-}{2}; \qquad t > 0, \tag{4.77}$$

$$\frac{\partial C_{pa}}{\partial t} = -\frac{\partial}{\partial x}\left(-D_{pa}\frac{\partial C_{pa}}{\partial x}\right), \qquad 0 \leqq x \leqq X_1^-; \qquad t \geqq 0. \tag{4.78}$$

The penetrant concentration at the $g - p$ interface is restricted to

$$C_{gc}(X_1^+, t) = C_{gc}^1 \tag{4.79}$$

and

$$C_{pc}(X_1^+, t) = C_{pc}^1 \equiv 0. \tag{4.80}$$

The simplifications made to obtain Eqs. (4.74a) and (4.74b) are: a) no penetrant flux exists in the glassy polymer region; b) only convective transport occurs in the gel region; c) $C_{pc}^1 = 0$; and d) $\alpha = 1$. Applying these simplifications to Eq. (4.57) for the penetrant, the relation

$$V_1(t) = v_g \tag{4.81}$$

is obtained. Substituting Eq. (4.81) along with Eqs. (4.74a) and (4.74b) into Eq. (4.57) for the active agent, the interface active agent flux continuity expression becomes

$$\left[-D_{pa}\frac{\partial C_{pa}}{\partial x}\right]_{X_1^-} - \left[-D_{ga}\frac{\partial C_{ga}}{\partial x}\right]_{X_1^+} = v_g C_{pa}^1. \tag{4.82}$$

iv. Case II transport with volume change: Derivation of the governing differential equations for this case is similar to that in the preceding case except that $\alpha \neq 1$. The appropriate simplifications are: a) no penetrant flux occurs in the polymer region; b) penetrant flux in the gel is only due to convective transport; and c) $C_{pc}^1 \equiv 0$. Therefore

$$V_g = v_g - (\alpha - 1)\, V_1(t) \tag{4.83a}$$

and

$$V_p = v_p \equiv 0. \tag{4.83b}$$

A relation between $V_1(t)$ and v_g has been derived by applying the simplifications presented in the preceding paragraph to Eq. (4.57). The resultant equation is

$$V_1(t) = \frac{v_g}{\alpha}. \tag{4.84}$$

The governing differential equations for this case are similar to those of the previous case and may be recovered by substituting (v_g/α) for v_g in Eqs. (4.75) through (4.78) and Eq. (4.82) (Singh and Fan, 1986).

v. Anomalous transport without volume change: The governing equations, Eqs. (4.50) and (4.51), represent the case of anomalous transport with volume change. Equations for the case without volume change may, therefore, be recovered by substituting $\alpha = 1$.

Singh and Fan (1986) have compared their equations for this case of $\alpha = 1$ with those presented by Peppas and Sinclair (1983); it has been shown that their penetrant flux continuity expression at the $g - p$ interface is not in agreement with that by Peppas and Sinclair (also see Singh, 1987).

The model of Singh and Fan (1986) has been simulated to examine the effect of the various parameters. The spatial coordinate in the model has been transformed to immobilize the moving boundaries and the simulation started from some small $t \approx 0^+$ in order to circumvent the singularity in the system at $t = 0$. Experimental data from Korsmeyer (1983), for the swelling and release behavior of P (HEMA-co-NVP) in water, have been fitted by this model through adjustment of the parameters. The results are illustrated in Figure 4.14a–c. The same set of parameters has been used to model the swelling and release data from one experiment. Agreement between experimental data and the model is good before t^*.

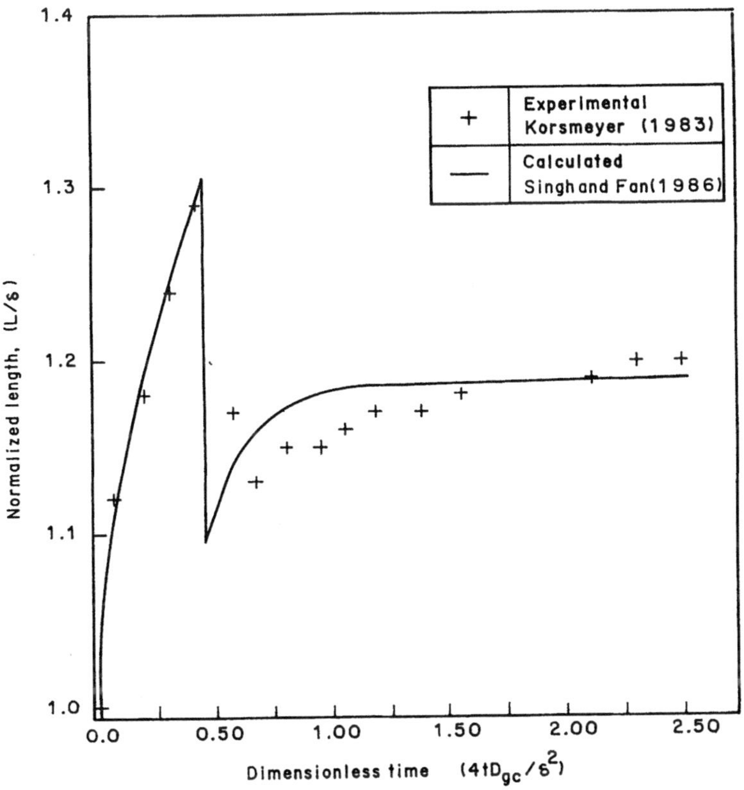

Fig. 4.14. a. Axial expansion of P (HEMA-co-NVP) with HEMA monomer fraction of 0.707, in water at 37 °C: Model parameters $\alpha = 1.31$, $D_{gc} = 3 \times 10^{-7}$ cm^2/s, $D_{pc} = 5 \times 10^{-8}$ cm^2/s, $v_g = v_p = 0$ (Singh and Fan, 1986)

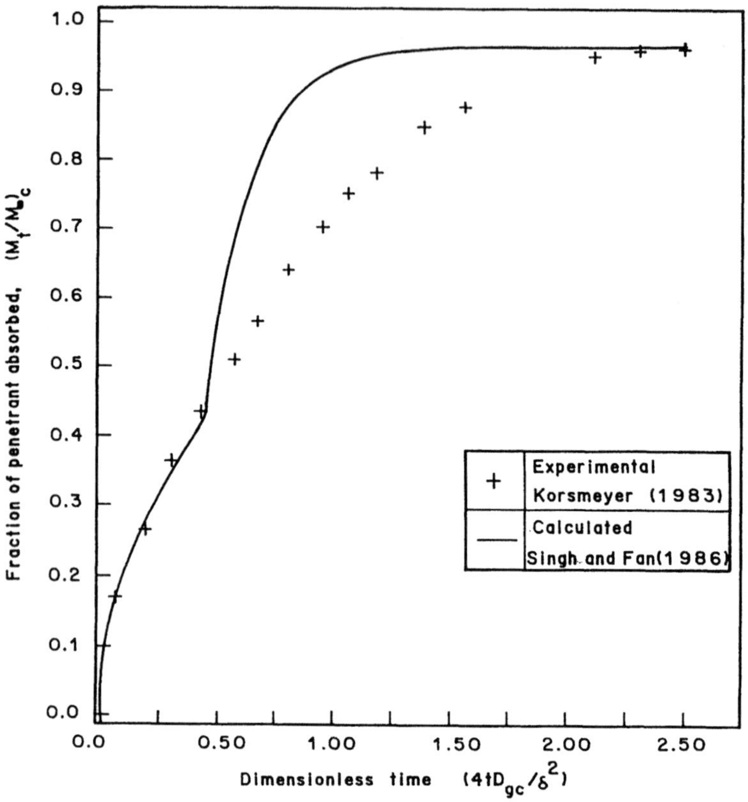

Fig. 4.14. b. Fraction of water absorbed by P(HEMA-co-NVP) with HEMA monomer fraction of 0.707 at 37 °C (Singh and Fan, 1986)

More recently, Lustig and Peppas (1987) have proposed a model for swelling-controlled systems which accounts for time-dependent volumetric expansion on the basis of a constitutive relation. The rate of polymer specific volume change is taken to be proportional to the difference between equilibrium and instantaneous specific volumes, $V_{eq,t}$ and V_t, i.e.,

$$\frac{dV_t}{dt} = \frac{V_{eq,t} - V_t}{\lambda} \tag{4.85}$$

where λ is the characteristic relaxation time, defined earlier in Eq. (4.5). The Fickian diffusion equation is applied to a rectangular slab kept with its faces exposed to the penetrant in an infinite well-stirred sink. The glassy polymer is uniformly loaded initially, and equilibrium is maintained at the outer surfaces at all times.

Diffusivities of penetrant, D_c, and active agent, D_a, are given in terms of functional forms derived from the free-volume theory

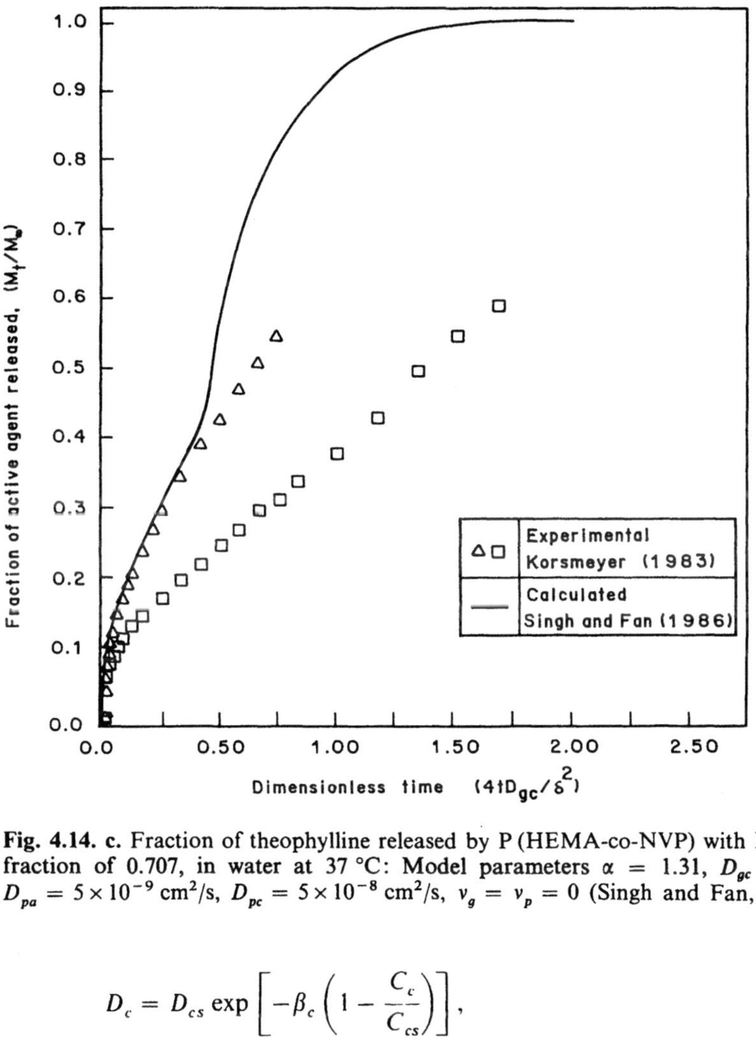

Fig. 4.14. c. Fraction of theophylline released by P (HEMA-co-NVP) with HEMA monomer fraction of 0.707, in water at 37 °C: Model parameters $\alpha = 1.31$, $D_{gc} = 3 \times 10^{-7}$ cm²/s, $D_{pa} = 5 \times 10^{-9}$ cm²/s, $D_{pc} = 5 \times 10^{-8}$ cm²/s, $v_g = v_p = 0$ (Singh and Fan, 1986)

$$D_c = D_{cs} \exp\left[-\beta_c\left(1 - \frac{C_c}{C_{cs}}\right)\right], \tag{4.86 a}$$

$$D_a = \frac{D_{as}}{C_{cs}} \exp\left[-\beta_a\left(1 - \frac{C_c}{C_{cs}}\right)\right], \tag{4.86 b}$$

where C_c is penetrant concentrant, C_{cs} the equilibrium penetrant concentration in swollen gel, D_{cs} and D_{as} are diffusivities of penetrant and active agent in equilibrium swollen gel, and β_c and β_a are material constants. Transport is restricted to the gel region and the position of the gel–glassy polymer interface is determined to be the point where local elongation is equal to the critical elongation required for the gelatinization phenomena, i.e., where local elongation is equal to the elongation at mechanical equilibrium swelling.

The model has been solved numerically using the "method of lines" and the seven independent parameters of the model analyzed (Lustig and Peppas, 1987). Deborah and swelling interface numbers are also obtained throughout the computation period. The main parameter influencing the performance of such a system is β_c, the material constant describing penetrant diffusivity as a function of penetrant concentration. The nature of solutions obtained in terms of penetrant concentration profiles, interface positions, and penetrant uptake is quite similar to that observed experimentally. The effect of active agent on penetrant sorption and polymer relaxation behavior is neglected due to an absence of coupling between β_a and penetrant sorption kinetics; this lack of interaction is true only for low active agent loadings.

Swelling-controlled release devices have great potential for controlled active agent delivery. Numerous experimental studies are being performed to determine the conditions leading to zero-order release from polymers undergoing relaxation and swelling. Mathematical modeling of these devices has been variously attempted by ideal Fickian diffusion (Peppas et al., 1980; Wang et al., 1980; Lee, 1981), non-ideal Fickian diffusion (Frisch, 1978; Rudolph, 1979, 1980; Vrentas et al., 1985; Korsmeyer et al., 1986a, 1986b; Lustig and Peppas, 1987), and anomalous transport (Singh and Fan, 1986). However, all these models isolate active agent transport from the penetrant transport and relaxational effects. Incorporation of this interaction by coupling of the theories of relaxation and transport with those for predicting diffusion coefficients should lead to an improved understanding of the behavior of swelling-controlled release devices, which will ultimately lead to better designs. Improved models may become intricate and cumbersome to use, but that is due to the complicated nature of the phenomenon itself.

References

Ahn, B. J., Smith, J. M., Parrish, W. R.: Non-Fickian Diffusion Through Polymer Films, AIChE J. *31*, 874–876 (1985)

Alfrey, T.: Glassy Polymer Diffusion is often Tractable. Chem. Eng. News *43* (41), 64–65 (1965)

Alfrey, T., Gurnee, E. F., Lloyd, W. G.: Diffusion in Glassy Polymers. J. Polym. Sci.: Part C, *12*, 249–261 (1966)

Astarita, G., Joshi, S.: Sample-Dimension Effects in the Sorption of Solvents in Polymers — A Mathematical Model, J. Membrane Sci. *4*, 165–182 (1978)

Astarita, G., Nicolais, L.: Physics and Mathematics of Heat and Mass Transfer in Polymers, Pure & Appl. Chem. *55*, 727–736 (1983)

Astarita, G., Sarti, G. C.: A Class of Mathematical Models for Sorption of Swelling Solvents in Glassy Polymers, Polym. Eng. Sci. *18*, 388–395 (1978)

Bagley, E., Long, F. A.: Two-Stage Sorption and Desorption of Organic Vapors in Cellulose Acetate, J. Am. Chem. Soc. *77*, 2172–2178 (1955)

Baird, B. R., Hopfenberg, H. B., Stannett, V. T.: The Effect of Molecular Weight and Orientation on the Sorption on n-Pentane by Glassy Polystyrene, Polym. Eng. Sci. *11*, 274–283 (1971)

Baker, R. W.: Controlled Release of Biologically Active Agents, pp. 179–184, John Wiley & Sons, New York (1987)

Bamba, M., Puisiuex, F., Marty, J.-P., Carstensen, J. T.: Release Mechanisms in Gel-forming Sustained Release Preparations, Int. J. Pharm. *2*, 307–315 (1979)

Bearman, R. J.: On the Linear Phenomenological Equations, J. Chem. Phys. *28*, 662–664 (1958)

Bearman, R. J., Kirkwood, J. G.: Statistical Mechanics of Transport Processes. XI. Equations of Transport in Multicomponent Systems, J. Chem. Phys. *28*, 136–145 (1958)

Berens, A. R., Hopfenberg, H. B.: Diffusion and Relaxation in Glassy Polymer Powders: 2. Separation of Diffusion and Relaxation Parameters, Polymer *19*, 489–496 (1978)

Bernier, G. A., Kambour, R. P.: The Role of Organic Agents in the Stress Crazing and Cracking of Poly(2,6-dimethyl-1,4-phenylene oxide), Macromolecules *1*, 393–400 (1968)

Bharucha-Reid, A. T.: Elements of the Theory of Markov Process and Their Applications, pp. 129–163, McGraw-Hill, New York (1960)

Cohen, D. S.: Theoretical Models for Diffusion in Glassy Polymers, J. Polym. Sci., Polym. Phys. Ed. *21*, 2057–2065 (1983)

Cohen, D. S.: Diffusive Fronts of Penetrants in Glassy Polymers, Physica *12D*, 369–374 (1984)

Cohen, D. S., Stanley, E. A.: Gaseous Diffusion in Glassy Polymers, SIAM J. Appl. Math. *43*, 949–970 (1983)

Crank, J.: Diffusion in Media with Variable Properties. Part III. Diffusion Coefficients which Vary Discontinuously with Concentration, Trans. Farad. Soc. *47*, 450–461 (1951)

Crank, J.: A Theoretical Investigation of the Influence of Molecular Relaxation and Internal Stress on Diffusion in Polymers, J. Polym. Sci. *XI*, 151–168 (1953)

Crank, J.: The Mathematics of Diffusion, pp. 290–293, Oxford University Press, London (1975)

Davidson III, G. W. R., Peppas, N. A.: Solute and Penetrant Diffusion in Swellable Polymers. V. Relaxation-Controlled Transport in P(HEMA-co-MMA) Copolymers, J. Controlled Release *3*, 243–258 (1986a)

Davidson III, G. W. R., Peppas, N. A.: Solute and Penetrant Diffusion in Swellable Polymers. VI. The Deborah and Swelling Interface Numbers as Indicators of the Order of Biomolecular Release, J. Controlled Release *3*, 259–271 (1986b)

Fan, L. T.: Mathematical Modeling and Process Optimization . In: Food Process Engineering, Vol. I, pp. 26–51, Food Processing Systems, P. Linko, Malkki, Y., Olkku, J. (Eds.), Applied Science Publishers, London (1979)

Feynman, R. P., Leighton, R. B., Sands, M.: The Feynman Lectures on Physics, pp. 43–44, Addison-Wesley Publishing Company, Reading, Massachusetts (1963)

Franson, N. M., Peppas, N. A.: Influence of Copolymer Composition on Non-Fickian Water Transport through Glassy Copolymers, J. Appl. Polym. Sci. *28*, 1299–1310 (1983)

Frisch, H. L.: Anomalous Polymer-Penetrant Permeation, J. Chem. Phys. *37*, 2408–2413 (1962)

Frisch, H. L.: Simultaneous Nonlinear Diffusion of a Solvent and Organic Penetrant in a Polymer, J. Polym. Sci. Polym. Phys. Ed. *16*, 1651–1664 (1978)

Frisch, H. L.: Sorption and Transport in Glassy Polymers — A Review, Polym. Eng. Sci. *20*, 2–13 (1980)

Frisch, H. L., Wang, T. T., Kwei, T. K.: Diffusion in Glassy Polymers II, J. Polym. Sci.: Part A-2, *7*, 879–887 (1969)

Good, W. R.: Diffusion of Water Soluble Drugs from Initially Dry Hydrogels, pp. 139–153. In: Polymeric Delivery Systems, Kostelnik, R. J. (Ed.), Gordon and Breach Science Publishers, New York (1978)

Gostoli, C., Sarti, G. C.: Diffusion and Localized Swelling Resistances in Glassy Polymers, Polym. Eng. Sci. *22*, 1018—1026 (1982)

Graham, N. B., McNeill, M. E.: Hydrogels for Controlled Drug Delivery, Biomaterials, *5*, 27—36 (1984)

Hopfenberg, H. B.: Anomalous Transport of Penetrants in Polymeric Membranes, pp. 16–32. In: Membrane Science and Technology: Industrial, Biological, and Waste Treatment Processes, J. E. Flinn (Ed.), Plenum Press, New York (1970)

Hopfenberg, H. B., Apicella, A., Saleeby, D. E.: Factors Affecting Water Sorption in and Solute Release from Glassy Ethylene-Vinyl Alcohol Copolymers, J. Membrane Sci. *8*, 273–282 (1981)

Hopfenberg, H. B., Frisch, H. L.: Transport of Organic Micromolecules in Amorphous Polymers, J. Polym. Sci. Polym. Letters, *B7*, 405–409 (1969)

Hopfenberg, H. B., Holley, R. H., Stannett, V.: The Effect of Penetrant Activity and Temperature on the Anomalous Diffusion of Hydrocarbons and Solvent Crazing in Polystyrene. Part I: Biaxially Oriented Polystyrene, Polym. Eng. Sci. *9*, 242–249 (1969)

Hopfenberg, H. B., Hsu, K. C.: Swelling-Controlled, Constant Rate Delivery Systems. Polym. Eng. Sci. *18*, 1186–1191 (1978)
Hopfenberg, H. B., Stannett, V. T., Folk, G. M.: Sorption Kinetics and Equilibria in Annealed Glassy Polyblends, Polym. Eng. Sci. *15*, 261–267 (1975)
Jacques, C. H. M., Hopfenberg, H. B., Stannett, V.: Super Case II Transport of Organic Vapors in Glassy Polymers, pp. 73–86. In: Permeability of Plastic Films and Coatings to Gases, Vapors, and Liquids, H. B. Hopfenberg (Ed.), Plenum Press, New York (1974)
Joshi, S., Astarita, G.: Diffusion-Relaxation Coupling in Polymers which Show Two-Stage Sorption Phenomena, Polymer *20*, 455–458 (1979)
Korsmeyer, R. W.: Solute and Penetrant Diffusion in Swellable Polymers, Ph. D. Dissertation, Purdue University, West Lafayette, Indiana, pp. 28–51 (1983)
Korsmeyer, R. W., Gurny, R., Doelker, E., Buri, P., Peppas, N. A.: Mechanisms of Solute Release from Porous Hydrophilic Polymers, Int. J. Pharm. *15*, 25–35 (1983)
Korsmeyer, R. W., Lustig, S. R., Peppas, N. A.: Solute and Penetrant Diffusion in Swellable Polymers. I. Mathematical Modeling, J. Polym. Sci., Polym. Phys. Ed. *24*, 395–408 (1986a)
Korsmeyer, R. W., Meerwall, E. V., Peppas, N. A.: Solute and Penetrant Diffusion in Swellable Polymers. II. Verification of Theoretical Models, J. Polym. Sci., Polym. Phys. Ed. *24*, 409–434 (1986b)
Korsmeyer, R. W., Peppas, N. A.: Effect of the Morphology of Hydrophilic Polymeric Matrixes on the Diffusion and Release of Water Soluble Drugs, J. Membrane Sci. *9*, 211–227 (1981)
Korsmeyer, R. W., Peppas, N. A.: Macromolecular and Modeling Aspects of Swelling-Controlled Systems, pp. 77–90. In: Controlled Release Delivery Systems, Roseman, T. J., Mansdorf, S. Z. (Eds.), Marcell Dekker, Inc., New York (1983)
Korsmeyer, R. W., Peppas, N. A.: Solute and Penetrant Diffusion in Swellable Polymers. III. Drug Release From Glassy Poly(HEMA-co-NVP) Copolymers, J. Controlled Release *1*, 89—98 (1984)
Kwei, T. K., Wang, T. T., Zupko, H. M.: Diffusion in Glassy Polymers. V. Combination of Fickian and Case II Mechanisms, Macromolecules *5*, 645–646 (1972)
Kwei, T. K., Zupko, H. M.: Diffusion in Glassy Polymers I, J. Polym. Sci: Part A-2, *7*, 867–877 (1969)
Lapidus, H., Lordi, N. G.: Drug Release from Compressed Hydrophilic Matrices, J. Pharm. Sci. *57*, 1292–1301 (1968)
Lee, P. I.: Controlled Drug Release from Polymeric Matrices Involving Moving Boundaries, pp. 39–48. In: Controlled Release of Pesticides and Pharmaceuticals, D. H. Lewis (Ed.), Plenum Press, New York (1981)
Lee, P. I.: Novel Approach to Zero-Order Drug Delivery via Immobilized Nonuniform Drug Distribution in Glassy Hydrogels, J. Pharm. Sci. *73*, 1344–1347 (1984a)
Lee, P. I.: Effect of Non-Uniform Initial Drug Concentration Distribution on the Kinetics of Drug Release from Glassy Hydrogel Matrices, Polymer *25*, 973–978 (1984b)
Long, F. A., Richman, D.: Concentration Gradients for Diffusion of Vapors in Glassy Polymers and their Relation to Time Dependent Diffusion Phenomena, J. Am. Chem. Soc. *82*, 513–519 (1960)
Lustig, S. R., Peppas, N. A.: Solute and Penetrant Diffusion in Swellable Polymers. VII. A Free Volume-Based Model with Mechanical Relaxation, J. Appl. Polym. Sci. *33*, 533—549 (1987)
McGregor, R.: Diffusion and Sorption in Fibers and Films, Vol. I, An Introduction with Particular Reference to Dyes, pp. 210–211, Academic Press, New York (1974)
Michaels, A. S., Bixler, H. J., Hopfenberg, H. B.: Controllably Crazed Polystyrene: Morphology and Permeability, J. Appl. Polym. Sci. *12*, 991–1007 (1968)
Nagy, M.: Some New Aspects of Research on Polymer Gels, Colloid and Polymer Sci. *263*, 245–265 (1985)
Neogi, P.: Anomalous Diffusion of Vapors Through Solid Polymers. Part I. Irreversible Thermodynamics of Diffusion and Solution Processes, AIChE J. *29*, 829–833 (1983)
Omar, M. A.: Elementary Solid State Physics: Principles and Applications, p. 143, Addison-Wesley Publishing Company, Reading, Massachusetts (1978).

Park, G. S.: The Glassy State and Slow Process Anomalies, pp. 141–163. In: Diffusion in Polymers, Crank, J., Parks, G. S. (Eds.), Academic-Press, London (1968)

Paul, D. R., Garcin, M., Garmon, W. E.: Solute Diffusion Through Swollen Polymer Membranes, J. Appl. Polym. Sci. *20*, 609–625 (1976)

Peppas, N. A.: Release of Bioactive Agents from Swellable Polymers: Theory and Experiments, pp. 279–289. In: Recent Advances in Drug Delivery Systems, Anderson, J. M., Kim, S. W. (Eds.), Plenum Publishing Corporation, New York (1984)

Peppas, N. A., Franson, N. M.: The Swelling Interface Number as a Criterion for Prediction of Diffusional Solute Release Mechanisms in Swellable Polymers, J. Polym. Sci., Polym. Phys. Ed. *21*, 983–997 (1983)

Peppas, N. A., Gurny, R., Doelker, E., Buri, P.: Modelling of Drug Diffusion through Swellable Polymeric Systems, J. Membrane Sci. *7*, 241–253 (1980)

Peppas, N. A., Meadows, D. L.: Macromolecular Structure and Solute Diffusion in Membranes: An Overview of Recent Theories, J. Membrane Sci. *16*, 361–377 (1983)

Peppas, N. A., Moynihan, H. J.: Solute Diffusion in Swollen Membranes. IV. Theories for Moderately Swollen Networks, J. Appl. Polym. Sci. *30*, 2589–2606 (1985)

Peppas, N. A., Reinhart, C. T.: Solute Diffusion in Swollen Membranes. Part I. A New Theory, J. Membrane Sci. *15*, 275–287 (1983)

Peppas, N. A., Sinclair, J. L.: Anomalous Transport of Penetrants in Glassy Polymers, Colloid and Polymer Sci. *261*, 404–408 (1983)

Peterlin, A.: Diffusion with Discontinuous Swelling. V. Type II Diffusion into Sheets and Spheres, J. Polym. Sci., Polym. Phys. Ed. *17*, 1741–1756 (1979)

Petropoulos, J. H., Roussis, P. P.: A Discussion of Theoretical Models of Anomalous Diffusion of Vapors in Polymers, pp. 219–232. In: Permeability of Plastic Films and Coatings to Gases, Vapors, and Liquids, Hopfenberg, H. B. (Ed.), Plenum Press, New York (1974)

Petropoulos, J. H., Roussis, P. P.: The Influence of Transverse Differential Swelling Stresses on the Kinetics of Sorption of Penetrants by Polymer Membranes, J. Membrane Sci. *3*, 343–356 (1978)

Rebenfeld, L., Makarewicz, P. J., Weignann, H.-D., Wilkes, G. L.: Interactions Between Solvents and Polymers in the Solid State, J. Macromol. Sci.-Rev. Macromol. Chem. *C15*, 279–393 (1976)

Reinhart, C. T., Peppas, N. A.: Solute Diffusion in Swollen Membranes. Part II. Influence of Crosslinking on Diffusive Properties, J. Membrane Sci. *18*, 227–239 (1984)

Roorda, W. E., Bodde, H. E., de Boer, A. G., Junginger, H. E.: Synthetic Hydrogels as Drug Delivery Systems, Pharmaceutisch Weekblad, Scientific Edition *8*, 165–189 (1986a)

Roorda, W. E., de Vries, M. A., Kosho, C., de Leede, L. G. T., de Boer, A. G., Junginger, H. E.: Zero-Order Release of Oxprenolol-HCl: A New Approach, pp. 206–207. In: Proceedings of the 13th International Symposium on Controlled Release of Bioactive Materials, I. A. Chaudry and C. Thies (Eds.), Norfolk, Virginia, August 3–6 (1986b)

Rudolph, F. B.: Diffusion in a Multicomponent Inhomogeneous System with Moving Boundaries. I. Swelling at Constant Volume, J. Polym. Sci., Polym. Phys. Ed. *17*, 1709–1718 (1979)

Rudolph, F. B.: Diffusion in a Multicomponent Inhomogeneous Multiphase System with Moving Boundaries. II. Increasing or Decreasing Volume (Swelling or Drying), J. Polym. Sci., Polym. Phys. Ed. *18*, 2323–2336 (1980)

Segot-Chicq, S., Peppas, N. A.: Solute and Penetrant Diffusion in Swellable Polymers. IV. Semicrystalline, Swelling-Controlled Release Systems of Poly(Ethylene-co-Vinyl Alcohol). J. Controlled Release *3*, 193–204 (1986)

Seinfeld, J. H., Lapidus, L.: Mathematical Models in Chemical Engineering, Vol. 3, pp. 225–229, Prentice Hall, Englewood Cliff, New Jersey (1974)

Sfirakis, A., Rogers, C. E.: Sorption and Diffusion of Alcohols in Amorphous Polymers, Polym. Eng. Sci. *21*, 542–547 (1981)

Singh, S. K.: Analysis and Development of Some Controlled Release Systems, Ph. D. Dissertation, Kansas State University, Manhattan, Kansas, pp. IV.19–IV.82 (1987)

Singh, S. K., Fan, L. T.: A Generalized Model for Swelling-Controlled Release Systems, Biotechnology Progress *2*, 145–156 (1986)

Thomas, N., Windle, A. H.: Discontinuous Shape Changes Associated with Case II Transport of Methanol in Thin Sheets of PMMA, Polymer *18*, 1195 (1977)
Thomas, N., Windle, A. H.: Case II Swelling of PMMA Sheet in Methanol, J. Membrane Sci. *3*, 337–342 (1978a)
Thomas, N., Windle, A. H.: Transport of Methanol in Poly(Methyl Methacrylate), Polymer *19*, 255–265 (1978b)
Thomas, N. L., Windle, A. H.: A Theory of Case II Diffusion, Polymer *23*, 529–542 (1982)
Tosun, I., Yilmaz, L.: On the Fickian Diffusion of Organic Penetrants in Glassy Polymers, Polym. Eng. Sci. *22*, 1043–1045 (1982)
Tu, Y.-O.: A Multi-Phase Stefan Problem Describing the Swelling and the Dissolution of Glassy Polymer, Quart. Appl. Math. *XXXV*, 269–285 (1977)
Tu, Y.-O., Ouano, A. C.: Model for the Kinematics of Polymer Dissolution, IBM J. Res. Develop. *12*, 197–208 (1977)
Urdahl, K. G., Peppas, N. A.: Anomalous Penetrant Transport in Glassy Polymers. V. Cyclohexane Transport in Polystyrene, J. Appl. Polym. Sci. *33*, 2669–2687 (1987)
van Kampen, N. G.: Stochastic Process in Physics and Chemistry, North Holland, Netherlands (1981)
Vieth, W. R., Howell, J. M., Hsieh, J. H.: Dual Sorption Theory, J. Membrane Sci. *1*, 177–220 (1976)
Vieth, W. R., Sladek, K. J.: A Model for Diffusion in a Glassy Polymer, J. Colloid Sci. *20*, 1014–1033 (1965)
Vrentas, J. S., Duda, J. L.: Diffusion in Polymer-Solvent Systems. III. Construction of Deborah Number Diagrams, J. Polym. Sci. Polym. Phys. Ed., *15*, 441–453 (1977)
Vrentas, J. S., Duda, J. L., Ling, H. C.: Enhancement of Impurity Removal from Polymer Films, J. Appl. Polym. Sci. *30*, 4499–4516 (1985)
Vrentas, J. S., Jarzebski, C. M., Duda, J. L.: A Deborah Number for Diffusion in Polymer-Solvent Systems, AIChE J. *21*, 894–901 (1975)
Wang, F. H. L., Duda, J. L., Vrentas, J. S.: Analysis of Impurity Migration in Plastic Containers, Polym. Eng. Sci. *20*, 120–127 (1980)
Wang, T. T., Kwei, T. K.: Diffusion in Glassy Polymers. Reexamination of Vapor Sorption Data, Macromolecules *6*, 919–921 (1973)
Wang, T. T., Kwei, T. K., Frisch, H. L.: Diffusion in Glassy Polymers. III, J. Polym. Sci. Part A-2, *7*, 2019–2028 (1969)
Windle, A. H.: Case II Sorption, pp. 75–118. In: Polymer Permeability, J. Comyn (Ed.), Elsevier Applied Science Publishers, London (1985)
Yariv, A.: An Introduction to Theory and Applications of Quantum Mechanics, pp. 253–259, John Wiley and Sons, New York (1982)
Yasuda, H., Lamaze, C. E.: Permeability of Solutes in Homogeneous Water-Swollen Polymer Membranes, J. Macromol. Sci. Phys. *85*, 111–134 (1971)
Yasuda, H., Peterlin, A., Colton, C. K., Smith, K. A., Merril, E. W.: Permeability of Solutes Through Hydrated Polymer Membranes. Part III. Theoretical Background for the Selectivity of Dialysis Membranes, Die Makromolekulare Chemie *126*, 177–186 (1969)
Yasuda, H., Stannett, V., Frisch, H. L., Peterlin, A.: The Permeability of Strained Polymer Films, Die Makromolekulare Chemie *73*, 188–202 (1964)
Yilmaz, L., Tosun, I., Gürkan, T., Gülcat, U.: Anomalous Diffusion of Liquids in Glassy Polymers, Mathematical Modelling *4*, 535–543 (1983)

5. Special Controlled-Release Systems

In this chapter, we shall briefly examine some controlled-release systems not widely modeled mathematically. Nevertheless, these systems are interesting since they offer a wide range of performance capabilities, and also constitute interesting modeling problems.

5.1 Osmotic Pumping Devices

Pumps of various configurations have been produced that exploit osmotic pressure as the driving force. Most of the research in this area has focused on drug delivery in the form of implantable or oral formulations. A simple osmotic pump consists of a core of compacted, solid, water-soluble active agent. The core is enclosed in a polymer membrane which is selectively water (or appropriate biological fluid) permeable but active agent impermeable, and in which a small orifice is drilled. In an aqueous environment, water is transported into the core by permeation through this membrane. If this membrane is non-extensible or is suitably supported, the hydrostatic pressure inside the core will build up, and a solution of active agent and water will be pumped out through the orifice (see, e.g., Bruck, 1983). The movement of water through the semipermeable membrane is governed by the difference between its chemical potential in the core, consisting of the saturated active agent solution when at steady state, and its chemical potential in the elution medium, consisting of pure solvent water when under infinite well-agitated sink conditions. If the core contains an excess of the solid active agent and the concentration of active agent in the elution medium does not change significantly, the chemical potential gradient of water will remain constant. Zero-order active agent delivery, therefore, is theoretically possible if molecular diffusional release of the active agent can be eliminated. A schematic representation of the functioning of an elementary osmotic pump is given in Fig. 5.1.

For an osmotic pump with a constant internal volume, the volumetric delivery rate of active agent solution is equal to the volumetric influx rate of water. The influx of water is governed by two opposing forces (Fig. 5.1). The driving force is the osmotic pressure difference, $\Delta\pi$, generated by the chemical potential gradient. The opposing force is the hydrostatic back-pressure, ΔP. The net volumetric flux of water into the pump is given by

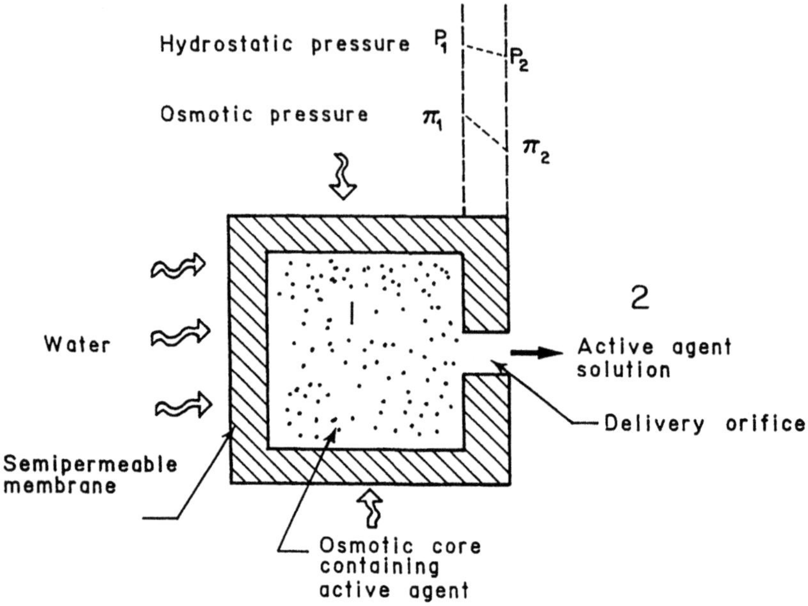

Fig. 5.1. Schematic representation of an elementary osmotic pump (Theeuwes, 1975)

$$J_w = \frac{L_p}{\delta}(\sigma \Delta\pi - \Delta P) \tag{5.1}$$

where L_P is the hydraulic permeability of the membrane, δ its thickness, and σ the reflection coefficient of the membrane. The reflection coefficient characterizes the leakiness of the membrane — it is equal to zero for a porous membrane and equal to one for an ideal semipermeable membrane. The positive sign with $\Delta\pi\,(= \pi_2 - \pi_1)$ indicates that water permeates up the osmotic pressure gradient, i.e., it flows from the phase in which it has a higher "escaping" tendency (dilute solution, high water vapor pressure, low osmotic pressure $= \pi_2$) to the phase with lower "escaping" tendency (concentrated solution, low water vapor pressure, high osmotic pressure $= \pi_1$) (Katchalsky and Curran, 1976; Theeuwes, 1980; Langer and Peppas, 1983). $\Delta P\,(= P_2 - P_1)$ is negative since the rigidity of the membrane causes the hydrostatic pressure inside the membrane, P_1, to increase; this hydrostatic head tends to oppose the inflow of water due to the osmotic pressure difference (Fig. 5.1).

Osmotic pressure can be related to the active agent concentration in the solution, C_a, through

$$\pi = C_a RT \tag{5.2}$$

where R is the gas constant and T is the temperature. It can also be determined from vapor pressure measurements by making use of the expression

$$\pi = \frac{RT}{v} \ln\left(\frac{p_0}{p}\right) \tag{5.3}$$

where v is the molar volume of the solvent (water), p_0 the vapor pressure of pure solvent, and p the vapor pressure of the solution. Substituting Eq. (5.2) into Eq. (5.1) yields

$$J_w = RT\sigma \frac{L_p}{\delta}(C_{a2} - C_{a1}) - \frac{L_p}{\delta}(P_2 - P_1) . \tag{5.4}$$

A device loaded with excess active agent generates a saturated solution of the active agent, C_{as}, such that

$$C_{a1} = C_{as} . \tag{5.5a}$$

Furthermore, if the elution medium is kept at a low concentration, then

$$C_{a2} \approx 0 . \tag{5.5b}$$

If the hydrostatic head, ΔP, is not allowed to build up because of the outflow from the delivery orifice, then

$$\Delta\pi \gg \Delta P . \tag{5.5c}$$

Thus, Eq. (5.4) reduces to

$$J_w = -RT\frac{L_p}{\delta}\sigma C_{as} . \tag{5.6}$$

The negative sign indicates that water flows up the active agent concentration gradient or down the *water* concentration gradient. J_w also represents the volumetric flux of solution out from the delivery orifice since the internal volume of the pump is a constant. The release rate of active agent, therefore, is

$$\left(\frac{dM_t}{dt}\right)_r = |J_w| \, AC_{as} \tag{5.7}$$

where A is the permeation area of the membrane. Substituting from Eq. (5.6) yields (Theeuwes, 1975)

$$\left(\frac{dM_t}{dt}\right)_r = \sigma ART \frac{L_p}{\delta} C_{as}^2 . \tag{5.8}$$

This expression is a simplified version of the zero-order release rate from a device operating at steady state without molecular diffusion of the active agent through the membrane or the delivery orifice. Further analysis of this device has been performed by Theeuwes (1975), for a non-zero-order release rate when the core does not contain excess active agent, and for the effect of molecular diffusion.

Various osmotic pump designs have been proposed. Among these is the Higuchi-Leeper pump (Higuchi and Leeper, 1976) illustrated in Fig. 5.2. The active agent in

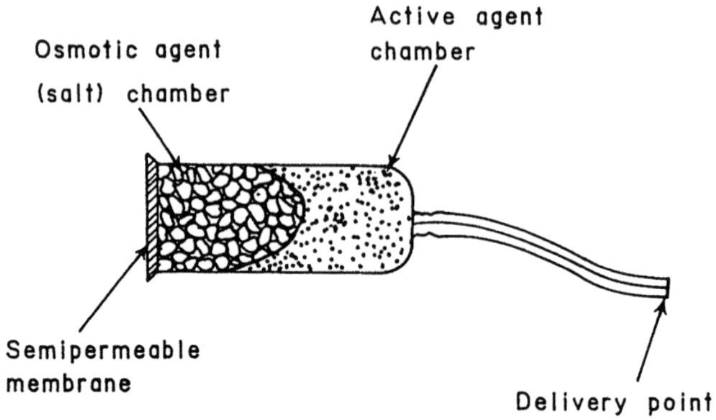

Fig. 5.2. Higuchi-Leeper osmotic pump (Higuchi and Leeper, 1976)

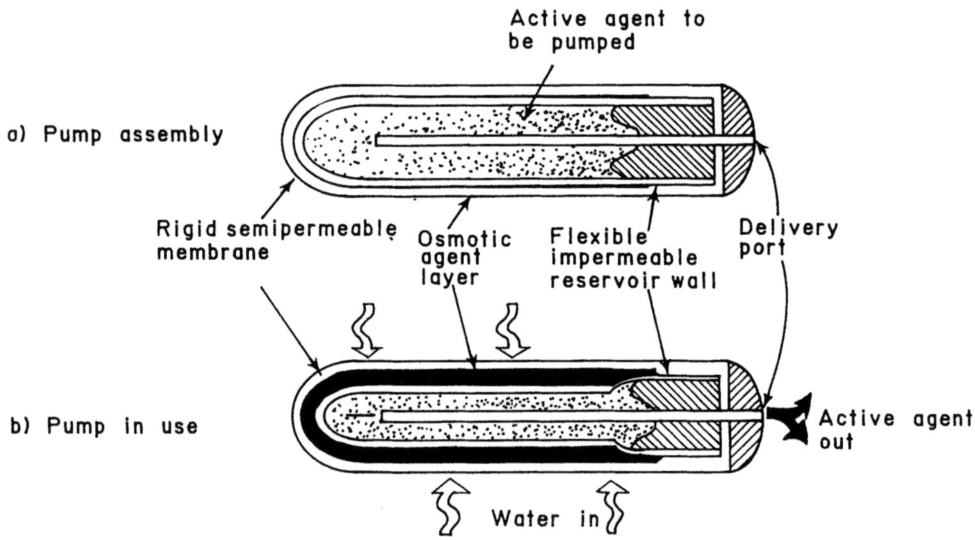

Fig. 5.3. Theeuwes miniature osmotic pump (Theeuwes, 1973)

the form of a liquid is separated from the osmotically active salt compartment by a moveable impermeable partition. A variant on this is the Theeuwes miniature osmotic pump (Theeuwes, 1973) in which the cellulose-based semipermeable membrane itself forms the outer rigid casing (Fig. 5.3). The performance of such a device, both *in vivo* and *in vitro*, is exhibited in Fig. 5.4 (Theeuwes and Yum, 1976). Zero-order release rates are attained; the *in vitro* rates correlate well with the *in vivo* availability. A detailed study of the performance of this device has been reported by Eckenhoff (1981).

The elementary osmotic pump (Fig. 5.1) is evidently a simplification of the miniature osmotic pump (Fig. 5.3). In the elementary osmotic pump, the active agent is either itself osmotically active or mixed with some suitable osmotically

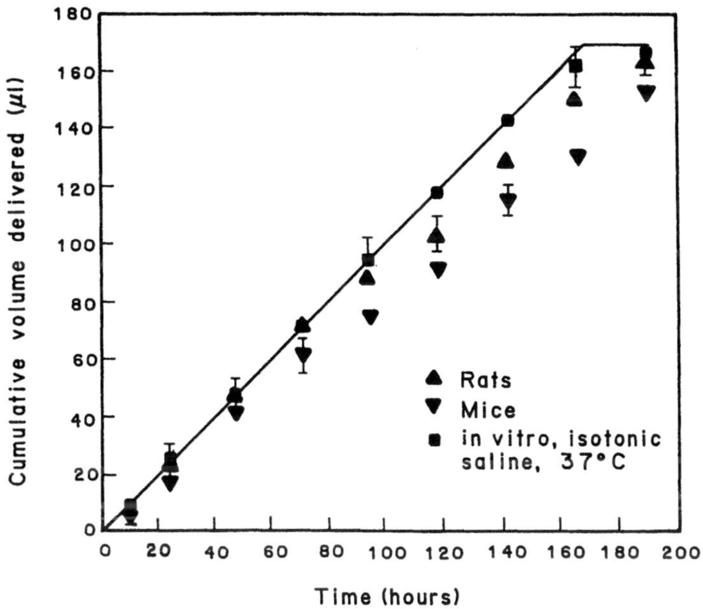

Fig. 5.4. Cumulative volume delivered from a Theeuwes miniature osmotic pump (Theeuwes and Yum, 1976)

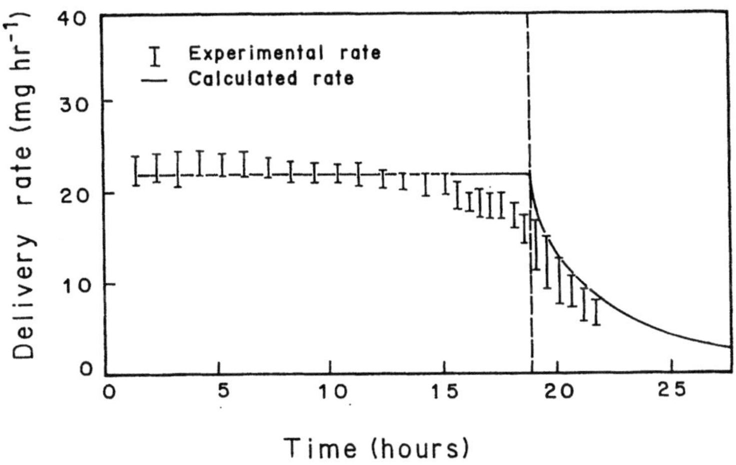

Fig. 5.5. Experimental and predicted release rates of potassium chloride from five elementary osmotic pumps in water at 37 °C (Theeuwes, 1975)

active compound to form the core. The core, formed in a tableting machine, is then coated with the semipermeable membrane and a hole drilled through the membrane. The *in vitro* release rate of potassium chloride into water from such a pump is illustrated in Fig. 5.5. The zero-order portion of the predicted delivery rate in this figure has been calculated from Eq. (5.8) (Theeuwes, 1975).

A different version of the elementary osmotic pump, called the controlled-porosity osmotic device has been described by Thombre (1985). In this device, the polymer coating contains a small amount of water-leachable and polymer-insoluble pore-former. The delivery orifice is not pre-formed; instead, pores are created in the semipermeable polymer by leaching-out the pore-former material by water. The active agent solution then flows out through these pores at the same volumetric rate as that of water influx osmosis.

5.2 Osmotically Activated Monolithic Devices

The devices described in the preceding section utilize osmotic pressure to pump the active agent. However, osmosis can also be used as a release mechanism for monolithic devices as indicated in Fig. 5.6 (Baker, 1987). The matrix is composed of dispersions of osmotically-active, active agents, having low solubility and diffusivity, within the polymer. When placed in an aqueous environment, water is imbibed osmotically by the active agent particles near the surface. Water uptake by these particles leads to a build-up of osmotic pressure which eventually becomes sufficient

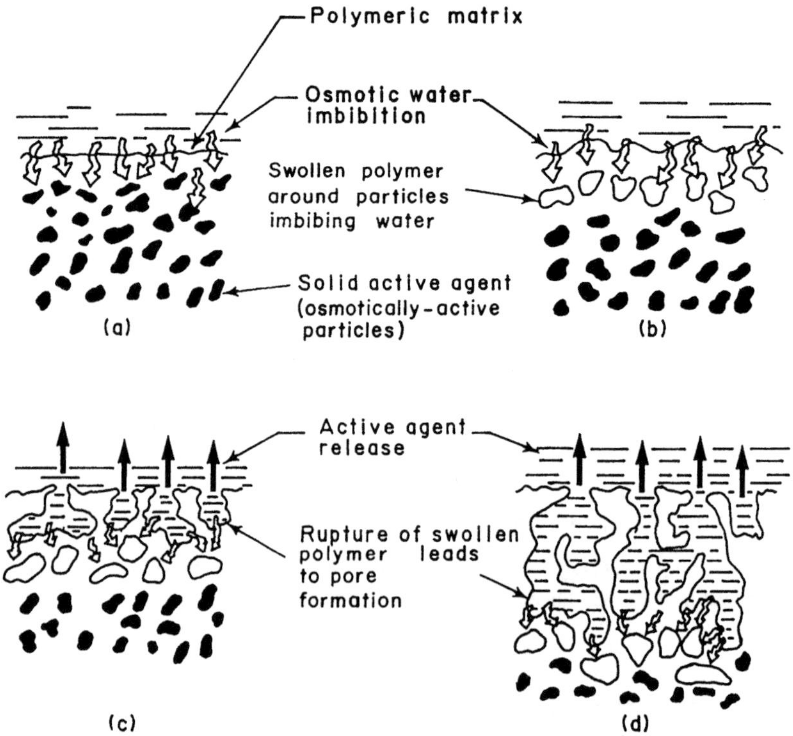

Fig. 5.6a–d. Mechanism of active agent release from monolithic devices by osmotic rupture (Baker, 1987)

to rupture the polymer matrix surrounding these particles. Subsequently, the active agent solution flows out through the fractures. Once this first layer of active agent particles is depleted below the saturation water-solubility level, the second layer of the particles becomes activated; eventually, the whole process is repeated. This proceeds in a serial fashion towards the interior of the matrix (Wright et al., 1981; Baker, 1987).

The serial rupturing mechanism allows the system to be broken into two regions: the zone of ruptured polymer and the zone of intact polymer. These zones are separated by a moving front of imbibing particles/rupturing polymer. The release rate from such a device is given by (Wright et al., 1981)

$$\left(\frac{dM_t}{dt}\right)_r = \frac{M_L}{t_b + t_s} \tag{5.9}$$

where M_L is the mass of active agent per layer, t_b the time for a particle in the active layer to imbibe water and rupture, and t_s the time required for saturated active agent solution to be pumped out. Both t_b and t_s are related to the osmotic pressure gradient across the matrix surface under the assumption that the flow of active agent solution between the rupture front and matrix surface is rapid. Wright et al. (1981) have performed a detailed analysis of the release rate from such a device. They have presented the results of experiments with devices fabricated from ethylene-vinyl acetate copolymers and various active agents. The results for pilocarpine nitrate release can be seen in Fig. 5.7.

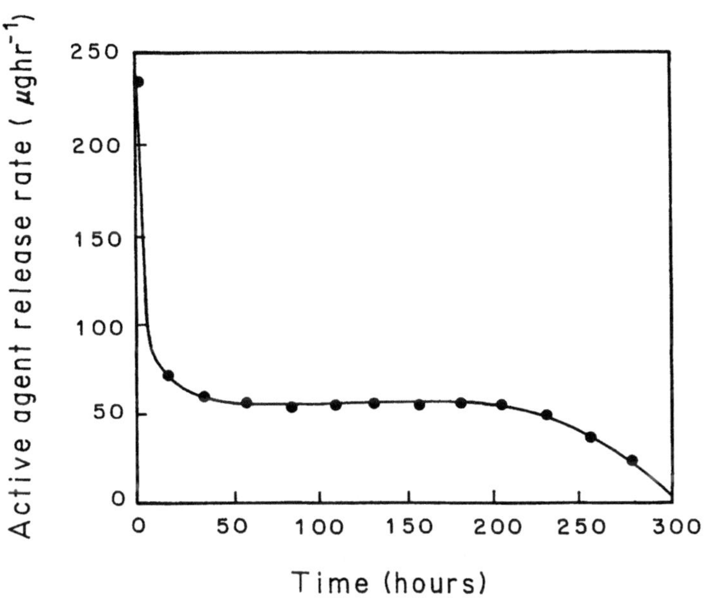

Fig. 5.7. Release of pilocarpine nitrate from an osmotically activated monolithic device containing 27% by volume of the active agent (Wright et al., 1981)

The osmotic release mechanism described above does not hold for loadings greater than 20–30% by volume of the active agent. At these loadings, the percolation threshold is approached, and the outcome is simple leaching of the active agent (Baker, 1987).

5.3 Externally Modulated Devices

If the dispersion of active agent and magnetic beads is produced within a polymer matrix and placed in an aqueous medium, release of the active agent occurs by diffusion in a manner typical of diffusion-controlled monolithic devices. However, when these matrices are exposed to an oscillating external magnetic field, the rate of release of active agent jumps to a higher value (Hsieh et al., 1981; Hsieh and Langer, 1983; Edelman et al., 1985; Kost et al., 1985). This externally modulated release mechanism has been extensively examined by Langer and his coworkers who have prepared the matrices from ethylene-vinyl acetate copolymers. The active agent and magnetic beads are incorporated uniformly by solvent-casting a mixture of polymer active agent at −80 °C on a glass mold and then placing beads uniformly on the cast. After two minutes, another layer of the mixture of polymer and active agent is cast over the beads, trapping them. Release rates with magnetic triggering have been found to be as much as thirty times higher than without triggering. While the mechanism responsible for magnetic modulation is not well understood, it has been suggested that the oscillating beads cause alternate compression and expansion of the pores, thus "pumping" the active agent. The

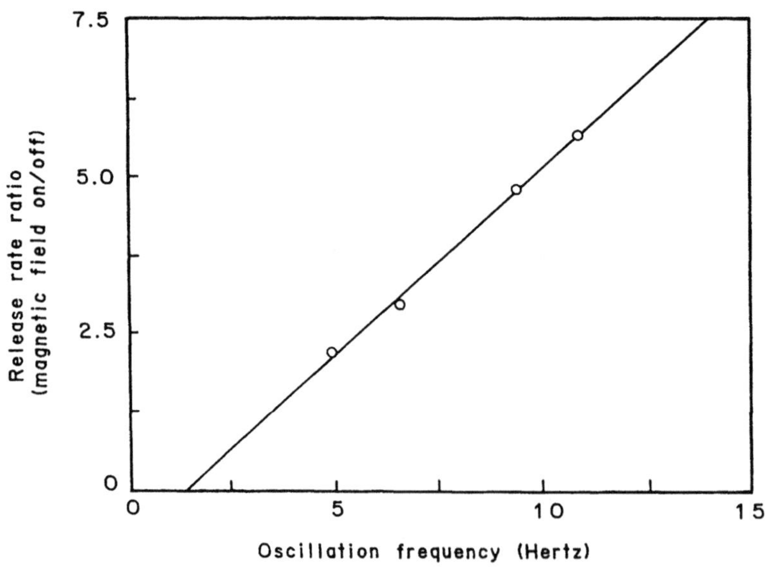

Fig. 5.8. Average ratio of the rates of bovine serum albumin release from matrices of ethylene-vinyl acetate copolymers containing magnetic beads as a function of magnetic oscillation frequency (Edelman et al., 1984)

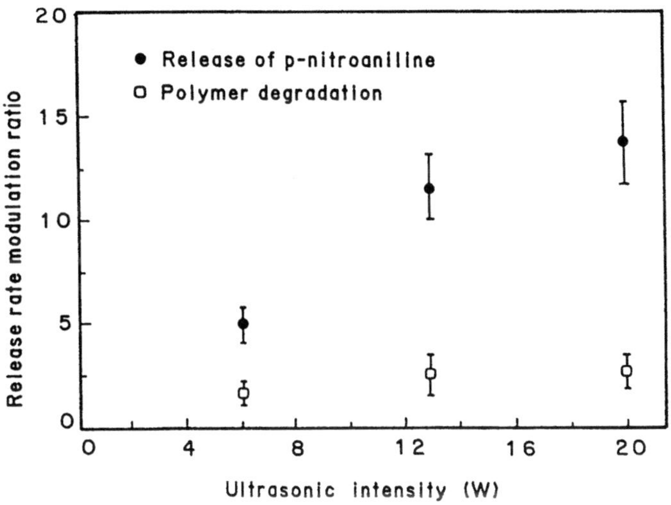

Fig. 5.9. Effect of ultrasound intensity on the modulation ratio for the release of *p*-nitroaniline from poly[bis (p-carboxy phenoxy) methane] (Kost et al., 1986)

effect of oscillation frequency on the release rate of borine serum albumin is demonstrated in Fig. 5.8 (Edelman et al., 1984).

Langer and his coworkers have also investigated ultrasound as a means of modulating the release of macromolecules from polymer matrices. Erodible and non-erodible polymers have been compression- or injection-molded with the active agent. When placed in water, the macromolecular agent diffuses out by a pore-diffusion release mechanism. Nevertheless, when subjected to ultrasound, the release rates are enhanced significantly from both types of polymers. Ultrasound appears to influence the biodegradability of the matrix as well as its permeability, as indicated by the enhanced rate of release from non-erodible matrices. Erosion of polymer surface is apparently intensified by cavitation and acoustic streaming. Figure 5.9 shows the effect of ultrasound intensity on the modulation ratio, defined as the ratio of the rate of release (or degradation) at a given period of ultrasound exposure to the average rate of release immediately preceding and following this exposure (Kost et al., 1986).

The two types of externally modulated systems described here offer the possibility of developing demand-responsive release systems, especially useful in medical therapeutics, where constant rate infusions of the drug must be periodically supplemented by increased dosages (e.g., extra insulin before meals for diabetics, anti-anginals for cardiovascular applications, etc.) (Kost et al., 1986). Compared to magnetic modulation, ultrasound modulation has the advantage that it does not require any special formulation techniques; it can be formed with erodible polymers, thus obviating the need for surgical removal of implants.

References

Baker, R. W.: Controlled Release of Biologically Active Agents, pp. 132–149, John Wiley and Sons, New York (1987)
Bruck, S. D.: Pharmacological Basis of Controlled Drug Delivery, pp. 1–13. In: Controlled Drug Delivery, Vol. I. Basic Concepts, Bruck, S. D. (Ed.), CRC Press, Boca Raton, Florida (1983)
Eckenhoff, B.: Osmotically Driven Pumps for Rate-Controlled Delivery of Solutions and Viscous Suspensions. In: Controlled Release Systems, Chandrasekaran, S. K. (Ed.), AIChE Symposium Series 206, *77*, 1–9 (1981)
Edelman, E., Kost, J., Bobeck, H., Langer, R.: Regulation of Drug Release from Porous Polymer Matrices by Oscillating Magnetic Fields, J. Biomed. Mater. Res. *19*, 67–83 (1985)
Edelman, E. R., Linhardt, R. J., Bobeck, H., Kost, J., Rosen, H. B., Langer, R.: Polymer Based Drug Delivery: Magnetically Modulated and Bioerodible Systems, pp. 279–292. In: Polymers as Biomaterials, Shalaby, S. W., Hoffman, A. S., Ratner, B. D., Horbett, T. A. (Eds.), Plenum Press, New York (1984)
Higuchi, T., Leeper, H. M.: Osmotic Dispenser with Means for Dispensing Active Agent Responsive to Osmotic Gradient, U.S. Patent 3,995,631 (1976)
Hsieh, D. S. T., Langer R.: Zero-Order Drug Delivery Systems with Magnetic Control, pp. 121–131. In: Controlled Release Delivery Systems, Roseman, T. J., Mansdorff, S. Z. (Eds.), Marcel Dekker, New York (1983)
Hsieh, D. S. T., Langer R., Folkman, J.: Magnetic Modulation of Release of Macromolecules from Polymers, Proc. Natl. Acad. Sci. *78*, 1863–1867 (1981)
Katchalsky, A., Curran, P. F.: Non-Equilibrium Thermodynamics in Biophysics, Harvard University Press, Cambridge, Massachusetts (1976)
Kost, J., Leong, K., Langer, R.: Ultrasonic Modulated Drug Delivery Systems, pp. 387–396. In: Polymers in Medicine. II, Chiellini, E., Giusti, P., Migliaresi, C., Nicolais, L. (Eds.), Plenum Press, New York (1986)
Kost, J., Noecker, R., Kunica, E., Langer, R.: Magnetically Controlled Release Systems: Effect of Polymer Composition, J. Biomed. Mater. Res. *19*, 935–940 (1985)
Langer, R., Peppas, N.: Chemical and Physical Structure of Polymers as Carriers for Controlled Release of Bioactive Agents: A Review, J. Macromol. Sci.-Rev. Macromol. Chem. Phys. *C23*, 61–126 (1983)
Theeuwes, F.: Osmotically Powered Active Agent Dispensing Device with Filling Means, U.S. Patent 3,760,984 (1973)
Theeuwes, F.: Elementary Osmotic Pump, J. Pharm. Sci. *64*, 1987–1991 (1975)
Theeuwes, F.: Delivery of Active Agents by Osmosis, pp. 195–205. In: Controlled Release Technologies: Methods, Theory, and Applications, Vol. II, Kydonieus, A. F. (Ed.), CRC Press, Boca Baton, Florida (1980)
Theeuwes, F., Yum, S. I.: Principles of the Design and Operation of Generic Osmotic Pumps for the Delivery of Semisolid or Liquid Drug Formulations, Ann. Biomed. Eng. *4*, 343–352 (1976)
Thombre, A. G.: Mechanism and Kinetics of Release from Two Controlled Drug Delivery Devices, pp. 107–205, Ph.D. Dissertation, University of Kansas, Lawrence, Kansas (1985)
Wright, J., Chandrasekaran, S. K., Gale, R., Swanson, D.: A Model for the Release of Osmotically Active Agents from Monolithic Polymeric Matrices. In: Controlled Release Systems, Chandrasekaran, S. K. (Ed.), AIChE Symposium Series 206 *77*, 62–68 (1981)

6. Stochastic Model for Diffusion in Porous or Heterogeneous Polymer Matrix

A number of new polymeric materials are being explored for possible utilization as carriers for the controlled release of biologically active agents. The morphological structure of most polymers is heterogeneous, consisting of two phases, namely crystalline and amorphous. These two phases are present in varying amounts depending on the composition and the thermal and processing history of the polymer, and are dispersed throughout the body of the polymer. Transport of active agents in such materials is influenced therefore by their molecular structure since the permeabilities of the two phases are different. Crystalline domains in a semi-crystalline polymer are virtually impermeable relative to the amorphous phase; consequently, the diffusing molecule has to follow a tortuous path through the permeable amorphous phase. This permeable phase is interspersed between the impermeable domains serving as boundaries or walls. It is evident that the characteristics of transport in the permeable-impermeable domains of a heterogeneous polymer are similar to those in a porous medium. Thus the release of an active agent from either the heterogeneous polymer or the porous matrix carrier can be analyzed analogously (see, e.g., Peterlin, 1979, Harland and Peppas, 1986).

Transport within a porous matrix is usually analyzed by the deterministic Fickian diffusion equation, representing a continuum approximation to the system. However, transport inside a pore network is a probabilistic phenomenon because of the random nature of the pores and the stochastic nature of the transport process. Each molecule diffuses by executing a random walk within the network along with collision with walls and with other molecules. The random nature of both the medium and the mechanism makes this phenomenon well suited for a stochastic analysis, as opposed to the conventional deterministic approach.

The deterministic approach does not provide any insight into the actual mechanism itself; correction for lack of "continuum" is made by utilization of "fudge factors" such as tortuosity, which are adjusted to make predicted values agree with experimental observations (Aris, 1975). A deep insight into the system will be afforded by a complete molecular level description based on the laws of motion; however, this approach is too complicated for computation. A stochastic approach provides more information than the deterministic approach, and yet is more tractable than a microscopic analysis. With the advent of increased and readily accessible computing power, the stochastic formulation becomes highly attractive because Monte Carlo simulations of these phenomena can be performed. The stochastic approach along with Monte Carlo simulations is applied here to the analysis of transport in heterogeneous polymer or porous matrices for illustration. Other phenomena, important in active agent delivery, that can be similarly analyzed include grinding and mixing (see, e.g., Fan and Shin, 1979; Duggirala, 1986), crystallization (see, e.g., Too et al., 1986), and polymerization and polymer degradation (see, e.g., Nassar et al., 1981; Chou et al., 1987).

Analysis of the phenomenon of concentration-gradient-driven mass transport in porous materials has received a great deal of attention in the chemical engineering literature. The equation usually applied to transport in porous media is

$$\varepsilon \frac{\partial(C)}{\partial t} = \nabla \cdot \varepsilon D_{\text{eff}} \, \nabla C . \tag{6.1}$$

This expression is rigorous only for a homogeneous medium (see, e.g., McGregor, 1974). However, due to its intuitive appeal and the existence of a large body of literature based on this equation, it has also been widely employed even for heterogeneous media. Equation (6.1) involves fluxes and concentrations defined in terms of the external dimensions and configuration of the whole solid mass, thereby implying averaging of the local properties in deriving it. Among the assumptions and constraints imposed if a point equation describing diffusion is to be integrated over an "averaging volume" to give Eq. (6.1) is

$$r', l' \ll \mathscr{L} \ll L$$

where r' and l' represent characteristic lengths for the structure of the porous media, $\mathscr{L}$ the characteristic length of the averaging volume, and L that of the macroscopic system (Whitaker, 1969). These constraints ensure that the volume and area averages for diffusion through a porous medium are identical. This justifies the use of the volume porosity, ε, in Eq. (6.1) to calculate the cross-sectional area open to diffusion in the porous medium (McGregor, 1974); a related requirement is that the pores be randomly oriented (Satterfield, 1970).

Various definitions of the effective diffusion coefficient or diffusivity are given to account for the complexity of the porous media structure (see, e.g., Akanni et al., 1987). The most common definition of the effective diffusivity is

$$D_{\text{eff}} = \frac{D_a}{\tau} \tag{6.2}$$

where τ is the tortuosity of the porous network, and D_a the average diffusion coefficient of the diffusing species in the fluid bathing the pores. D_a is variously based on the average pore radius (Satterfield, 1970) or defined as the pore-volume-averaged diffusivity in a pore (Johnson and Stewart, 1965; Feng and Stewart, 1973; Wang and Smith, 1983).

Transport in a porous medium can be portrayed as a probabilistic phenomenon, with a diffusing molecule having a choice among a number of paths to follow at any instant. The porous medium can itself be visualized as a network of randomly oriented, connected pore segments; the diffusing molecule undergoes a random walk in this medium (see, e.g., De Jong, 1958; Saffman, 1959; Scheidegger, 1965; Pismen, 1974).

Each step of the random walk, comprising passage through a single pore, is generally assumed to be statistically independent of the preceding one. Bhatia (1986) has argued that the above assumption of an uncorrelated random walk is not usually justified. He states that the fixed internal pore surface of the medium must provide a memory effect. Shortly after traversing a complete pore segment and arriving at a pore intersection, the diffusing molecule can, as a result of chance collisions, reverse direction. In such a case, the next full pore segment traversed may be precisely the last one and this no longer has a random orientation (Bhatia, 1985, 1986).

Directional autocorrelation between successive steps of a random walk is important for diffusion in an atomic lattice by the vacancy mechanism. When an atom jumps into a neighboring vacant lattice site, it creates a vacancy at the site it previously occupied. Consequently, the atom on its next jump has a greater than random probability of returning to its original site (Manning, 1961). However, the same reasoning cannot hold for diffusion in porous media. Each unit step in the random walk, consisting of the traversal of the length of a single pore, is actually made up of a number of smaller steps. The distance travelled by the molecule between collisions, with other molecules in the case of bulk diffusion or with pore walls in the case of Knudsen diffusion, is much smaller than the length of a pore segment. These collisions will result in a meandering motion of the molecule and will tend to erase any memory effect. The probability of returning through the previous pore will not be any different from that of exiting through any of the others at the intersection. Therefore, any unit step of the random walk will at most be weakly correlated to its preceding step (except in systems involving pressure-driven bulk flow).

Based on his description of a self-correlated random walk, Bhatia (1986) has derived a generalized form of the master equation for the diffusion process. The formulation has been solved analytically by seeking the continuum limit since individual steps of the random walk (the pore segments) are small compared to the macroscopic length scale (Bhatia, 1986). An expression for the effective diffusivity and a correction factor have been obtained, while the form of Eq. (6.1) has been justified for heterogeneous media assuming random (uniform) pore orientations. An equation of the form of Eq. (6.1) has also been obtained for spatially periodic media (Ryan et al., 1981) using the method of volume averaging. However, the specific case of non-uniform pore orientations has not been examined.

The objective of this chapter is to provide a general stochastic framework or model for transport in porous medium to illustrate the technique. A random-walk (uncorrelated) description of the diffusion phenomenon is adopted for this purpose. The molecules are considered capable of undergoing a first-order reaction in the pores; no *a priori* assumptions are made about pore orientation. The master equation for the phenomenon is derived; it is reduced to a generalized Fickian form of the diffusion equation consisting of a diffusive term and a drift (convective) term. An expression for the effective diffusivity is recovered from the analysis, and the drift velocity is defined. The process of transport in porous systems is also examined by an event-driven Monte Carlo simulation technique; both isotropic and anisotropic systems are considered.

It must be noted that a large number of models are available for transport in porous media including the spatially periodic network (of tubes) model (Fatt, 1956), the "turbulent" flow model (Scheidegger, 1959), the macropore-micropore model (Wakao and Smith, 1962), the convergent-divergent pore model (Foster and Butt, 1966), the dusty gas model (Mason et al. 1967), the random network model (Haring and Greenkorn, 1970), and the periodic capillary models for bidisperse media (Wakad and Nardse, 1974; Gavalas and Kim, 1981). Models for diffusion and reaction in ordered porous network catalysts have been presented by Theodorou and Wei (1983) and Mo and Wei (1986).

6.1 Derivation

A stochastic model for transport in porous media requires that all states that can possibly be occupied by a diffusing molecule, be identified. Transition rates between these states are also needed. A master equation can then be derived for the probability of a molecule being in any of the possible states. In what follows, the diffusing entity is regarded as a molecule; however, the same is valid (perhaps more so) for a multimolecular or mesoscopic diffusing entity (e.g., colloidal particles).

6.1.1 Statistical Model for the Porous/Heterogeneous-Polymer Matrix

The pore space may be considered as being composed of a network of intersecting cylindrical capillary segments, as shown in Fig. 6.1. The number of segments intersecting at any point is identical throughout the matrix and is equal to M. A pore segment is characterized by two parameter sets $\tilde{\omega}$ and $\tilde{\Theta}$; $\tilde{\omega}$ contains the

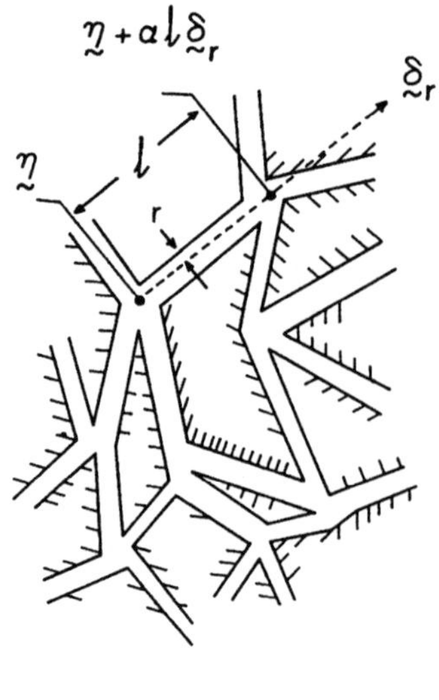

Fig. 6.1. Network of permeable domains in a semi-crystalline polymer or porous matrix

dimensionless pore radius ($r = r'/\bar{r}$) and the dimensionless length ($l = l'/\bar{l}$), while $\tilde{\Theta}$ contains the azimuthal angle (θ) and polar angle (Φ) specifying the orientation of the pore in the spherical coordinate system. Thus,

$$\tilde{\omega} \equiv (r, l) \tag{6.3a}$$

and

$$\tilde{\Theta} = (\theta, \Phi)\,. \tag{6.3b}$$

The unit vector along the axis of the pore segment, $\underset{\sim}{\delta}\boldsymbol{r}$, is related to the unit cartesian vector components $\underset{\sim}{\delta}x$, $\underset{\sim}{\delta}y$ and $\underset{\sim}{\delta}z$ by

$$\underset{\sim}{\delta}\boldsymbol{r} = \sin\theta\cos\Phi\,\underset{\sim}{\delta}x + \sin\theta\sin\Phi\,\underset{\sim}{\delta}y + \cos\theta\,\underset{\sim}{\delta}z\,. \tag{6.4}$$

The dimensions of the pore segments vary from pore to pore, but they are significantly smaller than the macroscopic dimensions of the matrix.

An intersection can be characterized by specifying the parameters of the M pore segments comprising the intersection. Thus, the set

$$\tilde{\tilde{\omega}} = [\tilde{\omega}_1, \tilde{\omega}_2, \ldots, \tilde{\omega}_M] \tag{6.5a}$$

characterizes the radii and length of the M pores, while

$$\tilde{\tilde{\Theta}} = [\tilde{\Theta}_1, \tilde{\Theta}_2, \ldots, \tilde{\Theta}_M] \tag{6.5b}$$

defines their orientations. If S such intersections exist in the matrix (based on unit volume), they may be characterized by the sets

$$\{\tilde{\tilde{\omega}}\} = [\tilde{\tilde{\omega}}_1, \tilde{\tilde{\omega}}_2, \ldots, \tilde{\tilde{\omega}}_S] \tag{6.6a}$$

and

$$\{\tilde{\tilde{\Theta}}\} = [\tilde{\tilde{\Theta}}_1, \tilde{\tilde{\Theta}}_2, \ldots, \tilde{\tilde{\Theta}}_S]\,. \tag{6.6b}$$

The dimensionless position coordinate in the matrix is given by $\tilde{\eta}$, defined as

$$\tilde{\eta} = \frac{\tilde{X}}{L} \tag{6.7}$$

where L is the macróscopic length scale, and $\tilde{X}$ is the vector defining the position coordinate with respect to a stationary frame of reference.

A probability density function for the characteristics of the matrix may be defined such that

$$p_S(\{\tilde{\tilde{\omega}}\}, \{\tilde{\tilde{\Theta}}\}; \alpha)\, d\{\tilde{\tilde{\omega}}\}\, d\{\tilde{\tilde{\Theta}}\}$$

is the probability that the parameters of the pore intersections in the matrix lie in the range of ($\{\tilde{\tilde{\omega}}\}$, $\{\tilde{\tilde{\omega}}\} + d\{\tilde{\tilde{\omega}}\}$) and ($\{\tilde{\tilde{\Theta}}\}$, $\{\tilde{\tilde{\Theta}}\} + d\{\tilde{\tilde{\Theta}}\}$), where

$$d\{\tilde{\tilde{\omega}}\} \equiv \prod_{j=1}^{S} d\tilde{\tilde{\omega}}_j \tag{6.8a}$$

and

$$d\{\tilde{\tilde{\Theta}}\} \equiv \prod_{j=1}^{S} d\tilde{\tilde{\Theta}}_j . \tag{6.8b}$$

(The density function, when integrated over all possible values of $\{\tilde{\tilde{\omega}}\}$ and $\{\tilde{\tilde{\Theta}}\}$, becomes equal to 1.) In the above definitions,

$$d\tilde{\tilde{\omega}} = \prod_{i=1}^{M} dr_i \, dl_i \left(\equiv \prod_{i=1}^{M} d\tilde{\omega}_i \right) \tag{6.9a}$$

and

$$d\tilde{\tilde{\Theta}} = \prod_{i=1}^{M} d\theta_i \, d\varphi_i \left(\equiv \prod_{i=1}^{M} d\tilde{\Theta}_i \right) . \tag{6.9b}$$

α is a characteristic parameter of the pore structure of the matrix; it is defined as the ratio of the pore length scale, $\bar{l}$, to the macroscopic length scale, L, i.e.,

$$\alpha = \frac{\bar{l}}{L} . \tag{6.10}$$

As $\alpha \to 0$, the continuum limit is approached.

The above formulation is adequate for describing the pore structure of the media being considered here. This description of porous media is similar to that employed by Bhatia (1986). [A much less detailed version was given by De Jong (1958)].

6.1.2 State of the Diffusing Molecule and State Transitions

The state of the diffusing molecule may, at any time, be defined by its position. The molecule is said to belong to (the neighborhood of) an intersection at $\tilde{\eta}$ as long as it is in the pore segments meeting at this intersection and has not exited from the end of any of these. Upon arriving at the next intersection, the molecule is considered to belong to (the neighborhood of) the new site even if, in the course of its meandering, it returns to the pore it has just traversed.

The procedure of deriving the master equation for the present system will involve following a particular "tagged" molecule as it moves in the matrix. The behavior of this molecule will be considered representative of all the molecules in the system. Thus, we define the probability distribution function, $P(\tilde{\eta}, t| \{\tilde{\tilde{\omega}}\}, \{\tilde{\tilde{\Theta}}\})$, as

$$P(\tilde{\eta}, t|\,\{\tilde{\tilde{\omega}}\}, \{\tilde{\tilde{\Theta}}\}) = \left[\begin{array}{l}\text{Probability of finding the molecule at (in the}\\ \text{neighborhood of) the intersection at } \tilde{\eta} \text{ at time } t,\\ \text{given that the state of the matrix is } (\{\tilde{\tilde{\omega}}\}, \{\tilde{\tilde{\Theta}}\}),\\ \text{and the position of the molecule at time zero is}\\ \text{known}\end{array}\right]$$

The "neighborhood" is regarded as a small region of radius $d\tilde{\eta}$ around the point $\tilde{\eta}$, i.e., formally

$$P(\tilde{\eta}, t|(\{\tilde{\tilde{\omega}}\}, \{\tilde{\tilde{\Theta}}\}) \equiv p(\tilde{\eta}, t|\,\{\tilde{\tilde{\omega}}\}, \{\tilde{\tilde{\Theta}}\})\, d\tilde{\eta}\,,$$

where $p(\tilde{\eta}, t|\,\{\tilde{\tilde{\omega}}\}, \{\tilde{\tilde{\Theta}}\})$ is the probability density function corresponding to $P(\tilde{\eta}, t|\,\{\tilde{\tilde{\omega}}\}, \{\tilde{\tilde{\Theta}}\})$.

A molecule can undergo a transition in its state by exiting from the intersection or by reacting in the pore system at the intersection. However, the time that elapses before it completes these transitions is also a random variable. We, therefore, define a probability distribution function for the residence time, $P_u(t,\tilde{\eta}|\,\{\tilde{\tilde{\omega}}\})$, such that

$$P_u(t, \tilde{\eta}|\,\{\tilde{\tilde{\omega}}\}) = \left[\begin{array}{l}\text{Probability of finding the molecule at (in the neighbor-}\\ \text{hood of) the intersection at } \tilde{\eta} \text{ at time } t, \text{ after it arrived}\\ \text{there at time } t = 0, \text{ given that the state of the matrix}\\ \text{is } \{\tilde{\tilde{\omega}}\}\end{array}\right].$$

After remaining in the intersection for a length of time t, the molecule may exit from one of the pores and arrive at the next intersection. The probability density function for this transition is given by $p_d(i, t, \tilde{\eta}|\,\{\tilde{\tilde{\omega}}\})$, such that

$$p_d(i, t, \tilde{\eta}|\,\{\tilde{\tilde{\omega}}\})\, dt = \left[\begin{array}{l}\text{Probability of the molecule exiting from the inter-}\\ \text{section at } \tilde{\eta} \text{ through the } i\text{th pore in the time inter-}\\ \text{val } (t, t + dt) \text{ having arrived there at time } t = 0,\\ \text{given that the state of the matrix is } \{\tilde{\tilde{\omega}}\}\end{array}\right]$$

Similarly, the molecule may "exit" or disappear from the intersection by reaction. The probability density function for this transition is given by $p_r(t, \tilde{\eta}|\,\{\tilde{\tilde{\omega}}\})$; in other words,

$$p_r(t, \tilde{\eta}|\,\{\tilde{\tilde{\omega}}\})\, dt = \left[\begin{array}{l}\text{Probability of the molecule undergoing reaction in the}\\ \text{pore system of the intersection at } \tilde{\eta} \text{ in the time}\\ \text{interval } (t, t + dt) \text{ having arrived there at time } t = 0,\\ \text{given that the state of the matrix is } \{\tilde{\tilde{\omega}}\}\end{array}\right]$$

These density functions are similar to those employed by Bhatia (1986).

The preceding developments may now be utilized to obtain the transition rates between the possible states of a diffusing molecule. It will be assumed that such transitions are Markovian. Thus, the probability of the molecule undergoing a transition at any point in time is independent of the length of time it has been in its present state. This implies that

$$\begin{bmatrix}\text{Probability of the molecule undergoing a transition (by exiting or}\\ \text{reacting) from the intersection at } \tilde{\eta} \text{ in the time interval } (t, t+dt)\\ \text{having arrived there at time } t=0\text{, given that the state of the matrix}\\ \text{is } \{\tilde{\bar{\omega}}\}\end{bmatrix}$$

$$= \begin{bmatrix}\text{Probability of finding the molecule at (in the neighborhood of)}\\ \text{the intersection at } \tilde{\eta} \text{ at time } t \text{ after it arrived there at time}\\ t=0\text{, given that the state of the matrix is } \{\tilde{\bar{\omega}}\}\end{bmatrix}$$

$$\times \begin{bmatrix}\text{Probability of the molecule undergoing a transition (by exiting or}\\ \text{reacting) from the intersection at } \tilde{\eta} \text{ in a time interval of length}\\ dt \text{ in } (t, t+dt)\text{, given that the state of the matrix is } \{\tilde{\bar{\omega}}\}\end{bmatrix}$$

For the transition involving the exodus from the end of the ith pore, we have

$$p_d(i, t, \tilde{\eta} \mid \{\tilde{\bar{\omega}}\})\, dt = P_u(t, \tilde{\eta} \mid \{\tilde{\bar{\omega}}\})\, \lambda_d(i, \tilde{\eta} \mid \{\tilde{\bar{\omega}}\})\, dt \tag{6.11}$$

where

$$\lambda_d(i, \tilde{\eta} \mid \{\tilde{\bar{\omega}}\})\, dt = \begin{bmatrix}\text{Probability of the molecule exiting from the inter-}\\ \text{section at } \tilde{\eta} \text{ through the } i\text{th pore in a time interval}\\ \text{of length } dt\text{, given that the state of the matrix is } \{\tilde{\bar{\omega}}\}\end{bmatrix}.$$

For reaction, similarly,

$$p_r(t, \tilde{\eta} \mid \{\tilde{\bar{\omega}}\})\, dt = P_u(t, \tilde{\eta} \mid \{\tilde{\bar{\omega}}\})\, \lambda_r(\tilde{\eta} \mid \{\tilde{\bar{\omega}}\})\, dt \tag{6.12}$$

where

$$\lambda_r(\tilde{\eta} \mid \{\tilde{\bar{\omega}}\})\, dt = \begin{bmatrix}\text{Probability of the molecule undergoing reaction in the}\\ \text{pore system of the intersection at } \tilde{\eta} \text{ in a time interval}\\ \text{of length } dt\text{, given that the state of the matrix is}\\ \{\tilde{\bar{\omega}}\}\end{bmatrix}.$$

λ_d and λ_r are, therefore, the actual transition intensity functions for the molecule.

After arriving at an intersection at time $t = 0$, the molecule may exit from it, react in it or remain in it at time t. Summing up these probabilities, we obtain

$$\sum_{i=1}^{M} \int_0^t p_d(i, \tau, \tilde{\eta} \mid \{\tilde{\bar{\omega}}\})\, d\tau + \int_0^t p_r(\tau, \tilde{\eta} \mid \{\tilde{\bar{\omega}}\})\, d\tau + P_u(t, \tilde{\eta} \mid \{\tilde{\bar{\omega}}\}) = 1\,. \tag{6.13}$$

6.1.3 Derivation of the Master Equation

The consideration of possible transitions experienced by a molecule in the matrix has not accounted for the transition by convective flow of bulk fluid (e.g., pressure-driven flow). Thus, the subsequent discussion is limited to constant density systems with no bulk fluid flow. The reaction is unimolecular, giving rise to a free moving inert product. This approach is also valid for dilute phase diffusion where the concentration of the diffusing species is small, and diffusion does not create a convective fluid flow. Furthermore, transitions into and out of the matrix have not been considered; these appear as boundary conditions.

The master equation for the present system may now be obtained by writing a probability balance on the state of a "tagged" molecule. This balance is given as

$$
\left[\begin{array}{l}\text{Probability of finding the molecule at (in the neighborhood of)}\\ \text{the intersection at } \tilde{\eta} \text{ at time } t+\Delta t\text{, given that the state of the}\\ \text{matrix is } (\{\tilde{\tilde{\omega}}\}, \{\tilde{\tilde{\Theta}}\})\end{array}\right]
$$

$$
-\left[\begin{array}{l}\text{Probability of the molecule arriving at the intersection at } \tilde{\eta} \text{ from}\\ \text{the surrounding intersections in the time interval } (t, t+\Delta t)\text{, given}\\ \text{that the state of the matrix is } \{\tilde{\tilde{\omega}}\}\end{array}\right]
$$

$$
\times\left[\begin{array}{l}\text{Probability of finding the molecule at (in the neighborhood of)}\\ \text{the surrounding intersections at time } t\text{, given that the state of the}\\ \text{matrix is } (\{\tilde{\tilde{\omega}}\}, \{\tilde{\tilde{\Theta}}\})\end{array}\right]
$$

$$
+\left\{1-\left[\begin{array}{l}\text{Probability of the molecule exiting from the intersection at}\\ \tilde{\eta} \text{ in the time interval } (t, t+\Delta t)\text{, given that the state of the}\\ \text{matrix is } \{\tilde{\tilde{\omega}}\}\end{array}\right]\right.
$$

$$
\left.-\left[\begin{array}{l}\text{Probability of the molecule reacting in the pore system of the}\\ \text{intersection at } \tilde{\eta} \text{ in the time interval } (t, t+\Delta t)\text{, given that the}\\ \text{state of the matrix is } \{\tilde{\tilde{\omega}}\}\end{array}\right]\right\}
$$

$$
\times\left[\begin{array}{l}\text{Probability of finding the molecule at (in the neighborhood of)}\\ \text{the intersection at } \tilde{\eta} \text{ at time } t\text{, given that the state of the}\\ \text{matrix is } (\{\tilde{\tilde{\omega}}\}, \{\tilde{\tilde{\Theta}}\})\end{array}\right]
$$

The first part in the right-hand side of this expression is the probability of the molecule arriving at the intersection from the surroundings in the stated time interval. The second part is equivalent to the probability that the molecule is at the intersection at $\tilde{\eta}$ at time t and maintains status-quo in the time interval $(t, t + \Delta t)$. The "surrounding intersections" include all the intersections at the ends of the M pores comprising the intersections under consideration at $\tilde{\eta}$. Each of these M intersections is at a distance $\alpha l_i \, \delta \boldsymbol{r}_i$ from $\tilde{\eta}$, i.e., at the position $\tilde{\eta} + \alpha l_i \, \delta \boldsymbol{r}_i$ (see Fig. 6.1).

Substituting the individual terms into the probability balance expression, we obtain,

$$P(\tilde{\eta}, t + \Delta t | \{\tilde{\omega}\}, \{\tilde{\Theta}\})$$
$$= \sum_{i=1}^{M} [\lambda_d(i, \tilde{\eta} + \alpha l_i \,\delta\mathbf{r}_i | \{\tilde{\omega}\}) \,\Delta t\, P(\tilde{\eta} + \alpha l_i \,\delta\mathbf{r}_i, t | \{\tilde{\omega}\}, \{\tilde{\Theta}\})]$$
$$+ \left[\left\{1 - \sum_{i=1}^{M} \lambda_d(i, \tilde{\eta} | \{\tilde{\omega}\}) \,\Delta t - \lambda_r(\tilde{\eta} | \{\tilde{\omega}\}) \,\Delta t\right\} P(\tilde{\eta}, t | \{\tilde{\omega}\}, \{\tilde{\Theta}\})\right]. \quad (6.14)$$

The summations arise since the transition into and out of the intersection at $\tilde{\eta}$ can occur through any of the M pores. Rearranging Eq. (6.14), dividing both sides by Δt, and subsequently taking the limit as $\Delta t \to 0$, yield

$$\frac{\partial P(\tilde{\eta}, t | \{\tilde{\omega}\}, \{\tilde{\Theta}\})}{\partial t}$$
$$= \sum_{i=1}^{M} \lambda_d(i, \tilde{\eta} + \alpha l_i \,\delta\mathbf{r}_i | \{\tilde{\omega}\}) \, P(\tilde{\eta} + \alpha l_i \,\delta\mathbf{r}_i, t | \{\tilde{\omega}\}, \{\tilde{\Theta}\})$$
$$- \left[\sum_{i=1}^{M} \lambda_d(i, \tilde{\eta} | \{\tilde{\omega}\}) + \lambda_r(\tilde{\eta} | \{\tilde{\omega}\})\right] P(\tilde{\eta}, t | \{\tilde{\omega}\}, \{\tilde{\Theta}\}). \quad (6.15)$$

This equation is the master equation for the probability distribution for the location of a tagged diffusing molecule in the matrix.

Equation (6.15) may be simplified by recognizing that the lengths of the pore segments are small compared to the macroscopic length scale. Thus, the first term on the right-hand side may be expanded as a Taylor series around $\tilde{\eta}$, giving

$$\lambda_d(i, \tilde{\eta} + \alpha l_i \,\delta\mathbf{r}_i | \{\tilde{\omega}\}) \, P(\tilde{\eta} + \alpha l_i \,\delta\mathbf{r}, t | \{\tilde{\omega}\}, \{\tilde{\Theta}\})$$
$$= \lambda_d(i, \tilde{\eta} | \{\tilde{\omega}\}) \, P(\tilde{\eta}, t | \{\tilde{\omega}\}, \{\tilde{\Theta}\})$$
$$+ \alpha l_i \,\delta\mathbf{r}_i \cdot \nabla[\lambda_d(i, \tilde{\eta} | \{\tilde{\omega}\}) \, P(\tilde{\eta}, t | \{\tilde{\omega}\}, \{\tilde{\Theta}\})]$$
$$+ \frac{1}{2} \alpha^2 l_i^2 \,\delta\mathbf{r}_i \,\delta\mathbf{r}_i : \nabla\nabla[\lambda_d(i, \tilde{\eta} | \{\tilde{\omega}\}) \, P(\tilde{\eta} | \{\tilde{\omega}\}, \{\tilde{\Theta}\})]$$
$$+ O[(\alpha l_i)^3]. \quad (6.16)$$

Substituting Eq. (6.16) into Eq. (6.15) and dropping the higher order terms yield

$$\frac{\partial P(\tilde{\eta}, t | \{\tilde{\omega}\}, \{\tilde{\theta}\})}{\partial t}$$
$$= \sum_{i=1}^{M} \frac{1}{2} \alpha^2 l_i^2 \,\delta\mathbf{r}_i \,\delta\mathbf{r}_i : \nabla\nabla[\lambda_d(i, \tilde{\eta} | \{\tilde{\omega}\}) \, P(\tilde{\eta}, t | \{\tilde{\omega}\}, \{\tilde{\Theta}\})]$$
$$+ \sum_{i=1}^{M} \alpha l_i \,\delta\mathbf{r}_i \cdot \nabla[\lambda_d(i, \tilde{\eta} | \{\tilde{\omega}\}) \, P(\tilde{\eta}, t | \{\tilde{\omega}\}, \{\tilde{\Theta}\})]$$
$$- \lambda_r(\tilde{\eta} | \{\tilde{\omega}\}) \, P(\tilde{\eta}, t | \{\tilde{\omega}\}, \{\tilde{\Theta}\}). \quad (6.17)$$

This equation is the master equation for the probability of finding the tagged molecule at an intersection, in terms of the parameters of the matrix. Note that by conditioning on the matrix characteristics, we imply that the motions of this tagged molecule are followed in a well-defined system.

6.1.4 Recovery of the Diffusion Equation

To transform the probability balance equation, Eq. (6.17), into a mass balance equation, the transition intensities need to be defined in terms of the characteristics of the matrix. From the analysis, originally presented by Pismen (1974), and extended by Bhatia (1986) to reacting systems, we have

$$\lambda_d(i, \tilde{\eta} \mid \{\tilde{\tilde{\omega}}_j\}) = \frac{1}{\alpha^2}\left(\frac{2D_e}{L^2}\right)\frac{r_i^2 D_i}{l_i \sum_{i=1}^{M} r_i^2 l_i} \tag{6.18a}$$

$$\equiv \frac{1}{\alpha^2} k_{1i}(\tilde{\omega}_i; \tilde{\tilde{\omega}}) \tag{6.18b}$$

and

$$\lambda_r(\tilde{\eta} \mid \{\tilde{\tilde{\omega}}_j\}) = \left(\frac{2k_r}{\bar{r}}\right)\frac{\sum_{i=1}^{M} r_i l_i}{\sum_{i=1}^{M} r_i^2 l_i} \tag{6.19a}$$

$$\equiv k_2(\tilde{\tilde{\omega}})\,. \tag{6.19b}$$

These expressions have been obtained by solving the diffusion equations simultaneously for each pore connected to an intersection, with common boundary conditions at the intersection. The molecules are initially placed at the intersection, and the flux at any pore outlet is taken as the exit intensity. D_e is an effective diffusivity utilized for de-dimensionalization, D_i the dimensionless diffusivity in the ith pore, and k_r the first-order surface reaction rate constant (Bhatia, 1986). It is evident that the above transition intensities are functions of the characteristics of the intersection under consideration only, and not of the whole system even though they have been defined as conditional on $\{\tilde{\tilde{\omega}}\}$.

For the present purpose, D_i is the significant parameter. In a porous matrix where the pores are bathed with the environmental fluid, it is representative of the diffusivity in the pore fluid. If the pores are large enough, this may be taken to be the diffusivity in the bulk fluid itself (both liquids and gases). In gaseous systems where the radius of the pores becomes small and diffusion falls in the Knudsen region, it is representative of the Knudsen diffusivity. Analogous definitions may be employed for heterogeneous polymeric systems where D_i is the diffusivity within the amorphous polymer. Various theories are available for predicting this diffusivity, e.g., statistical mechanical theories (Pace and Datyner, 1979) and free-volume theories (Vrentas and Duda, 1979) [see also McGregor (1974) for a brief review of this topic].

Substituting Eqs. (6.18b) and (6.19b) into Eq. (6.17) gives

$$\frac{\partial P(\tilde{\eta}, t \mid \{\tilde{\bar{\omega}}\}, \{\tilde{\bar{\Theta}}\})}{\partial t}$$

$$= \sum_{i=1}^{M} \frac{1}{2} l_i^2 \,\underset{\sim}{\delta r}_i \,\underset{\sim}{\delta r}_i : \nabla\nabla[k_{1i}(\tilde{\omega}_i; \tilde{\bar{\omega}})\; P(\tilde{\eta}, t \mid \{\tilde{\bar{\omega}}\}, \{\tilde{\bar{\Theta}}\})]$$

$$+ \sum_{i=1}^{M} \alpha^{-1} l_i \,\underset{\sim}{\delta r}_i \cdot \nabla[k_{1i}(\tilde{\omega}_i; \tilde{\bar{\omega}})\; P(\tilde{\eta}, t \mid \{\tilde{\bar{\omega}}\}, \{\tilde{\bar{\Theta}}\})]$$

$$- k_2(\tilde{\bar{\omega}})\, P(\tilde{\eta}, t \mid \{\tilde{\bar{\omega}}\}, \{\tilde{\bar{\Theta}}\})\,. \tag{6.20}$$

The quantity of most interest is the concentration of the diffusing entities at any point. Towards this end, in the absence of interaction between the diffusing molecules, we have (Appendix 6A)

$$E[U(\tilde{\eta}, t \mid \{\tilde{\bar{\omega}}\}, \{\tilde{\bar{\Theta}}\})] = NP(\tilde{\eta}, t \mid \{\tilde{\bar{\omega}}\}, \{\tilde{\bar{\Theta}}\}) \tag{6.21}$$

where $U(\tilde{\eta}, t \mid \{\omega\}, \{\Theta\})$ is the number of molecules at (in the neighborhood of) the position $\tilde{\eta}$ at time t, given that the state of the matrix is $(\{\tilde{\bar{\omega}}\}, \{\tilde{\bar{\Theta}}\}$, and N is the total number of molecules in the matrix at any instant t. $E[X]$ denotes the expectation value of random variable X. Molecules can only be found at $\tilde{\eta}$ values corresponding to intersections. The volume of an intersection comprises the volumes of all the M pores comprising the intersection. Denoting this volume by $V(\tilde{\eta})$, we have

$$W(\tilde{\eta}, t \mid \{\tilde{\bar{\omega}}\}, \{\tilde{\bar{\Theta}}\}) = \frac{E[U(\tilde{\eta}, t \mid \{\tilde{\bar{\omega}}\}, \{\tilde{\bar{\Theta}}\})]}{V(\tilde{\eta})}\,. \tag{6.22}$$

This expression represents the number concentration of molecules at the intersection, based on the local intersection volume; transformation to the system-volume basis requires multiplication by the local porosity. Division of this number concentration by the Avogadro number converts it into the molar concentration. Combining Eqs. (6.21) and (6.22) gives

$$P(\tilde{\eta}, t \mid \{\tilde{\bar{\omega}}\}, \{\tilde{\bar{\Theta}}\}) = \frac{W(\tilde{\eta}, t \mid \{\tilde{\bar{\omega}}), \{\tilde{\bar{\Theta}}\})\, V(\tilde{\eta})}{N}\,. \tag{6.23}$$

This may be substituted into Eq. (6.17) to obtain the master equation in terms of the number concentration (based on local pore volume), as given below:

$$V(\tilde{\eta}) \frac{\partial W(\tilde{\eta}, t \mid \{\tilde{\bar{\omega}}\}, \tilde{\bar{\Theta}}\})}{\partial t}$$

$$= \sum_{i=1}^{M} \frac{1}{2} l_i^2 \,\underset{\sim}{\delta r}_i \,\underset{\sim}{\delta r}_i : \nabla\nabla[k_{1i}(\tilde{\omega}_i; \tilde{\bar{\omega}})\, V(\tilde{\eta})\;\; W(\tilde{\eta}, t \mid \{\tilde{\bar{\omega}}\}, \{\tilde{\bar{\Theta}}\})]$$

$$+ \sum_{i=1}^{M} \alpha^{-1} l_i \,\underset{\sim}{\delta r}_i \cdot \nabla[k_{1i}(\tilde{\omega}_i; \tilde{\bar{\omega}})\, V(\tilde{\eta})\;\; W(\tilde{\eta}, t \mid \{\tilde{\bar{\omega}}\}, \{\tilde{\bar{\Theta}}\})]$$

$$- k_2(\tilde{\bar{\omega}})\, V(\tilde{\eta})\, W(\tilde{\eta}, t \mid \{\tilde{\bar{\omega}}\}, \{\tilde{\bar{\Theta}}\})\,. \tag{6.24}$$

The quantity N has been cancelled from both sides. Note that by not differentiating N with respect to t, we have restricted the model to a closed system where N can be taken to be a constant. However, transitions into and out of the matrix will appear in the form of boundary conditions when solving the master equation, and therefore the derivation is applicable to both closed and open systems. Application of the ∇ operator to $V(\tilde{\eta})$ in Eq. (6.24) and division of the resultant expression by $V(\tilde{\eta})$ lead to

$$\begin{aligned}\frac{\partial W}{\partial t} = \sum_{i=1}^{M} \frac{1}{2} l_i^2\, \underset{\sim}{\delta \boldsymbol{r}}_i\, \underset{\sim}{\delta \boldsymbol{r}}_i : \left[\nabla\nabla(k_{1i}W) + 2(\nabla k_{1i}W)\cdot\left(\frac{\nabla V}{V}\right) + k_{1i}W\,\frac{\nabla\nabla V}{V}\right] \\ + \sum_{i=1}^{M} \alpha^{-1} l_i\, \underset{\sim}{\delta \boldsymbol{r}}_i \cdot \left[\nabla k_{1i}W + k_{1i}W\,\frac{\nabla V}{V}\right] \\ - k_2 W \end{aligned} \tag{6.25}$$

where the parameters have been dropped for simplicity of notation. Noting that

$$\nabla k_{1i}W = k_{1i}\, \nabla W + W\, \nabla k_{1i} \tag{6.26a}$$

and

$$\nabla\nabla k_{1i}W = \nabla\cdot(k_{1i}\,\nabla W) + (\nabla k_{1i})\cdot(\nabla W) + W\, \nabla\nabla k_{1i} \tag{6.26b}$$

we obtain from Eq. (6.25)

$$\begin{aligned}\frac{\partial W}{\partial t} = \frac{1}{2}\sum_{i=1}^{M} l_i^2\, \underset{\sim}{\delta \boldsymbol{r}}_i\, \underset{\sim}{\delta \boldsymbol{r}}_i : [\nabla\cdot(k_{1i}\,\nabla W) + (\nabla k_{1i})\cdot(\nabla W) + W\,\nabla\nabla k_{1i}] \\ + \sum_{i=1}^{M}\left[\alpha^{-1} l_i + l_i^2\, \underset{\sim}{\delta \boldsymbol{r}}_i \cdot \frac{\nabla V}{V}\right] \underset{\sim}{\delta \boldsymbol{r}}_i \cdot [k_{1i}\nabla W + W\nabla k_{1i}] \\ + \sum_{i=1}^{M}\left[\alpha^{-1} l_i\, \underset{\sim}{\delta \boldsymbol{r}}_i \cdot \frac{\nabla V}{V} + \frac{1}{2} l_i^2\, \underset{\sim}{\delta \boldsymbol{r}}_i\, \underset{\sim}{\delta \boldsymbol{r}}_i : \frac{\nabla\nabla V}{V}\right] k_{1i}W \\ - k_2 W \end{aligned} \tag{6.27}$$

or

$$\begin{aligned}\frac{\partial W}{\partial t} = \frac{1}{2}\sum_{i=1}^{M} l_i^2\, \underset{\sim}{\delta \boldsymbol{r}}_i\, \underset{\sim}{\delta \boldsymbol{r}}_i : \nabla\cdot(k_{1i}\,\nabla W) \\ + \sum_{i=1}^{M}\left[\frac{1}{2} l_i^2\, \underset{\sim}{\delta \boldsymbol{r}}_i \cdot \nabla k_{1i} + \alpha^{-1} l_i k_{1i} + l_i^2 k_{1i}\, \underset{\sim}{\delta \boldsymbol{r}}_i \cdot \frac{\nabla V}{V}\right] \underset{\sim}{\delta \boldsymbol{r}}_i \cdot \nabla W \\ + \sum_{i=1}^{M}\Bigg[\frac{1}{2} l_i^2\, \underset{\sim}{\delta \boldsymbol{r}}_i\, \underset{\sim}{\delta \boldsymbol{r}}_i : \nabla\nabla k_{1i} + \alpha^{-1} l_i\, \underset{\sim}{\delta \boldsymbol{r}}_i \cdot \nabla k_{1i} \\ + l_i^2\left(\underset{\sim}{\delta \boldsymbol{r}}_i \cdot \frac{\nabla V}{V}\right)(\underset{\sim}{\delta \boldsymbol{r}}_i \cdot \nabla k_{1i}) + \alpha^{-1} k_{1i} l_i\, \underset{\sim}{\delta \boldsymbol{r}}_i \cdot \frac{\nabla V}{V} \\ + \frac{1}{2} l_i^2 k_{1i}\, \underset{\sim}{\delta \boldsymbol{r}}_i\, \underset{\sim}{\delta \boldsymbol{r}}_i : \frac{\nabla\nabla V}{V}\Bigg] W \\ - k_2 W\,. \end{aligned} \tag{6.28}$$

This equation should be valid for all systems with or without reaction. A non-reacting system at thermodynamic equilibrium requires that

$$\frac{\partial W}{\partial t} = 0 \tag{6.29a}$$

and

$$\nabla W = 0\,. \tag{6.29b}$$

For these conditions to be satisfied in a non-reacting system, the third part of the right-hand side of Eq. (6.28) must be identically zero at equilibrium, i.e.,

$$\sum_{i=1}^{M}\Bigg[\frac{1}{2}l_i^2\,\underset{\sim}{\delta r}_i\,\underset{\sim}{\delta r}_i : \nabla\nabla k_{1i} + \alpha^{-1}l_i\,\underset{\sim}{\delta r}_i\cdot\nabla k_{1i}$$
$$+\,l_i^2\left(\underset{\sim}{\delta r}_i\cdot\frac{\nabla V}{V}\right)(\underset{\sim}{\delta r}_i\cdot\nabla k_{1i}) + \alpha^{-1}k_{1i}l_i\,\underset{\sim}{\delta r}_i\,\frac{\nabla V}{V}$$
$$+\,\frac{1}{2}l_i^2k_{1i}\,\underset{\sim}{\delta r}_i\,\underset{\sim}{\delta r}_i : \frac{\nabla\nabla V}{V}\Bigg] \equiv 0\,. \tag{6.30}$$

Since all terms in the above expression are constants representing parameters of the matrix, they will be identically zero, irrespective of the thermodynamic state of the system (in or out of equilibrium). The simplified version of Eq. (6.28) is, therefore,

$$\frac{\partial W}{\partial t} = \frac{1}{2}\sum_{i=1}^{M} l_i^2\,\underset{\sim}{\delta r}_i\,\underset{\sim}{\delta r}_i : \nabla\cdot(k_{1i}\nabla W)$$
$$+\sum_{i=1}^{M}\left[\frac{1}{2}l_i^2\,\underset{\sim}{\delta r}_i\cdot\nabla k_{1i} + \alpha^{-1}l_ik_{1i} + l_i^2k_{1i}\,\underset{\sim}{\delta r}_i\cdot\frac{\nabla V}{V}\right]\underset{\sim}{\delta r}_i\cdot\nabla W$$
$$-\,k_2W\,. \tag{6.31}$$

Note that starting from the behavior of a single molecule, we have obtained an equation for the population of the molecules under the assumption that all molecules will behave in a similar fashion.

The preceding development has been based on the matrix characteristics. However, since these characteristics are seldom well defined, the equation will now be averaged over the probability density function for these characteristics. This will give us an "averaged" or expected-concentration equation. We denote an expectation operator by $\langle\cdot\rangle$, such that for any function $g(\cdot)$

$$\langle g\rangle = \int_{\{\tilde{\tilde{\omega}}\}}\int_{\{\tilde{\tilde{\Theta}}\}} g(\tilde{\tilde{\omega}}, \tilde{\tilde{\Theta}}, \tilde{\eta}, t\,|\,\{\tilde{\tilde{\omega}}\}, \{\tilde{\tilde{\Theta}}\})\, p_S(\{\tilde{\tilde{\omega}}\}, \{\tilde{\tilde{\Theta}}\}; \alpha)\, d\{\tilde{\tilde{\omega}}\}\, d\{\tilde{\tilde{\Theta}}\}\,. \tag{6.32}$$

Multiplying both sides of Eq. (6.31) by $p_S(\{\tilde{\tilde{\omega}}\}, \{\tilde{\tilde{\Theta}}\}; \alpha)\, d\{\tilde{\tilde{\omega}}\}\, d\{\tilde{\tilde{\Theta}}\}$ and performing the above mentioned integration yield, formally

$$\frac{\partial \langle W \rangle}{\partial t} = \frac{1}{2} \left\langle \sum_{i=1}^{M} l_i^2 \, \underset{\sim}{\delta \boldsymbol{r}}_i \, \underset{\sim}{\delta \boldsymbol{r}}_i : \nabla \cdot \{k_{1i}[\nabla W]\} \right\rangle$$
$$+ \left\langle \sum_{i=1}^{M} \left[\frac{1}{2} l_i^2 \, \underset{\sim}{\delta \boldsymbol{r}}_i \cdot \nabla k_{1i} + \alpha^{-1} l_i k_{1i} + l_i^2 k_{1i} \underset{\sim}{\delta \boldsymbol{r}}_i \cdot \frac{\nabla V}{V} \right] \underset{\sim}{\delta \boldsymbol{r}}_i \cdot \nabla W \right\rangle$$
$$- \langle k_2 W \rangle . \tag{6.33}$$

In attempting to evaluate the expectations on the right-hand side of Eq. (6.33), it must be noted that the intersections in the matrix are not truly independent of each other. While the pore characteristics within each intersection may be independently and identically distributed, neighboring intersections share a common pore segment and their distributions, therefore, are correlated. Nevertheless, since α is a small parameter, we may write

$$p_S(\{\tilde{\tilde{\omega}}\}, \{\tilde{\tilde{\Theta}}\}; \alpha) = p_{S0}(\{\tilde{\tilde{\omega}}\}, \{\tilde{\tilde{\Theta}}\}) + \alpha p_{S1}(\{\tilde{\tilde{\omega}}\}, \{\tilde{\tilde{\Theta}}\}; \alpha) + O(\alpha^2) \tag{6.34}$$

where $p_{S0}(\cdot)$ is the zero-order approximation to $p_S(\cdot)$, representing the probability density function applicable if all intersections are independent of and identical to each other; $p_{S1}(\cdot)$ is the first-order correction term containing the effect of nearest-neighbor interactions, while the higher-order interactions are neglected. We shall elaborate on p_{S0} elsewhere. Meanwhile, note that from the definition of p_S and the requirement that density functions integrate to 1, p_{S0} and p_{S1} are constrained, respectively, by

$$\int_{\{\tilde{\tilde{\omega}}\}} \int_{\{\tilde{\tilde{\Theta}}\}} p_{S0}(\{\tilde{\tilde{\omega}}\}, \{\tilde{\tilde{\Theta}}\})\, d\{\tilde{\tilde{\omega}}\}\, d\{\tilde{\tilde{\Theta}}\} = 1 \tag{6.35 a}$$

and

$$\int_{\{\tilde{\tilde{\omega}}\}} \int_{\{\tilde{\tilde{\Theta}}\}} p_{S1}(\{\tilde{\tilde{\omega}}\}, \{\tilde{\tilde{\Theta}}\}; \alpha)\, d\{\tilde{\tilde{\omega}}\}\, d\{\tilde{\tilde{\Theta}}\} = 0 . \tag{6.35 b}$$

Substituting Eq. (6.34) into Eq. (6.32) yields

$$\langle g \rangle = \langle g \rangle_0 + \alpha \langle g \rangle_1 \tag{6.36 a}$$

where

$$\langle g \rangle_0 = \int_{\{\tilde{\tilde{\omega}}\}} \int_{\{\tilde{\tilde{\Theta}}\}} g(\tilde{\tilde{\omega}}, \tilde{\tilde{\Theta}}, \tilde{\eta}, t \mid \{\tilde{\tilde{\omega}}\}, \{\tilde{\tilde{\Theta}}\})\, p_{S0}(\{\tilde{\tilde{\omega}}\}, \{\tilde{\tilde{\Theta}}\})\, d\{\tilde{\tilde{\omega}}\}\, d\{\tilde{\tilde{\Theta}}\} \tag{6.36 b}$$

and

$$\langle g \rangle_1 = \int_{\{\tilde{\tilde{\omega}}\}} \int_{\{\tilde{\tilde{\Theta}}\}} g(\tilde{\tilde{\omega}}, \tilde{\tilde{\Theta}}, \tilde{\eta}, t \mid \{\tilde{\tilde{\omega}}\}, \{\tilde{\tilde{\Theta}}\})\, p_{S1}(\{\tilde{\tilde{\omega}}\}, \{\tilde{\tilde{\Theta}}\}; \alpha)\, d\{\tilde{\tilde{\omega}}\}\, d\{\tilde{\tilde{\Theta}}\} . \tag{6.36 c}$$

Utilizing the above definitions in Eq. (6.33), and equating terms of similar orders of approximation, we obtain

$$\frac{\partial\langle W(\tilde{\eta}, t\,|\,\{\tilde{\omega}\}, \{\tilde{\tilde{\Theta}}\})\rangle_0}{\partial t}$$

$$= \frac{1}{2}\left\langle \sum_{i=1}^{M} l_i^2\ \underset{\sim}{\delta \boldsymbol{r}_i}\ \underset{\sim}{\delta \boldsymbol{r}_i} : \nabla \cdot \{k_{1i}(\tilde{\omega}_i;\ \tilde{\tilde{\omega}})\ [\nabla W(\tilde{\eta}, t\,|\,\{\tilde{\tilde{\omega}}\}, \{\tilde{\tilde{\Theta}}\})]\} \right\rangle_0$$

$$+ \left\langle \sum_{i=1}^{M} \left[\frac{1}{2}\ l_i^2\ \underset{\sim}{\delta \boldsymbol{r}_i} \cdot \nabla k_{1i}(\tilde{\omega}_i;\ \tilde{\tilde{\omega}}) + \alpha^{-1}\ l_i k_{1i}(\tilde{\omega}_i;\ \tilde{\tilde{\omega}}) + l_i^2 k_{1i}(\tilde{\omega}_i;\ \tilde{\tilde{\omega}})\ \underset{\sim}{\delta \boldsymbol{r}_i} \cdot \frac{\nabla V(\tilde{\eta})}{V(\tilde{\eta})}\right] \underset{\sim}{\delta \boldsymbol{r}_i} \cdot \nabla W(\tilde{\eta}, t\,|\,\{\tilde{\tilde{\omega}}\}, \{\tilde{\tilde{\Theta}}\}) \right\rangle_0$$

$$- \langle k_2(\tilde{\tilde{\omega}})\ W(\tilde{\eta}, t\,|\,\{\tilde{\tilde{\omega}}\}, \{\tilde{\tilde{\Theta}}\})\rangle_0\,, \tag{6.37}$$

and

$$\frac{\partial\langle W(\tilde{\eta}, t\,|\,\{\tilde{\tilde{\omega}}\}, \{\tilde{\tilde{\Theta}}\})\rangle_1}{\partial t}$$

$$= \frac{1}{2}\left\langle \sum_{i=1}^{M} l_i^2\ \underset{\sim}{\delta \boldsymbol{r}_i}\ \underset{\sim}{\delta \boldsymbol{r}_i} : \nabla \cdot \{k_{1i}(\tilde{\omega}_i;\ \tilde{\tilde{\omega}})\ [\nabla W(\tilde{\eta}, t\,|\,\{\tilde{\tilde{\omega}}\}, \{\tilde{\tilde{\Theta}}\})]\} \right\rangle_1$$

$$+ \left\langle \sum_{i=1}^{M} \left[\frac{1}{2}\ l_i^2\ \underset{\sim}{\delta \boldsymbol{r}_i} \cdot \nabla k_{1i}(\tilde{\omega}_i;\ \tilde{\tilde{\omega}}) + \alpha^{-1}\ l_i k_{1i}(\tilde{\omega}_i;\ \tilde{\tilde{\omega}}) + l_i^2 k_{1i}(\tilde{\omega}_i;\ \tilde{\tilde{\omega}})\ \underset{\sim}{\delta \boldsymbol{r}_i} \cdot \frac{\nabla V(\tilde{\eta})}{V(\tilde{\eta})}\right] \underset{\sim}{\delta \boldsymbol{r}_i} \cdot \nabla W(\tilde{\eta}, t\,|\,\{\tilde{\tilde{\omega}}\}, \{\tilde{\tilde{\Theta}}\}) \right\rangle_1$$

$$- \langle k_2(\tilde{\tilde{\omega}})\ W(\tilde{\eta}, t\,|\,\{\tilde{\tilde{\omega}}\}, \{\tilde{\tilde{\Theta}}\})\rangle_1 \tag{6.38}$$

after re-introduction of the parameters. The term $\langle W(\cdot)\rangle_0$ may be looked upon as the mean or zero-order concentration observed at any point. $\langle W(\cdot)\rangle_1$ is the first-order correction to this mean, arising when a better approximation to the system statistics, $p_S(\cdot)$, is adopted in place of $p_{S0}(\cdot)$. For the purpose of the present work, this factor will not be considered further.

The mean or zero-order equation, Eq. (6.37), may be simplified to (Appendix 6B)

$$\frac{\partial\langle C(\tilde{\eta}, t\rangle_0}{\partial t}$$

$$= \frac{1}{2}\left\langle \sum_{i=1}^{M} l_i^2\ \underset{\sim}{\delta \boldsymbol{r}_i}\ \underset{\sim}{\delta \boldsymbol{r}_i}\ k_{1i}(\tilde{\omega}_i;\ \tilde{\tilde{\omega}}) \right\rangle_0 : \nabla\nabla\langle C(\tilde{\eta}, t)\rangle_0$$

$$+ \left\langle \sum_{i=1}^{M} \left[\alpha^{-1} l_i k_{1i}(\tilde{\omega}_i;\ \tilde{\tilde{\omega}}) + l_i^2 k_{1i}(\tilde{\omega}_i;\ \tilde{\tilde{\omega}})\ \underset{\sim}{\delta \boldsymbol{r}_i} \cdot \frac{\nabla V}{V}\right] \underset{\sim}{\delta \boldsymbol{r}_i} \right\rangle_0 \cdot \nabla\langle C(\tilde{\eta}, t)\rangle_0$$

$$- \langle k_2(\tilde{\tilde{\omega}})\rangle_0\ \langle C(\tilde{\eta}, t)\rangle_0 \tag{6.39}$$

where,

$$C \equiv C(\tilde{\eta}, t) = \int_{\{\tilde{\tilde{\omega}}\}} \int_{\{\tilde{\tilde{\Theta}}\}} W(\tilde{\eta}, t \,|\, \{\tilde{\tilde{\omega}}\}, \{\tilde{\tilde{\Theta}}\})\, p_{SO}(\{\tilde{\tilde{\omega}}\}, \{\tilde{\tilde{\Theta}}\})\, d\{\tilde{\tilde{\omega}}\}\, d\{\tilde{\tilde{\Theta}}\}\,. \tag{6.40}$$

The conditionality in the definition of $W(\cdot)$ is integrated out by averaging over all possible intersection characteristics. Thus $C(\cdot)$ represents the actual number concentration that would be observed at any point and corresponds to the deterministic concentration in Eq. (6.1).

Equation (6.39) is of the form

$$\frac{\partial \langle C \rangle_0}{\partial t} = D\colon \nabla\nabla \langle C \rangle_0 - v \cdot \nabla \langle C \rangle_0 - K_R \langle C \rangle_0\,.$$

The definitions of D and v in this expression need be delineated further; nevertheless, it is worth noting that the stochastic model or the Fokker–Planck representation of the diffusion process naturally contains the drift (convective) term (see, e.g., Bharucha-Reid, 1960; Seinfeld and Lapidus, 1974; Fan and Shin, 1979; van Kampen, 1981). Moreover, this "convective" term is independent of any bulk fluid-flow term within the pores induced by an external pressure difference across them. The significance of this drift velocity may be understood in terms of the tendency of molecules to make jumps preferably in one direction or the other. In an isotropic system (randomly oriented pores) with no concentration gradient, molecules at an intersection have an equal probability of jumping in all directions. If a concentration gradient is imposed on this system, a larger number of jumps are made from the higher to the lower concentration than vice versa. This leads to the diffusive flux, arising solely because a larger number of molecules exist in the direction of a higher concentration; the individual molecules themselves have no preferred direction of jump.

In an anisotropic system (with non-uniform pore orientations), a larger number of pores in one direction will "force" the individual molecules at the intersection to jump more frequently in this direction. This may be visualized by considering transport in the x-y plane. If every intersection has one pore pointing in the negative-x direction and two in the positive-x direction, the probability of molecules moving to the positive-x side is twice that of movement in the opposite direction. The pore structure thus renders the jump frequency of an individual molecule in one direction across the diffusion plane different from that in the opposite direction; hence, the drift velocity results. [An analogous definition of drift in an atomic lattice is given by Manning (1965).] The drift velocity by itself cannot create a flux in the absence of a concentration gradient since any drift velocity-induced flux in this situation will produce a gradient and, thereby, a diffusive flux that will tend to negate this drift velocity-induced flux.

To summarize, a non-uniform (non-random) pore orientation in a matrix, leading to a spatial gradient in the microstructure, gives rise to a drift velocity; there are more pores at one end of such a matrix than the other.

6.2 Transport Coefficients from the Diffusion Equation

In the preceding section, it has been shown that the master equation formulation for transport leads to a mean or zero-order diffusion equation, Eq. (6.39), of the form

$$\frac{\partial \langle C \rangle_0}{\partial t} = D_{\text{eff}} : \nabla\nabla \langle C \rangle_0 - v \cdot \nabla \langle C \rangle_0 - K_R \langle C \rangle_0 \tag{6.41}$$

along with the parallel first-order correction equation of the form

$$\frac{\partial \langle C \rangle_1}{\partial t} = \langle D^+ : \nabla\nabla C \rangle_1 - \langle v^+ \cdot \nabla C \rangle_1 - \langle K_R^+ C \rangle_1 \, .$$

Note that from the definition of $\langle \cdot \rangle_0$ and $\langle \cdot \rangle_1$ in Eq. (6.36a), a quantity, $\langle C \rangle$, that can be expressed as

$$\langle C \rangle = \langle C \rangle_0 + \alpha \langle C \rangle_1$$

is obtained. This is the experimentally observable or measurable concentration.

Conventionally, we resort to the zero-order equation, Eq. (6.41), for analysis of transport in a porous or heterogeneous polymeric matrix. In this section, we shall recover the mean or effective diffusion coefficient, D_{eff}, and the drift or convective velocity term, v, from this zero-order diffusion equation. The definition of p_{SO}, as stated in the context of Eq. (6.34), will be exploited for this purpose. p_{SO} is the probability density function of the matrix characteristics if each intersection were taken independent of and identical to the others, i.e., in isolation. The following definition is the outcome,

$$p_{SO}(\{\tilde{\tilde{\omega}}\}, \{\tilde{\tilde{\Theta}}\}) = \prod_{m=1}^{S} p_O(\tilde{\tilde{\omega}}_m, \tilde{\tilde{\Theta}}_m) \tag{6.42}$$

where $p_0(\tilde{\tilde{\omega}}, \tilde{\tilde{\Theta}})$ is the probability density function for the characteristics of an intersection, i.e., of the set of pores comprising the intersection.

In certain cases, we shall simplify further by assuming that all pores within an intersection are independently and identically distributed. This leads to

$$p_O(\tilde{\tilde{\omega}}, \tilde{\tilde{\Theta}}) = \prod_{j=1}^{M} f_O(\tilde{\omega}_j, \tilde{\Theta}_j) \, . \tag{6.43a}$$

Thus,

$$p_{SO}(\{\tilde{\tilde{\omega}}\}, \{\tilde{\tilde{\Theta}}\}) = \prod_{m=1}^{S} \left[\prod_{j=1}^{M} f_{Om}(\tilde{\omega}_j, \tilde{\Theta}_j) \right] \tag{6.43b}$$

where $f_0(\tilde{\omega}, \tilde{\Theta})$ is the probability density function for the characteristics of an individual pore within an intersection.

6.2.1 Mean or Effective Diffusion Coefficient

The effective diffusivity will be extracted from the first term on the right-hand side of Eq. (6.39). Comparing this equation with Eq. (6.41), we obtain

$$D_{\text{eff}} = \frac{1}{2}\left\langle \sum_{i=1}^{M} l_i^2 \, \underset{\sim}{\delta \boldsymbol{r}}_i \, \underset{\sim}{\delta \boldsymbol{r}}_i \, k_{1i}(\tilde{\omega}_i;\, \tilde{\tilde{\omega}}) \right\rangle_0 \tag{6.44}$$

or, from the definition of $\langle \cdot \rangle_0$ in Eq. (6.36b),

$$D_{\text{eff}} = \frac{1}{2} \int_{\{\tilde{\tilde{\omega}}\}} \int_{\{\tilde{\tilde{\Theta}}\}} \sum_{i=1}^{M} [l_i^2 \, \underset{\sim}{\delta \boldsymbol{r}}_i \, \underset{\sim}{\delta \boldsymbol{r}}_i \, k_{1i}(\tilde{\omega}_i;\, \tilde{\tilde{\omega}}) \, p_{SO}(\{\tilde{\tilde{\omega}}\}, \{\tilde{\tilde{\Theta}}\}) \, d\{\tilde{\tilde{\omega}}\} \, d\{\tilde{\tilde{\Theta}}\}] . \tag{6.45}$$

Substituting for p_{SO} from Eq. (6.42),

$$D_{\text{eff}} = \frac{1}{2} \int_{\{\tilde{\tilde{\omega}}\}} \int_{\{\tilde{\tilde{\Theta}}\}} \sum_{i=1}^{M} \left[l_i^2 \, \underset{\sim}{\delta \boldsymbol{r}}_i \, \underset{\sim}{\delta \boldsymbol{r}}_i \, k_{1i}(\tilde{\omega}_i;\, \tilde{\tilde{\omega}}) \prod_{j=1}^{S} [p_0(\tilde{\tilde{\omega}}_j, \tilde{\tilde{\Theta}}_j) \, d\tilde{\tilde{\omega}}_j \, d\tilde{\tilde{\Theta}}_j] \right] . \tag{6.46}$$

Integrating over all intersections, except the one under consideration at $\tilde{\eta}$, results in

$$D_{\text{eff}} = \frac{1}{2} \int_{\tilde{\tilde{\omega}}} \int_{\tilde{\tilde{\Theta}}} \sum_{i=1}^{M} [l_i^2 \, \underset{\sim}{\delta \boldsymbol{r}}_i \, \underset{\sim}{\delta \boldsymbol{r}}_i \, k_{1i}(\tilde{\omega}_i;\, \tilde{\tilde{\omega}}) \, p_0(\tilde{\tilde{\omega}}_i, \tilde{\tilde{\Theta}}_i) \, d\tilde{\tilde{\omega}}_i \, d\tilde{\tilde{\Theta}}_i] . \tag{6.47}$$

Substituting for $k_1(\tilde{\omega};\, \tilde{\tilde{\omega}})$ from Eq. (18b) into this expression gives

$$D_{\text{eff}} = \left(\frac{D_e}{L^2}\right) \int_{\tilde{\tilde{\omega}}} \int_{\tilde{\tilde{\Theta}}} \sum_{i=1}^{M} \left[l_i \, \underset{\sim}{\delta \boldsymbol{r}}_i \, \underset{\sim}{\delta \boldsymbol{r}}_i \, \frac{r_i^2 D_i}{\sum_{i=1}^{M} r_i^2 l_i} \, p_0(\tilde{\tilde{\omega}}_i, \tilde{\tilde{\Theta}}_i) \, d\tilde{\tilde{\omega}}_i \, d\tilde{\tilde{\Theta}}_i \right] . \tag{6.48}$$

Integrating formally, we have

$$D_{\text{eff}} = \left(\frac{D_e}{L^2}\right) \left\langle \sum_{i=1}^{M} \frac{l_i r_i^2 D_i \, \underset{\sim}{\delta \boldsymbol{r}}_i \, \underset{\sim}{\delta \boldsymbol{r}}_i}{\sum_{i=1}^{M} r_i^2 l_i} \right\rangle_0^{\approx} \tag{6.49}$$

where $\langle \cdot \rangle_0^{\approx}$ denotes the average over all possible intersection characteristics, utilizing the probability density function, $p_0(\cdot)$, for this purpose, i.e.,

$$\langle g \rangle_0^{\approx} = \int_{\tilde{\tilde{\omega}}} \int_{\tilde{\tilde{\Theta}}} g(\tilde{\tilde{\omega}}, \tilde{\tilde{\Theta}}, \tilde{\eta}, t) \, p_0(\tilde{\tilde{\omega}}, \tilde{\tilde{\Theta}}) \, d\tilde{\tilde{\omega}} \, d\tilde{\tilde{\Theta}} . \tag{6.50}$$

If all M pores at an intersection are identical and independent of each other, the simplification in Eq. (6.43a) may be utilized in Eq. (6.49), leading to

$$D_{\text{eff}} = \left(\frac{D_e}{L^2}\right) M \left\langle \frac{l_i r_i^2 D_i \, \delta\mathbf{r}_i \, \delta\mathbf{r}_i}{\sum_{i=1}^{M} r_i^2 l_i} \right\rangle_0^{\approx} . \tag{6.51}$$

For an approximation, we let

$$E\left[\frac{X}{Y}\right] \approx \frac{E[X]}{E[Y]} .$$

This approximation is reasonable when the pores are taken to be independent of each other. The numerator and denominator on the right-hand side of Eq. (6.51) may then be integrated separately to give

$$\langle l_i r_i^2 D_i \, \delta\mathbf{r}_i \, \delta\mathbf{r}_i \rangle_0^{\approx} \equiv \langle l r^2 D \, \delta\mathbf{r} \, \delta\mathbf{r} \rangle_0^{\sim}$$

and

$$\left\langle \sum_{i=1}^{M} r_i^2 l_i \right\rangle_0^{\approx} \equiv M \langle r^2 l \rangle_0^{\sim}$$

whereby Eq. (6.51) yields

$$D_{\text{eff}} = \left(\frac{D_e}{L^2}\right) \frac{\langle r^2 l D \, \delta\mathbf{r} \, \delta\mathbf{r} \rangle_0^{\sim}}{\langle r^2 l \rangle_0^{\sim}} . \tag{6.52}$$

The operator $\langle \cdot \rangle_0^{\sim}$ indicates that the averaging has been performed over all possible characteristics of an individual pore through the density function $f_0(\cdot)$, i.e.,

$$\langle g \rangle_0^{\sim} = \int_{\tilde{\omega}} \int_{\tilde{\Theta}} g(\tilde{\omega}, \tilde{\Theta}, \tilde{\eta}, t) \, f_0(\tilde{\omega}, \tilde{\Theta}) \, d\tilde{\omega} \, d\tilde{\Theta} . \tag{6.53}$$

D_{eff}, as obtained above, is a second-order tensor since the derivation is general and valid for anisotropic systems also. The diagonal elements represent diffusivities along the principal axes, while the off-diagonal elements arise due to the fact that in anisotropic systems, the flux vector may not be normal to the surfaces of constant concentration; this flux vector may point in a direction different from the concentration gradient vector. For an isotropic system (i.e., a system with random pore orientations), the diagonal elements will be equal and the off-diagonal elements will reduce to zero.

6.2.2 Drift or Convective Velocity

The drift velocity, v, will be extracted from the second term on the right-hand side of Eq. (6.39). Comparing this equation with Eq. (6.41),

$$v = -\left\langle \sum_{i=1}^{M} \left\{ \alpha^{-1} l_i k_{1i}(\tilde{\omega}_i;\, \tilde{\tilde{\omega}}) + l_i^2 k_{1i}(\tilde{\omega}_i;\, \tilde{\tilde{\omega}}) \left[\underset{\sim}{\delta \boldsymbol{r}}_i \cdot \frac{\nabla V}{V} \right] \right\} \underset{\sim}{\delta \boldsymbol{r}}_i \right\rangle_0 \tag{6.54}$$

or, from Eq. (6.36b),

$$v = -\int_{\{\tilde{\tilde{\omega}}\}} \int_{\{\tilde{\tilde{\Theta}}\}} \sum_{i=1}^{M} \left[\left\{ \alpha^{-1} l_i k_{1i}(\tilde{\omega}_i;\, \tilde{\tilde{\omega}}) + l_i^2 k_{1i}(\tilde{\omega}_i,\, \tilde{\tilde{\omega}}) \left[\underset{\sim}{\delta \boldsymbol{r}}_i \cdot \frac{\nabla V}{V} \right] \right\} \underset{\sim}{\delta \boldsymbol{r}}_i \right.$$
$$\left. \times\, p_{S0}(\{\tilde{\tilde{\omega}}\}, \{\tilde{\tilde{\Theta}}\})\, d\{\tilde{\tilde{\omega}}\}\, d\{\tilde{\tilde{\Theta}}\} \right]. \tag{6.55}$$

Following a development similar to that in Eqs. (6.46) and (6.47), we obtain

$$v = -\int_{\tilde{\tilde{\omega}}} \int_{\tilde{\tilde{\Theta}}} \sum_{i=1}^{M} \left[\alpha^{-1} l_i k_{1i}(\tilde{\omega}_i;\, \tilde{\tilde{\omega}})\, \underset{\sim}{\delta \boldsymbol{r}}_i\, p_0(\tilde{\tilde{\omega}}_i,\, \tilde{\tilde{\Theta}}_i)\, d\tilde{\tilde{\omega}}_i\, d\tilde{\tilde{\Theta}}_i \right]$$
$$- \int_{\tilde{\tilde{\omega}}} \int_{\tilde{\tilde{\Theta}}} \sum_{i=1}^{M} \left[l_i^2 k_{1i}(\tilde{\omega}_i;\, \tilde{\tilde{\omega}}) \left[\underset{\sim}{\delta \boldsymbol{r}}_i \cdot \frac{\nabla V}{V} \right] \underset{\sim}{\delta \boldsymbol{r}}_i\, p_0(\tilde{\tilde{\omega}}_i,\, \tilde{\tilde{\Theta}}_i)\, d\tilde{\tilde{\omega}}_i\, d\tilde{\tilde{\Theta}}_i \right]. \tag{6.56}$$

Substituting for $k_1(\cdot)$ and integrating yield

$$v = -2\alpha^{-1} \left(\frac{D_e}{L^2} \right) \left\langle \sum_{i=1}^{M} \frac{r_i^2 D_i\, \underset{\sim}{\delta \boldsymbol{r}}_i}{\sum_{i=1}^{M} r_i^2 l_i} \right\rangle_0^{\approx}$$
$$- 2 \left(\frac{D_e}{L^2} \right) \left\langle \sum_{i=1}^{M} \frac{r_i^2 l_i D_i \left[\underset{\sim}{\delta \boldsymbol{r}}_i \cdot \frac{\nabla V}{V} \right] \underset{\sim}{\delta \boldsymbol{r}}_i}{\sum_{i=1}^{M} r_i^2 l_i} \right\rangle_0^{\approx}. \tag{6.57}$$

The above expression represents the drift velocity obtained by averaging over all possible intersection characteristics, utilizing the density function $p_0(\cdot)$ for this purpose [see Eq. (6.50)]. A further simplification can result if the pore characteristics within each intersection are identically and independently distributed. The procedure is similar to that employed to derive Eq. (6.52) from Eq. (6.49); the result is

$$v = -\left(\frac{D_e}{L_2} \right) 2\alpha^{-1} \frac{\langle r^2 D\, \underset{\sim}{\delta \boldsymbol{r}} \rangle_0^{\sim}}{\langle r^2 l \rangle_0^{\sim}} - \left(\frac{D_e}{L^2} \right) 2 \frac{\left\langle r^2 l D \left[\underset{\sim}{\delta \boldsymbol{r}} \cdot \frac{\nabla V}{V} \right] \underset{\sim}{\delta \boldsymbol{r}} \right\rangle_0^{\sim}}{\langle r^2 l \rangle_0^{\sim}} \tag{6.58}$$

with the operator $\langle\cdot\rangle_{\tilde{0}}$ as defined in Eq. (6.53), based on pore characteristics density $f_0(\cdot)$.

Equation (6.58) indicates that the drift velocity is composed of two terms, the first being directly related to the distribution of pore orientations, and the second to the intersection void-volume (or porosity) distribution.

6.3 Transport Coefficients in Specific Systems

In the preceding section, general expressions have been obtained for the transport coefficients in porous or heterogeneous polymeric matrices, in terms of the distribution of intersection characteristics [Eqs. (6.49) and (6.57)]. The general expressions are simplified under the assumption of independently and identically distributed pore characteristics. The resultant simplified expressions, Eqs. (6.52) and (6.58), will be further simplified to specific cases in this section. Note that for Eq. (6.42), all intersections have been taken to be identically distributed. This would imply isotropy in the usual sense, i.e., each intersection is similar to the others. However, in the present work, the term "isotropic" is further limited to systems in which the pores within an intersection are also randomly oriented, i.e., each intersection, apart from being similar to all other intersections, also appears the same in all directions.

6.3.1 Mean or Effective Diffusion Coefficient

The factor (D_e/L^2) is required in the right-hand side of Eq. (6.52), defining the mean or effective diffusivity, to balance the dimensions. From our definition, D_{eff} has the dimension of $[T^{-1}]$ while D is dimensionless. Absorbing this factor on the left-hand side, we have

$$D^*_{\text{eff}} = \frac{\langle r^2 l D \,\underset{\sim}{\delta r}\, \underset{\sim}{\delta r} \rangle_{\tilde{0}}}{\langle r^2 l \rangle_{\tilde{0}}} \tag{6.59}$$

where

$$D^*_{\text{eff}} = D_{\text{eff}} \left(\frac{L^2}{D_e} \right).$$

D^*_{eff} is thus the dimensionless effective diffusivity that may be observed in a porous matrix over a number of experimental determinations. D is the corresponding diffusivity in a single pore; it can be considered as the molecular diffusivity, D_{mol}, if the mean free path of the molecules is sufficiently small compared to the pore radii (liquid and gaseous systems). On the other hand, if the mean free path is larger than or of similar magnitude to the pore radii (dilute gaseous systems only), D can be considered to be the Knudsen diffusivity, D_{knud}.

Further simplifications of the expression for effective diffusivity, Eq. (6.59), will depend on the nature of the pore characteristics, given by the density function $f_0(\tilde{\omega}, \tilde{\Theta})$.

a) Uncorrelated pore orientation and dimensions. For a matrix in which the pore orientations are independent of their dimensions, we may write

$$f_0(\tilde{\omega}, \tilde{\Theta}) = h_0(\tilde{\omega})\, g_0(\tilde{\Theta}) \,. \tag{6.60}$$

Under this condition, the expression for D^*_{eff} reduces to

$$D^*_{\text{eff}\,1} = \frac{\langle r^2 l D \rangle_0^{\sim}}{\langle r^2 l \rangle_0^{\sim}} \langle \underset{\sim}{\delta r}\, \underset{\sim}{\delta r} \rangle_0^{\sim} \tag{6.61}$$

since D, in general, does not depend on pore orientation. Rearranging Eq. (6.61) leads to the definition of D^*_{eff} in terms of the pore-volume-averaged diffusivity $\bar{D}$, i.e.,

$$D^*_{\text{eff}\,1} = \frac{\bar{D}}{\tau} \tag{6.62}$$

where

$$\tau = [\langle \underset{\sim}{\delta r}\, \delta r \rangle]^{-1} \tag{6.63a}$$

and

$$\bar{D} = \frac{\langle r^2 l D \rangle_0^{\sim}}{\langle r^2 l \rangle_0^{\sim}} \,. \tag{6.63b}$$

This definition of $\bar{D}$ was originally derived by Johnson and Stewart (1965), by considering the diffusion flux per unit of void plus non-void area. The random walk model derived here has naturally given rise to this definition of $\bar{D}$. Wang and Smith (1983) have demonstrated that this expression is the most suitable when attempting to measure the tortuosity factors in matrices with a wide range of pore sizes.

b) Constant pore length, and uncorrelated pore orientation and radius. This case leads to the simplified expression

$$D^*_{\text{eff}\,2} = \frac{\langle r^2 D \rangle_0^{\sim}}{\langle r^2 \rangle_0^{\sim}} \langle \underset{\sim}{\delta r}\, \underset{\sim}{\delta r} \rangle \,. \tag{6.64}$$

c) Randomly oriented pores, and uncorrelated orientation and dimensions. In such a system, the pore orientations are uniformly distributed, i.e., the angular distribution of θ is sinusoidal while that of Φ is uniform (De Jong, 1958; Bhatia, 1986). Thus, the density function for orientations becomes

$$g(\tilde{\Theta}) = \frac{\sin\theta}{4\pi}, \qquad 0 \leqq \theta \leqq \pi \,; \qquad 0 \leqq \Phi \leqq 2\pi \,. \tag{6.65}$$

The above distribution may be employed to evaluate $\langle \underset{\sim}{\delta r}\, \underset{\sim}{\delta r} \rangle$, thereby giving

$$\langle \underset{\sim}{\delta r}\, \underset{\sim}{\delta r} \rangle = \frac{\bar{\delta}}{3} \tag{6.66}$$

where $\bar{\delta}$ is the identity tensor (diagonal elements equal to 1, and off-diagonal elements equal to zero). The symmetric nature of this tensor indicates the isotropy of the system due to the randomness of the pore orientation. Substituting Eq. (6.66) into Eq. (6.61) leads to

$$D^*_{\text{eff}\,3} = \frac{1}{3} \frac{\langle r^2 l D \rangle_0^{\sim}}{\langle r^2 l \rangle_0^{\sim}} \bar{\delta}\,. \tag{6.67}$$

Moreover, from Eq. (6.63a) we have

$$\tau = 3\,. \tag{6.68}$$

The tortuosity value predicted by this model for an isotropic system is identical to that calculated by Johnson and Stewart (1965) and Feng and Stewart (1973) employing the mass-flux model.

In all specific cases considered above, D can be the molecular diffusivity, D_{mol}, or the Knudsen diffusivity; D_{knud}, or their combination, depending upon the nature of the diffusion phenomenon in a single pore. Each of these is a function of the pore radius. However, if the pore radii are large enough for the diffusivity in the pore to be equal to bulk fluid diffusivity (i.e., in the absence of pore walls),

$$D = D_{\text{mol}} \equiv D_{\text{bulk}}\,. \tag{6.69}$$

Then, from Eq. (6.65), we obtain

$$D^*_{\text{eff}\,4} = \frac{D_{\text{bulk}}}{3} \bar{\delta}\,. \tag{6.70}$$

This corresponds to the well-known result for an isotropic system (Satterfield, 1970).

For the case of small pore radii, in the Knudsen regime for gaseous systems

$$D = D_{\text{knud}} \equiv \mu r \tag{6.71}$$

where μ is a constant that depends on the mass of the diffusing molecule and temperature of the system only (Smith, 1970). Thus, from Eq. (6.67),

$$D^*_{\text{eff}\,5} = \mu \frac{1}{3} \frac{\langle r^3 l D \rangle_0^{\sim}}{\langle r^2 l \rangle_0^{\sim}} \bar{\delta}\,. \tag{6.72}$$

In a matrix system where the fluid filling the pores is a gas, and the radii of the permeable regions are in the range between those for the bulk and Knudsen diffusivities, D in Eq. (6.67) can be written as an average of the form (Satterfield, 1970)

$$D = [D_{\text{bulk}}^{-1} + D_{\text{knud}}^{-1}]^{-1} \equiv [D_{\text{bulk}}^{-1} + (\mu r)^{-1}]^{-1} \tag{6.73}$$

and used to evaluate the effective diffusivity.

6.3.2 Drift or Convective Velocity

The zero-order drift velocity coefficient has been obtained, as expressed in Eq. (6.58), in which v has the dimensions of $[T^{-1}]$. Absorbing the factor (D_e/L^2) into the left-hand side of the equation, the dimensionless drift velocity, v^*, becomes

$$v^* = -2\alpha^{-1}\frac{\langle r^2 D\,\delta\mathbf{r}\rangle_0^{\sim}}{\langle r^2 l\rangle_0^{\sim}} - 2\frac{\left\langle r^2 lD\left[\delta\mathbf{r}\cdot\dfrac{\nabla V}{V}\right]\delta\mathbf{r}\right\rangle_0^{\sim}}{\langle r^2 l\rangle_0^{\sim}}\,. \tag{6.74}$$

Different pore characteristics, as reflected in the function $f_0(\tilde{\omega}, \tilde{\Theta})$, result in various forms of v^*.

a) Constant porosity. If the pore volume is constant at all points in the matrix, i.e., $V(\tilde{\eta})$ is not a function of the position,

$$v_1^* = -2\alpha^{-1}\frac{\langle r^2 D\,\delta\mathbf{r}\rangle_0^{\sim}}{\langle r^2 l\rangle_0^{\sim}}\,. \tag{6.75}$$

b) Uncorrelated pore orientation and dimensions. If the dimensions of the pores are distributed independent of their orientations [Eq. (6.60)], we have

$$v_2^* = -2\alpha^{-1}\frac{\langle r^2 D\rangle_0^{\sim}}{\langle r^2 l\rangle_0^{\sim}}\langle\delta\mathbf{r}\rangle - 2\frac{\left\langle r^2 lD\left[\delta\mathbf{r}\cdot\dfrac{\nabla V}{V}\right]\right\rangle_0^{\sim}}{\langle r^2 l\rangle_0^{\sim}}\langle\delta\mathbf{r}\rangle\,. \tag{6.76}$$

c) Constant pore length, and uncorrelated pore orientation and radius. This case results in

$$v_3^* = -2\alpha^{-1}\frac{\langle r^2 D\rangle_0^{\sim}}{\langle r^2\rangle_0^{\sim}}\langle\delta\mathbf{r}\rangle - 2\frac{\left\langle r^2 D\left[\delta\mathbf{r}\cdot\dfrac{\nabla V}{V}\right]\right\rangle_0^{\sim}}{\langle r^2\rangle_0^{\sim}}\langle\delta\mathbf{r}\rangle\,. \tag{6.77}$$

d) Randomly oriented pores, and uncorrelated orientation and dimensions. Randomly oriented pores have the density function given in equation (6.65), thereby leading to

$$\langle\delta\mathbf{r}\rangle = 0 \tag{6.78}$$

since all directions are equally probable. In such a system (isotropic by definition), from Eq. (6.76),

$$v_4^* \equiv 0 . \tag{6.79}$$

The drift velocity is zero in an isotropic matrix since the frequencies of molecular jumps are identical in all directions; in the absence of a concentration gradient, the geometry of the pores does not favor any particular direction. Hence, in the usual analyses of porous systems, the absence of the drift velocity indicates that the more restrictive assumption of isotropy in terms of pore orientations has been implicitly invoked..

e) Constant porosity, uncorrelated orientation and dimensions, and non-uniform orientation. Under these conditions

$$v_5^* = -2\alpha^{-1} \frac{\langle r^2 D \rangle_0^{\sim}}{\langle r^2 l \rangle_0^{\sim}} \langle \underset{\sim}{\delta r} \rangle . \tag{6.80}$$

For a constant pore length ($l_i = \bar{l}$)

$$v_6^* = -\frac{2\alpha^{-1}}{\bar{l}} \frac{\langle r^2 D \rangle_0^{\sim}}{\langle r^2 \rangle_0^{\sim}} \langle \underset{\sim}{\delta r} \rangle \tag{6.81}$$

or

$$v_6^* = -\frac{2\alpha^{-1}}{\bar{l}} D_{\mathrm{eff}\,2}^* \frac{\langle \underset{\sim}{\delta r} \rangle}{\langle \underset{\sim}{\delta r}\, \underset{\sim}{\delta r} \rangle} . \tag{6.82}$$

This equation implies that in a system with non-uniform pore orientations, the drift term could be significant; a preferred pore orientation generates a greater probability for movement in a particular direction, which in turn, generates the drift or convective flux over and above the diffusive flux.

6.4 Monte Carlo Simulations of Transport in Porous Media

The effects of the pore orientation distribution on transport in porous media have been examined numerically by Monte Carlo simulations. Estimates of transport coefficients in different networks have been recovered. Two varieties of networks have been generated for this purpose. One variety includes uniform networks in which the positions of the intersections are uniformly distributed, and the other includes skewed networks in which certain quadrants are more probable than others. Reaction in the pores is not considered, and no attempt has been made to obtain complete concentration profiles in the system. The pores are assumed to have constant pore radii, but variable lengths and orientations. (For a detailed discussion of Monte Carlo simulations of the master equation with examples, see Fox, 1987; also see, e.g., Rajamani et al., 1986.)

6.4.1 Procedure

The simulation procedure essentially involves the generation of a pore network, calculation of the transition intensities and the associated probabilities, determination of the event (jump of a molecule from one intersection to another), and its inter-arrival time (time elapsed between two events). The probabilities, event, and times are determined repeatedly until the requisite number of events have occurred or the desired time elapsed. The procedure is detailed in Singh (1987).

The following three types of networks are created here in a $1 \times 1 \times 1$ cube for the purpose of simulation.

a) Uniform in all three directions:
 {intersection coords. $(u_1 - 0.5, u_2 - 0.5, u_3 - 0.5)$; $u_1, u_2, u_3 \in U[0, 1]$}
b) Skewed to the right in "x", and uniform in the other two:
 {intersection coords. $(2u_1 - u_1^2 - 0.5, u_2 - 0.5, u_3 - 0.5)$; $u_1, u_2, u_3 \in U[0, 1]$}
c) Skewed to the right in all three directions:
 {intersection coords. $(2u_1 - u_1^2 - 0.5, 2u_2 - u_2^2 - 0.5, 2u_3 - u_3^2 - 0.5)$; $u_1, u_2, u_3 \in U[0, 1]$}.

The uniform network approximates an isotropic network most closely; the other two types have preferential pore orientations. The pore network in each case was formed by first determining the intersection positions according to the type of network desired. These intersections were then connected to M of their nearest neighbors, starting from the intersection at the origin (Singh, 1987). Different networks within each type were created by employing different sets of random variables. A coordination number, M, of 5 was employed.

The transition intensities within each pore were obtained from Eq. (6.18a), under the assumptions of constant pore radii, and

$$D_e = 1\,(L^2 T^{-1}),$$

$$D = 1\,(-),$$

and

$$L = 1\,(L)$$

from the dimensions of the cube. At time $t = 0$, the molecules were distributed among the intersections such that the intersection at the origin had 3000, intersections within a radius of 0.1 units of the origin had 1000, intersections at a distance between 0.1 and 0.2 units from the origin had 500, those between 0.2 and 0.3 units 300, those between 0.3 and 0.4 units 100, and those at a distance greater than 0.4 units no molecules. The transitions made by the molecules were followed along with their inter-arrival times, providing information about the number of molecules at each intersection at various times.

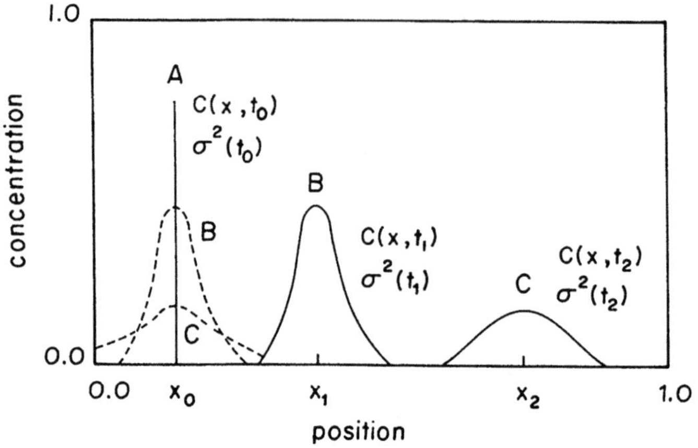

Fig. 6.2. Schematic diagram illustrating significance of the diffusivity and drift velocity

Effective diffusivities and drift velocities in the networks have been recovered from the results of simulation. The effective diffusivity can be estimated from the mean square displacement of the molecules around the center of mass per unit time, while the rate of movement of this center of mass is the drift velocity (see, e.g., Fan and Shin, 1979). This is illustrated in Fig. 6.2 in one dimension. "A" is the initial distribution of molecules at time t_0, assumed to be a delta-function. Curves B and C represent the distributions at times t_1 and t_2, respectively. The means (centers of mass) of distributions B and C are at x_1 and x_2, respectively, while their variances are $\sigma_x^2(t_1)$ and $\sigma_x^2(t_2)$, respectively. The mean of the distribution A is at x_0, while its variance is zero. The average diffusivity between points x_0 and x_1, x_0 and x_2, or a general set x_0 and x_j, is then given by (Fan and Shin, 1979)

$$D(x_0 \rightarrow x_j) \approx \frac{1}{t_j - t_0} \frac{1}{2} \sigma_x^2(t_j) , \qquad j = 1, 2, \ldots .$$

Similarly, the drift velocity is

$$v(x_0 \rightarrow x_j) \approx \frac{x_j - x_0}{t_j - t_0} , \qquad j = 1, 2, \ldots .$$

(When the drift velocity is negligible, the initial concentration distribution A undergoes a transition to B′ and C′ as shown in Fig. 6.2.) The above relations were employed to obtain estimates of diffusivities and drift velocities in all three coordinate axes, using the variances and centers of mass corresponding to the axis under consideration.

(The simulations were performed on the Cray–XMP supercomputer at the National Center for Supercomputing Applications at the University of Illinois, Urbana-Champaign. The program-runs were time intensive; each run was limited, therefore, to 30 minutes CPU time; this resulted in approximately 20,000 transition events.

The steady state was not approached in any of the runs. The algorithm and FORTRAN code are provided in APPENDIX 6C.)

A set of simulation was also performed starting with all molecules placed at the origin. In this case, however, the vast majority of transitions were from the intersection at the origin to the intersections connected to it. A preponderance of these transitions biased the estimates of the various coefficients. This problem could be avoided by running the program until a near-steady-state is achieved, but would require a large amount of CPU time. On the other hand, with the molecules distributed among a larger number of intersections initially, as detailed in a previous paragraph, the transition will be well distributed and provide better estimates of the coefficients.

The relations derived earlier in this chapter can be employed to estimate values of the transport coefficients in the networks generated. This requires creation of the pore network only; knowledge of the positions of the intersections and pore lengths allows the pore orientation vectors to be readily obtained, and thence the averages in Eqs. (6.44) and (6.54) over the complete matrix. Only the diagonal elements in the diffusivity tensor were estimated, although the other elements could also be evaluated with this procedure. Due to the difficulty in calculating the term

$$\left[\underset{\sim}{\delta \boldsymbol{r}} \cdot \frac{\nabla V}{V}\right] \underset{\sim}{\delta \boldsymbol{r}}$$

for the network generated, the drift velocities were estimated only with the first term on the right-hand side of Eq. (6.54). As with the others, these networks were generated on the Cray – XMP.

6.4.2 Results

The results of the simulations are presented in Table 6.1. The parameter values obtained from the Monte Carlo runs are given in the table in the column labelled "simulated". The diffusivities range between 0.280 and 0.340; the skewness of the network does not show any particular effect on these values. The type of network is more clearly reflected in the drift velocities with those for uniform networks being approximately equal to zero. This is to be expected since the uniform network generated here corresponds most closely to an isotropic system [Eq. (6.79)].

In the x-skewed network, the drift velocity in the x-direction is greater than zero while those in the y- and z-directions are almost zero (Table 6.1). The pores in this type of network are preferentially oriented towards the positive-x axis, resulting in a higher probability of movement or "drift" towards this direction.

In the third type of network considered here, the preferred orientation of the pores is towards the quadrant with positive x, y, and z. This is also reflected in the values of the drift velocities (Table 6.1).

Table 6.1. Comparison of the transport coefficients obtained from Monte Carlo simulations with those estimated from Eqs. (6.44) and (6.54).

Network	Coefficient	Simulated[a] [Monte Carlo]	Estimated[b] [Eqs. (6.44), (6.54)]
uniform	D_{xx}	0.30376	0.33625
	D_{yy}	0.32304	0.33480
	D_{zz}	0.30951	0.32894
	v_x	0.10549	−0.00117
	v_y	−0.07225	−0.00233
	v_z	−0.05545	−0.00843
x-skewed[c]	D_{xx}	0.34006	0.31742
	D_{yy}	0.30282	0.34373
	D_{zz}	0.30621	0.33884
	v_x	0.51780	0.20464
	v_y	−0.03823	−0.00687
	v_z	−0.13722	−0.00774
x,y,z-skewed[c]	D_{xx}	0.33518	0.33572
	D_{yy}	0.28938	0.33307
	D_{zz}	0.32420	0.33120
	v_x	0.27787	0.24974
	v_y	0.55472	0.21122
	v_z	0.26858	0.20518

[a] average over 3 Monte Carlo runs per type of network
[b] average over 8 pore networks generated per type of network
[c] skewed by taking the appropriate coordinates $= (2u - u^2 - 0.5)$; $u \in U\,[0, 1]$

The transport coefficients estimated with Eqs. (6.44) and (6.54) are listed in Table 6.1 in the column labelled "estimated". These estimated values correspond closely to those expected from a consideration of the nature of the network. The agreement between the estimated and simulated values is very good in most cases. The discrepancies between the simulated and estimated values for the drift velocities may be a consequence of the fact that only three Monte Carlo simulations were performed for each type of network; better averages will require a large number of runs due to the wide variations in the estimates from each run. Furthermore, only part of Eq. (6.54) was employed to estimate these values for the skewed networks. Nevertheless, the difference between the simulated and estimated values is not large. Thus, the first part of Eq. (6.54) may suffice to obtain a reasonable estimate of this transport coefficient in most cases.

The relatively large magnitude of the drift velocities is due to the large average pore lengths ($\approx$0.12 units) in the networks created, thus resulting in large average displacements and, therefore, large v's. A denser network will lead to smaller pores and smaller v's; however, the diffusivities will not be influenced.

6.5 Concluding Remarks

To summarize the conclusions of the preceding sections of this chapter, the master equation for the position of a diffusing molecule has been derived and reduced to the form

$$\frac{\partial\langle C\rangle_0}{\partial t} = D_{\mathrm{eff}} : \nabla\nabla\langle C\rangle_0 - v \cdot \nabla\langle C\rangle_0 - K_R\langle C\rangle_0 \,. \tag{6.41}$$

The subscript 0 indicates the mean value. From Eq. (6.41), expressions have been obtained for the mean diffusion coefficient and the drift or convective velocity in terms of the distribution of the intersection of pore characteristics.

The resultant expression for the mean or effective diffusion coefficient in terms of pore characteristics distribution, $f_0(\tilde{\omega}, \tilde{\Theta})$, is given in Eq. (6.59). While the expression is not very different from the definition provided by Johnson and Stewart (1965) [see Eq. (6.62)], it is somewhat more general, having been derived for anisotropic systems. More important, it has been arrived at from a mechanistic consideration of the diffusion phenomenon without resorting to the Fickian formulation. In this sense, the present analysis provides a rigorous basis for the use of a pore-volume based effective diffusivity in porous media, furthermore it reduces to the conventional form for isotropic media [see Eq. (6.67)].

The drift velocity, given in Eq. (6.74) in terms of the pore characteristics distribution, is a consequence of the pore structure of the matrix that makes the probability of jumps of the individual molecules in one direction higher than in others. As stated earlier, it should not be confused with convective transport caused by the flow of bulk fluid within the pores.

An isotropic system, defined here in a restrictive sense (identical pore intersections along with random pore orientations within an intersection) to will reduce the convective term to zero [Eq. (6.79)]. Most analyses of transport data from porous or heterogeneous polymers do not take this factor into account, thereby implicitly assuming such a restrictive system.

As $\alpha \to 0$, the continuum limit is approached. From the definition of the continuum approximation, we know that a diffusing molecule is able to move in all directions in such a medium. Thus, in the limit of $\alpha \to 0$, we have $\langle \delta \mathbf{r} \rangle \to 0$. The drift velocity, therefore, does not exist in a homogeneous system.

The differential equation obtained by Bhatia (1986) from a consideration of correlated random walk in a porous system is of the form

$$\frac{\partial C'}{\partial t} = \nabla_x \cdot (D'_{\mathrm{eff}} - D_{\mathrm{corr}})\,\nabla_x C' - K'C'$$

where D_{corr} is a correction factor for the effective diffusivity, D'_{eff}, arising as a result of the correlations between successive random walk steps (Bhatia, 1986). The analysis of Bhatia (1986) includes the *a priori* assumption of random or uniform pore orientation; because of this assumption, the drift velocity is not recovered. The general expression obtained here for the mean or effective diffusivity reduces to that obtained by Bhatia (1986), under this assumption of randomness of pore orientation [see Eq. (6.67)]. As a consequence of neglecting correlations

Fig. 6.3. A cross-linked pore network

between intersections, the present analysis does not have a term corresponding to the correction factor. Nevertheless, this term is embodied in the first-order correction equation, Eq. (6.38), obtained here.

Applying the two models [that of Bhatia (1986) and the present] to a cross-linked pore network (Fig. 6.3) also leads to identical results. In such a network, the correction factor of Bhatia (1986) is equal to zero, while the convective term of the present model drops out.

Monte Carlo simulations of transport in porous networks prove the existence of the convective term when analyzing systems with preferred pore orientations. The relations derived here may be utilized to estimate the drift velocities, given the knowledge of the pore parameters. It is apparent from the present analysis that the drift velocity can affect the observed flux in certain systems; when not accounted for, it will affect the calculated diffusion coefficient and tortuosity. Further work, especially Monte Carlo simulation, is needed to elucidate the detailed nature of this factor and to analyze its dependence on system parameters.

The use of Eq. (6.1) instead of Eq. (6.41) (without the reaction term), to analyze diffusion in porous/heterogeneous polymeric matrices, is particularly inappropriate since these matrices are usually anisotropic. The presence of a convective term can lead to differences between expected and observed concentrations. These differences can be particularly critical if the matrix is employed to deliver a pharmaceutically active agent. Depending on the distribution of the permeable regions, the concentration obtained may be lower or higher than desired, with the concomitant loss of performance of the device.

Some of the polymeric systems being employed/examined for drug release satisfy this anisotropy criterion. Among the systems reported are those made by mixing hydrophobic polymers (usually in powder form) with a drug (usually solid) and compressing the mixture below their melting points. This produces tablets or sustained release devices with irregular pores. When placed in an appropriate medium (e.g., water for water-soluble drugs), release occurs by dissolution of the drug in the water-filled pores followed by diffusion (Gurny et al., 1982). As dissolution progresses, the size and shape of the pores change continuously. In attempting to predict the release of KCl from ethyl cellulose tablets with a diffusion–

dissolution model, Gurny et al. (1982) have noted that the deviation from prediction becomes large as the tablet proceeds from diffusion to dissolution control. They have attributed the deviation to the change in void fraction and porosity caused by depletion of the drug in polymer. It is also worth noting that the deviation appears to increase as the system becomes dilute (due to slow dissolution rate).

Successes have been achieved in delivering macromolecular drugs from hydrophilic polymers (see, e.g., Langer and Folkman, 1978). Langer and his coworkers have employed solvent casting techniques with ethylene-vinyl acetate copolymers to prepare their matrices. This involves the dissolution of drug and polymer in a suitable solvent which is subsequently evaporated, producing a formation with microchannels. When placed in water, these polymers undergo swelling, thereby causing the size and shapes of the microchannels to change continuously. Since the swelling process progresses inwards, there is a gradient in the extent of swelling undergone by a pore depending on the position of the pore. Lapidus and Lordi (1968), have obtained a similar system by compressing mixtures of drug and hydrophilic polymer granules to obtain tablets. Four out of the 15 tortuosity values calculated by them lie in the range 4–6. A possible explanation for these high tortuosity values can be given in terms of the analysis presented here. Leaching of the active agent from these matrices will give rise to a pore structure with fine finger-like pore projections on the inside, connected to a few pores on the surface (somewhat like a tree with the trunk connecting the surface to the roots in the interior). The drift velocity in these matrices will be towards the interior and will thus slow the release rate. The calculated effective diffusivity, when data are analyzed based on the Fickian equation, will be small and will lead to high estimated tortuosities.

The structure of gels (e.g., hydrogels used in swelling-controlled release systems) has also been found to be heterogeneous. These gels are characterized by the existence of extensively organized (crystalline) regions containing a large amount of the gel substance, between which a small number of long coiling polymer chains establish the connections and hold the gel together topologically. The organized regions act as effective crosslinks and the chain segments between them as effective network segments (Silberberg, 1976). While the network segment distribution can be assumed to be isotropic inside the gel, this will not be true at the surface. The surface layers will be much more diffuse than the interior of the gel. After a study of three model gel systems, Pines and Prins (1973) have concluded that a structural model for gels, based on randomly crosslinked Gaussian chains, is often inadequate.

Gels formed on the surface of swelling-controlled release systems during the release process are further constrained by the inner glassy core. The above mentioned heterogeneity, between the surface and the interior, will be even more pronounced in these gels than in gels in isolation.

The stochastic formulation of concentration-gradient-driven transport in porous media presented in this chapter has permitted a rigorous justification for using an equation of the diffusive–convective type for analysis. The assumptions and restrictions involved in reducing this equation to the Fickian diffusion form have been elaborated. General relations between the transport coefficients and pore structure parameters have been recovered. The expression for the effective diffusivity reduces to that appearing in the available literature, under certain simplifying assumptions. Similarly, the expression for the drift velocity enables us to elucidate the conditions under which this term can be neglected, as is usually the case. The expressions and analysis presented here, therefore, provide a method of estimating, qualitatively and quantitatively, the role of various structural factors in the phenomenon of transport in such media.

References

Akanni, K. A., Evans, J. W., Abramson, I. S.: Effective Transport Coefficients in Heterogeneous Media, Chem. Eng. Sci. *42*, 1945–1954 (1987)

Aris, R.: The Mathematical Theory of Diffusion and Reaction in Permeable Catalysts, Vol. I, The Theory of the Steady State, pp. 23–25, Clarendon Press, Oxford (1975)

Bharucha-Reid, A. T.: Elements of the Theory of Markov Processes and Their Applications, pp. 129–163, McGraw-Hill, New York (1960)

Bhatia, S. K.: Directional Autocorrelation and the Diffusional Tortuosity of Capillary Porous Media, J. Catalysis *93*, 192–196 (1985)

Bhatia, S. K.: Stochastic Theory of Transport in Inhomogeneous Media, Chem. Eng. Sci. *41*, 1311–1324 (1986)

Chou, S. T., Fan, L. T., Nassar, R.: Stochastic Analysis of Stepwise Cellulose Degradation Data, Presented at the AIChE 1987 Summer National Meeting, Minneapolis, Minnesota, August 16–19 (1987).

De Jong, G. De J.: Longitudinal and Transverse Diffusion in Granular Deposits, Trans. Am. Geophys. Union *39*, 67–74 (1958)

Duggirala, S. K.: Stochastic Analysis of Particulate Processes-A Study of Attrition, Sieving and Grinding, M. S. Thesis, Department of Chemical Engineering, Kansas State University, Manhattan, Kansas (1986)

Fan, L. T., Shin, S. H.: Stochastic Diffusion Model of Non-Ideal Mixing in a Horizontal Drum Mixer, Chem. Eng. Sci. *34*, 811–820 (1979)

Fatt, I.: The Network Model of Porous Media, Parts I, II, III, Petroleum Transactions, AIME *207*, 144–181 (1956)

Feng, C., Stewart, W. E.: Practical Models for Isothermal Diffusion and Flow of Gases in Porous Solids, Ind. Eng. Chem. Fundam. *12*, 143–147 (1973)

Foster, R. N., Butt, J. B.: A Computational Model for the Structure of Porous Materials Employed in Catalysis, AIChE J. *12*, 180–185 (1966)

Fox, R. O.: Master Equation Formulation and Monte Carlo Simulation of Some Multinomial Stochastic Processes from Chemical Engineering, PhD Dissertation, Kansas State University, Manhattan, Kansas (1987)

Gavalas, G. R., Kim, S.: Periodic Capillary Models of Diffusion in Porous Solids, Chem. Eng. Sci. *36*, 1111–1122 (1981)

Gurny, R., Doelker, E., Peppas, N. A.: Modelling of Sustained Release of Water-Soluble Drugs from Porous, Hydrophobic Polymers, Biomaterials *3*, 27–32 (1982)

Harland, R. S., Peppas, N. A.: Drug Diffusion in Heterogeneous Polymeric Systems with Impermeable Domains, pp. 153–154. In: Proceedings of the 13th International Symposium on Controlled Release of Bioactive Materials, Chaudry, I., Theis, C. (Eds.), August 3–6 Norfolk, Virginia (1986)

Haring, R. E., Greenkorn, R. A.: A Statistical Model of a Porous Medium with Non-uniform Pores, AIChE J. *16*, 477–483 (1970)

Johnson, M. F. L., Stewart, W. E.: Pore Structure and Gaseous Diffusion in Solid Catalysts, J. Catalysis *4*, 248–252 (1965)

Langer, R., Folkman, J.: Sustained Release of Macromolecules from Polymers, pp. 175–196. In: Polymeric Delivery Systems, Kostelnik, R. J. (Ed.), Gordon and Breach Science Publishers, New York (1978)

Lapidus, H., Lordi, N. G.: Drug Release from Compressed Hydrophilic Matrices, J. Pharm. Sci. *57*, 1293–1301 (1968)

Manning, J. R.: Diffusion in a Chemical Concentration Gradient, Phys. Rev. *124*, 470–482 (1961)
Manning, J. R.: Correlated Walk and Diffusion Equations in a Driving Force, Phys. Rev. *139*, A 126–A 135 (1965)
Mason, E. A., Malinauskas, A. P., Evans III, R. B.: Flow and Diffusion of Gases in Porous Media, J. Chem. Phys. *46*, 3199–3216 (1967)
McGregor, R.: Diffusion and Sorption in Fibers and Thin Films, Vol. I., pp. 193–218, Academic Press, New York (1974)
Mo, W. T., Wei, J.: Effective Diffusivity in Partially Blocked Zeolite Catalyst, Chem. Eng. Sci. *41*, 703–710 (1986)
Nassar, R., Too, J. R., Fan, L. T.: Stochastic Modeling of Polymerization in a Continuous Flow Reactor, J. Appl. Polym. Sci. *26*, 3745–3759 (1981)
Pace, R. J., Datyner, A.: Statistical Mechanical Model for Diffusion of Simple Penetrants in Polymers. I. Theory, J. Polym. Sci., Polym. Phys. Ed. *17*, 437–451 (1979)
Peterlin, A.: Transport Phenomena and Polymer Morphology, Makromolekulare Chem. Suppl. *3*, 215–232 (1979)
Pines, E., Prins, W.: Structure-Property Relations of Thermoreversible Macromolecular Hydrogels, Macromolecules *6*, 888–895 (1973)
Pismen, I. M.: Diffusion in Porous Media of a Random Structure, Chem. Eng. Sci. *29*, 1227–1236 (1974)
Rajamani, K., Pate, W. T., Kinneberg, D. J.: Time-Driven and Event-Driven Monte Carlo Simulations of Liquid-Liquid Dispersions: A Comparison, Ind. Eng. Chem. Fundam. *25*, 746–752 (1986)
Ryan, D., Carbonell, R. G., Whitaker, S.: A Theory of Diffusion and Reaction in Porous Media. In: Transport with Chemical Reactions, AIChE Symposium Series 202 *77*, 46–62 (1981)
Saffman, P. G.: A Theory of Dispersion in a Porous Medium, J. Fluid Mech. *6*, 321–349 (1959)
Satterfield, C. N.: Mass Transfer in Heterogeneous Catalysis, pp. 1–77, MIT Press, Cambridge, Massachusetts (1970)
Scheidegger, A. E.: Statistical Approach to Miscible Displacement in Porous Media, The Canadian Mining and Metallurgical Bulletin, pp. 26–30, January (1959)
Scheidegger, A. E.: Statistical Theory of Flow Through Porous Media. Trans. Soc. Rheol. *9*, 313–319 (1965)
Seinfeld, J. H., Lapidus, L.: Mathematical Models in Chemical Engineering, Vol. 3, pp. 225–229, Prentice Hall, Englewood Cliff, New Jersey (1974)
Silberberg, A.: The Hydrogel-Water Interface. In: Hydrogels for Medical and Related Applications, J. D. Andrade (Ed.), ACS Symposium Series *31*, 198–205 (1976)
Singh, S. K.: Analysis and Development of Some Controlled Release Systems, pp. III.48–III.56, PhD Dissertation, Kansas State University, Manhattan, Kansas (1987)
Smith, J. M.: Chemical Engineering Kinetics, pp. 268–269, McGraw-Hill, New York (1970)
Theodorou, D., Wei, J.: Diffusion and Reaction in Blocked and High Occupancy Zeolite Catalysts, J. Catalysis *83*, 205–224 (1983)
Too, J. R., Nassar, R., Chou, S. T., Fan, L. T.: Stochastic Analysis of Crystallization in an Open Flow System, J. Chinese I. Ch. E. *17*, 303–313 (1986)
van Kampen, N. G.: Stochastic Processes in Physics and Chemistry, North-Holland, Amsterdam, Netherlands (1981)
Vrentas, J. S., Duda, J. L.: Molecular Diffusion in Polymer Solutions, AIChE J. *25*, 1–24 (1979)
Wakad, N., Nardse, Y.: Effective Diffusivities and Deadend Pores, Chem. Eng. Sci. *29*, 1304–1306 (1974)
Wakao, N., Smith, J. M.: Diffusion in Catalyst Pellets, Chem. Eng. Sci. *17*, 825–834 (1962)
Wang, C.-T., Smith, J. M.: Tortuosity Factors for Diffusion in Catalyst Pellets, AIChE J. *29*, 132–136 (1983)
Whitaker, S.: Advances in Theory of Fluid Motion in Porous Media, Ind. Eng. Chem. *61* (12), 14–28 (1969)

Appendix 6A. Conversion of Probability to Number Concentration

Let us define

$U(\tilde{\eta}, t \mid \{\tilde{\omega}\}, \{\tilde{\Theta}\})$

= Number of molecules at (in the neighborhood of) the position $\tilde{\eta}$ at time t, given that the state of the matrix is $(\{\tilde{\omega}\}, \{\tilde{\Theta}\})$.

If N is the number of molecules in the matrix at time t, then we have

$$U(\tilde{\eta}, t \mid \{\tilde{\omega}\}, \{\tilde{\Theta}\}) = \sum_{i=1}^{N} I_{(\tilde{\eta}, \tilde{\eta} + d\tilde{\eta})}(\tilde{x}_i) \qquad (6A.1)$$

where $I(\cdot)$ is the indicator function and $\tilde{x}_i$ represents the position of the ith molecule

$$I_{(\tilde{\eta}, \tilde{\eta} + d\tilde{\eta})}(\tilde{x}_i) = \begin{cases} 1 & \text{if } \tilde{x}_i \in (\tilde{\eta}, \tilde{\eta} + d\tilde{\eta}) \\ 0 & \text{otherwise} \end{cases}. \qquad (6A.2)$$

Equation (6A.1) thus represents a count of the number of molecules in the neighborhood of $\tilde{\eta}$. The expected value of U is

$$E[U(\tilde{\eta}, t \mid \{\tilde{\omega}\}, \{\tilde{\Theta}\})] = \int_{\{\tilde{\chi}\}} \left[\sum_{i=1}^{N} I_{(\tilde{\eta}, \tilde{\eta} + d\tilde{\eta})}(\tilde{x}_i) \right] g(\{\tilde{x}\}, t \mid \{\tilde{\omega}\}, \{\tilde{\Theta}\})\, d\{\tilde{x}\} \qquad (6A.3)$$

where

$\{\tilde{x}\}$ = set of position vectors of the N molecules in the matrix at time t

and

$g(\{\tilde{x}\}, t \mid \{\tilde{\omega}\}, \{\tilde{\Theta}\})$

= Probability density function for the set of positions of the N molecules in the matrix at time t, $\{\tilde{x}\}$, given that the state of the matrix is $(\{\tilde{\omega}\}, \{\tilde{\Theta}\})$.

The integration is therefore over all possible values of $\{\tilde{x}\}$ denoted by $\{\tilde{\chi}\}$.

An expression for $g(\cdot)$ may be obtained by reverting back to our definition of $P(\tilde{x}, t \mid \{\tilde{\omega}\}, \{\tilde{\Theta}\})$, which states that

$P(\tilde{x}, t \mid \{\tilde{\omega}\}, \{\tilde{\Theta}\})$

= Probability of finding a "tagged" molecule at (in the neighborhood of) the intersection at $\tilde{x}$ at time t, given that the state of the matrix is $(\{\tilde{\omega}\}, \{\tilde{\Theta}\})$.

The same $P(\cdot)$ may also be interpreted as

$P(\tilde{x}, t \mid \{\tilde{\omega}\}, \{\tilde{\Theta}\})$

= Probability that the position of the molecule at time t is at (in the neighborhood of) the point $\tilde{x}$, given that the state of the matrix is $(\{\tilde{\omega}\}, \{\tilde{\Theta}\})$.

The notion that the molecule is in the neighborhood of $\tilde{x}$ corresponds to the statement that the position of the molecule is in the range $(\tilde{x}, \tilde{x} + d\tilde{x})$. Thus

$$P(\tilde{x}, t \mid \{\tilde{\tilde{\omega}}\}, \{\tilde{\tilde{\Theta}}\}) = p(\tilde{x}, t \mid \{\tilde{\tilde{\omega}}\}, \{\tilde{\tilde{\Theta}}\})\, d\tilde{x} \tag{6A.4}$$

where $p(\cdot)$ is the probability density function for $P(\cdot)$. In the absence of interaction among molecules, the behavior of a single molecule may be considered representative of all molecules. Therefore,

$$g(\{\tilde{x}\}, t \mid \{\tilde{\tilde{\omega}}\}, \{\tilde{\tilde{\Theta}}\}) = \prod_{j=1}^{N} p(\tilde{x}_j, t \mid \{\tilde{\tilde{\omega}}\}, \{\tilde{\tilde{\Theta}}\}) \,. \tag{6A.5}$$

Substituting Eq. (6A.5) into Eq. (6A.3) and reversing the order of integration and summation, we have

$$E[U(\cdot)] = \sum_{i=1}^{N} \left[\int_{\{\tilde{x}\}} I_{(\tilde{\eta},\, \tilde{\eta}+d\tilde{\eta})}(x_i) \prod_{i=1}^{N} [p(\tilde{x}_j, t \mid \{\tilde{\tilde{\omega}}\}, \{\tilde{\tilde{\Theta}}\})\, d\tilde{x}_j] \right] . \tag{6A.6}$$

Expanding,

$$E[U(\cdot)] = \sum_{i=1}^{N} \left[\int_{\tilde{x}_1} \int_{\tilde{x}_2} \cdots \int_{\tilde{x}_i} \cdots \int_{\tilde{x}_N} I_{(\tilde{\eta},\, \tilde{\eta}+d\tilde{\eta})}(x_i) \cdot \prod_{j=1}^{N} [p(\tilde{x}_j, t \mid \{\tilde{\tilde{\omega}}\}, \{\tilde{\tilde{\Theta}}\})\, d\tilde{x}_j] \right] . \tag{6A.7}$$

Rearranging the integrals,

$$E[U(\cdot)] = \sum_{i=1}^{N} \left[\left[\prod_{\substack{j=1\\ j\neq i}}^{N} \int_{\tilde{x}_j} p(\tilde{x}_j, t \mid \{\tilde{\tilde{\omega}}\}, \{\tilde{\tilde{\Theta}}\})\, d\tilde{x}_j \right] \times \left[\int_{\tilde{x}_i} I_{(\tilde{\eta},\, \tilde{\eta}+d\tilde{\eta})}(\tilde{x}_i)\, p(\tilde{x}_i, t \mid \{\tilde{\tilde{\omega}}\}, \{\tilde{\tilde{\Theta}}\})\, d\tilde{x}_i \right] \right] . \tag{6A.8}$$

On evaluating the integrals, the only term that remains is that corresponding to the ith molecule, the rest integrates to 1. Thus,

$$E[U(\cdot)] = \sum_{i=1}^{N} \left[\int_{\tilde{x}_i} I_{(\tilde{\eta},\, \tilde{\eta}+d\tilde{\eta})}(\tilde{x}_i)\, p(\tilde{x}_i, t \mid \{\tilde{\tilde{\omega}}\}, \{\tilde{\tilde{\Theta}}\})\, d\tilde{x}_i \right] . \tag{6A.9}$$

From the definition of the indicator function in Eq. (6A.2), we obtain

$$E[U(\cdot)] = \sum_{i=1}^{N} p(\tilde{\eta}, t \mid \{\tilde{\tilde{\omega}}\}, \{\tilde{\tilde{\Theta}}\})\, d\tilde{\eta} \tag{6A.10}$$

or

$$E[U(\cdot)] = \sum_{i=1}^{N} P(\tilde{\eta}, t \mid \{\tilde{\tilde{\omega}}\}, \{\tilde{\tilde{\Theta}}\}) \,. \tag{6A.11}$$

Equation (6A.11) yields, for molecules behaving identically and independently,

$$E[U(\tilde{\eta}, t \mid \{\tilde{\tilde{\omega}}\}, \{\tilde{\tilde{\Theta}}\})] = NP(\tilde{\eta}, t \mid \{\tilde{\tilde{\omega}}\}, \{\tilde{\tilde{\Theta}}\}) \tag{6A.12}$$

which corresponds to Eq. (6.21) in the text.

Appendix 6B. Derivation of Equation (6.39) from Equation (6.37)

Equation (6.37) in the text is reiterated here.

$$\begin{aligned}
&\frac{\partial \langle W(\tilde{\eta}, t \mid \{\tilde{\tilde{\omega}}\}, \{\tilde{\tilde{\Theta}}\})\rangle_0}{\partial t} \\
&= \frac{1}{2}\left\langle \sum_{i=1}^{M} l_i^2\, \underset{\sim}{\delta \boldsymbol{r}}_i\, \underset{\sim}{\delta \boldsymbol{r}}_i : \nabla \cdot \{k_{1i}(\tilde{\omega}_i; \tilde{\tilde{\omega}})\, [\nabla W(\tilde{\eta}, t \mid \{\tilde{\tilde{\omega}}\}, \{\tilde{\tilde{\Theta}}\})]\} \right\rangle_0 \\
&\quad + \left\langle \sum_{i=1}^{M} \left[l_i^2 \underset{\sim}{\delta \boldsymbol{r}}_i \cdot \nabla k_{1i}(\tilde{\omega}_i; \tilde{\tilde{\omega}}) + \alpha^{-1} l_i k_{i1}(\tilde{\omega}; \tilde{\tilde{\omega}}) \right.\right. \\
&\qquad \left.\left. + \, l_i^2 k_{i1}(\tilde{\omega}; \tilde{\tilde{\omega}})\, \underset{\sim}{\delta \boldsymbol{r}}_i \cdot \frac{\nabla V(\tilde{\eta})}{V(\tilde{\eta})} \right] \delta \boldsymbol{r}_i \cdot \nabla W(\tilde{\eta}, t \mid \{\tilde{\tilde{\omega}}\}, \{\tilde{\tilde{\Theta}}\}) \right\rangle_0 \\
&\quad - \langle k_2(\tilde{\tilde{\omega}})\, W(\tilde{\eta}, t \mid \{\tilde{\tilde{\omega}}\}, \{\tilde{\tilde{\Theta}}\})\rangle_0
\end{aligned} \tag{6B.1}$$

where the operator $\langle \cdot \rangle_0$ represents

$$\langle q \rangle_0 = \int_{\{\tilde{\tilde{\omega}}\}} \int_{\{\tilde{\tilde{\Theta}}\}} q(\tilde{\tilde{\omega}}, \tilde{\tilde{\Theta}}, \tilde{\eta}, t \mid \{\tilde{\tilde{\omega}}\}, \{\tilde{\tilde{\Theta}}\})\, p_{S0}(\{\tilde{\tilde{\omega}}\}, \{\tilde{\tilde{\Theta}}\})\, d\{\tilde{\tilde{\omega}}\}\, d\{\tilde{\tilde{\Theta}}\} . \tag{6B.2}$$

Note that from Eq. (6.22) in the text

$$W(\tilde{\eta}, t \mid \{\tilde{\tilde{\omega}}\}, \{\tilde{\tilde{\Theta}}\}) = \frac{E[U(\tilde{\eta}, t \mid \{\tilde{\tilde{\omega}}\}, \{\tilde{\tilde{\Theta}}\})]}{V(\tilde{\eta})} . \tag{6B.3}$$

The $\langle \cdot \rangle_0$ operations in Eq. (6B.1) can be generalized and written as follows,

$$\langle \cdot \rangle_0 = \left\langle \frac{h(\{\tilde{\tilde{\omega}}\}, \{\tilde{\tilde{\Theta}}\}, \tilde{\eta}, t)}{V(\tilde{\eta})}\, E[U(\tilde{\eta}, t \mid \{\tilde{\tilde{\omega}}\}, \{\tilde{\tilde{\Theta}}\})] \right\rangle_0 . \tag{6B.4}$$

This equation may be expanded to yield

$$\begin{aligned}
\langle \cdot \rangle_0 = &\int_{\{\tilde{\tilde{\omega}}\}} \int_{\{\tilde{\tilde{\Theta}}\}} \frac{h(\{\tilde{\tilde{\omega}}\}, \{\tilde{\tilde{\Theta}}\}, \tilde{\eta}, t)}{V(\tilde{\eta})}\, E[U(\tilde{\eta}, t \mid \{\tilde{\tilde{\omega}}\}, \{\tilde{\tilde{\Theta}}\})] \\
&\times p_{S0}(\{\tilde{\tilde{\omega}}\}, \{\tilde{\tilde{\Theta}}\})\, d\{\tilde{\tilde{\omega}}\}\, d\{\tilde{\tilde{\Theta}}\} .
\end{aligned} \tag{6B.5}$$

In the above equation, $\tilde{\eta}$ and t are parameters while $(\{\tilde{\tilde{\omega}}\}, \{\tilde{\tilde{\Theta}}\})$ are the random variables. $E[U(\tilde{\eta}, t \mid \{\tilde{\tilde{\omega}}\}, \{\tilde{\tilde{\Theta}}\})]$ is thus only a function of $(\{\tilde{\tilde{\omega}}\}, \{\tilde{\tilde{\Theta}}\})$. Integration over the distribution function of $(\{\tilde{\tilde{\omega}}\}, \{\tilde{\tilde{\Theta}}\})$ will yield, formally

$$\langle \cdot \rangle_0 = \left\langle \frac{h(\{\tilde{\tilde{\omega}}\}, \{\tilde{\tilde{\Theta}}\}, \tilde{\eta}, t)\, U(\tilde{\eta}, t)}{V(\tilde{\eta})} \right\rangle_0 \tag{6B.6}$$

The conditionality has thus been removed by integrating over all possible intersection characteristics. Neglecting the covariance, equation (6B.6) may be written as

$$\langle \cdot \rangle_0 = \langle h(\{\tilde{\tilde{\omega}}\}, \{\tilde{\tilde{\Theta}}\}, \tilde{\eta}, t) \rangle_0 \left\langle \frac{U(\tilde{\eta}, t)}{V(\tilde{\eta})} \right\rangle_0 . \tag{6B.7}$$

Defining

$$\langle C(\tilde{\eta}, t) \rangle_0 = \left\langle \frac{U(\tilde{\eta}, t)}{V(\tilde{\eta})} \right\rangle_0 \tag{6 B.8a}$$

or

$$\langle C(\tilde{\eta}, t) \rangle_0 = \frac{1}{V(\tilde{\eta})} \int\limits_{\{\tilde{\tilde{\omega}}\}} \int\limits_{\{\tilde{\tilde{\Theta}}\}} E[U(\tilde{\eta}, t \mid \{\tilde{\tilde{\omega}}\}, \{\tilde{\tilde{\Theta}}\})]\, p_{S0}(\{\tilde{\tilde{\omega}}\}, \{\tilde{\tilde{\Theta}}\})\, d\{\tilde{\tilde{\omega}}\}\, d\{\tilde{\tilde{\Theta}}\} \tag{6 B.8 b}$$

we recover the average number concentration at any point in the matrix.

From Eq. (6.18 b) in the text, we have

$$\nabla \cdot \{k_1(\tilde{\omega}; \tilde{\tilde{\omega}})\, [\nabla C(\tilde{\eta}, t)]\} = k_1(\tilde{\omega}; \tilde{\tilde{\omega}})\, [\nabla \nabla C(\tilde{\eta}, t)] \tag{6B.9}$$

since

$$\nabla k_1(\tilde{\omega}; \tilde{\tilde{\omega}}) \equiv 0 . \tag{6B.10}$$

Substituting Eqs. (6 B.7), (6 B.8 a), (6 B.9), and (6 B.10) into Eq. (6 B-1) yields

$$\begin{aligned} \frac{\partial \langle C(\tilde{\eta}, t) \rangle_0}{\partial t} &= \frac{1}{2} \left\langle \sum_{i=1}^{M} l_i^2 k_{1i}(\tilde{\omega}_i; \tilde{\tilde{\omega}})\, \delta \boldsymbol{r}_i\, \delta \boldsymbol{r}_i \right\rangle_0 : \nabla \nabla \langle C(\tilde{\eta}, t) \rangle_0 \\ &+ \left\langle \sum_{i=1}^{M} \left[\alpha^{-1} l_i k_{i1}(\tilde{\omega}; \tilde{\tilde{\omega}}) + l_i^2 k_{i1}(\tilde{\omega}; \tilde{\tilde{\omega}})\, \delta \boldsymbol{r}_i \cdot \frac{\nabla V}{V} \right] \delta \boldsymbol{r}_i \right\rangle_0 \cdot \nabla \langle C(\tilde{\eta}, t) \rangle_0 \\ &- \langle k_2(\tilde{\tilde{\omega}}) \rangle_0 \langle C(\tilde{\eta}, t) \rangle_0 . \end{aligned} \tag{6 B.11}$$

The derivatives on the right-hand side can be taken outside the expectation since this operator does not act on $\tilde{\eta}$ as it is not a random variable.

Equation (6B.11) corresponds to Eq. (6.39) in the text.

Appendix 6C. Algorithm and FORTRAN Code for Monte Carlo Simulation of Transport in Porous Networks

An event-driven Monte-Carlo simulation (see, e.g., Rajamani et al., 1986) was employed in this work. The steps involved in the process are as follows:

1. Generate intersections in a $1 \times 1 \times 1$ cube
   a. Pore intersection no. 1 is placed at the origin.
   b. Generate 3 uniform random variables

$$(u_1, u_2, u_3) \in U[0, 1] .$$

   c. If *uniform* network is to be generated, assign $(u_1 - 0.5, u_2 - 0.5, u_3 - 0.5)$ as coordinates of the next pore intersection.
   If skewed network is to be created:
   — for *x-skewed* network, assign $(2u_1 - u_1^2 - 0.5, u_2 - 0.5, u_3 - 0.5)$ as coordinates of the next pore intersection.
   — for *x, y, z-skewed* network, assign $(2u_1 - u_1^2 - 0.5, 2u_2 - u_2^3 - 0.5, 2u_3 - u_3^2 - 0.5)$ as coordinates of the next pore intersection.
   d. Repeat steps 1b, 1c till all required (S) intersections have been created and stored in POS(I, 1/2/3).

Note: Schemes other than $2u - u^2$ can be utilized to generate a non-uniform network.

2. Generate pore network
   a. Find the intersection (J), closest to the one under consideration (I).
   b. Does this intersection (J) already have M interconnections?
   c. If yes, find the next closest intersection and repeat from step 2b onwards.
   If no, connect intersections (I) and (J) and increment the numbers of interconnections of both by 1.
   d. Repeat steps 2a–2c until M interconnections have been generated for (I).
   e. Repeat steps 2a–2d for all S intersections.

Note: Near the end of the above procedure, there may be some intersections left which are unable to achieve a full complement of M pores, depending on the relative values of M and S.

3. Determine transition intensities
   a. Determine length of each interconnection (IJ).
   b. Evaluate the transition intensity from intersection (I) to (J) by Eq. (6.18), i.e.,

$$\lambda_{ij} = \left(\frac{D_e}{L^2}\right) \frac{2r_{ij}^2 D}{l_{ij} \sum_{k=1}^{M} r_{ik}^2 l_{ik}}$$

   where l_{ik} and r_{ik} are the length and radius of the pore going from (I) to (K).

Note: All pore radii were taken to be equal; D_e, D, were taken equal to 1; L is equal to 1. Transition intensities were stored in TRN(I, J)

4. Generate number-of-molecules vector from initial conditions.
   a. Based on initial conditions, place appropriate number of molecules at each intersection.

Note: Number of molecules at intersection (I) were stored in NUM(I). For starting the present simulations, 3000 molecules were placed at the origin, 1000 were placed at intersections within a radius of 0.1 units from the origin, 500 were placed at intersections between 0.1 and 0.2 units distant from the origin, 300 at those between 0.2 and 0.3 units from the origin, 100 at those between 0.3 and 0.4 units from the origin, with none at those intersections farther than 0.4 units from the origin.

5. Determine inter-arrival time and transition event

   Inter-arrival time
   a. Generate uniform random variable $u_1 \in U[0, 1]$.
   b. Under the condition that inter-arrival time is exponentially distributed, obtain the inter-arrival time from

$$t = -\ln(u_1)/\text{RATE}$$

   Transition event
   c. Generate uniform random number $u_2 \in U[0, 1]$.
   d. Determine probability of transition by multiplying each row (I) of the transition intensity matrix TRN(I, J) by the number of molecules at that intersection NUM(I), and then dividing each element of the resultant matrix by the sum of these elements (= RATE), i.e.,

$$\text{PTRN(I, J)} = \text{TRN(I, J)} * \text{NUM(I)/RATE}$$

   where

$$\text{RATE} = \sum_i \sum_j \text{TRN(I, J)}^* \ \text{NUM(I)}.$$

   RATE is the total transition rate in (no. of events/time)
   e. Determine the event that occurs by adding the individual elements of PTRN from left to right, top row down, until the sum exceeds u_2 at element (I, J). The event is then designated to be a jump of one molecule from intersection (I) to intersection (J).
   f. Update NUM(·) vector by subtracting 1 from NUM(I) and adding 1 to NUM(J).
   g. Accumulate inter-arrival time to obtain clock-time.
   h. Repeat steps 5a–5g until desired number of events have occurred or desired clock-time has elapsed.

6. Determine diffusivities and velocities at time t_m corresponding to the occurrence of event No. m
   a. Determine center-of-mass of system, $\langle \cdot \rangle_m$

$$\langle x \rangle_m = \frac{\sum_{i=1}^{S} \text{NUM(I)}^* \ \text{POS(I, 1)}}{\sum_{i=1}^{S} \text{NUM(I)}}$$

$$\langle y\rangle_m = \frac{\sum_{i=1}^{S} \text{NUM(I)* POS(I, 2)}}{\sum_{i=1}^{S} \text{NUM(I)}}$$

$$\langle z\rangle_m = \frac{\sum_{i=1}^{S} \text{NUM(I)* POS(I, 3)}}{\sum_{i=1}^{S} \text{NUM(I)}}$$

b. Determine mean square distances, $\langle\!\langle \cdot \rangle\!\rangle_m$

$$\langle\langle x\rangle\rangle_m = \frac{\sum_{i=1}^{S} \text{NUM(I)* (POS(I, 1)} - \langle x\rangle_m)^2}{\sum_{i=1}^{S} \text{NUM(I)}}$$

$$\langle\langle y\rangle\rangle_m = \frac{\sum_{i=1}^{S} \text{NUM(I)* (POS(I, 2)} - \langle y\rangle_m)^2}{\sum_{i=1}^{S} \text{NUM(I)}}$$

$$\langle\langle z\rangle\rangle_m = \frac{\sum_{i=1}^{S} \text{NUM(I)* (POS(I, 3)} - \langle z\rangle_m)^2}{\sum_{i=1}^{S} \text{NUM(I)}}$$

c. Set velocities at time t_m as

$$v_x = \frac{\langle x\rangle_m - \langle x\rangle_{m-1}}{t_m - t_{m-1}}$$

$$v_y = \frac{\langle y\rangle_m - \langle y\rangle_{m-1}}{t_m - t_{m-1}}$$

$$v_z = \frac{\langle z\rangle_m - \langle z\rangle_{m-1}}{t_m - t_{m-1}}$$

d. Set effective diffusivities at time t_m as

$$D_{xx} = \frac{\langle\!\langle x\rangle\!\rangle_m - \langle\!\langle x\rangle\!\rangle_{m-1}}{2(t_m - t_{m-1})}$$

$$D_{yy} = \frac{\langle\!\langle y\rangle\!\rangle_m - \langle\!\langle y\rangle\!\rangle_{m-1}}{2(t_m - t_{m-1})}$$

$$D_{zz} = \frac{\langle\!\langle z\rangle\!\rangle_m - \langle\!\langle z\rangle\!\rangle_{m-1}}{2(t_m - t_{m-1})}$$

Note: The numerators and denominators in the above expressions are very small since they represent differences created by one event only. Dividing a small number by another leads to wildly fluctuating estimates of the parameters. This problem was avoided by taking differences every 50 or 100 events, i.e., between *m* and (*m* — 50) or (*m* — 100).

Listing Program I: NETWORK

```
C ************************************************************************
C OBJECTIVE: CREATE NETWORK OF PORES AND SIMULATE
C            DIFFUSION.
C ITYPE: 1 = UNIFORM NETWORK; 2 = SKEWED NETWORK.
C ICORD: 199 = X-SKEWED, X COORD. 2U—U*U—0.5
C        999 & ITYPE=1 UNIFORM
C        999 & ITYPE=2 X,Y,Z-SKEWED, ALL COORDS.
C              2U—U*U—0.5
C NPORE = NUMBER OF PORES PER INTERSECTION
C IMAX = NUMBER OF INTERSECTIONS IN CUBE
C NMOLES = NUMBER OF MOLECULES, TOTAL
C POS(I,1/2/3) = COORDS. OF INTSXN. #I
C ICN(I,J) = INTERCONNXN. MATRIX BETWEEN INTSXN. I & J
C            0 = NO CONNXN, 1 = CONNXN.
C TRN (I,J) = TRANSITION INTENSITY BETWEEN INTSXN. I & J
C PLN(I,J) = PORE LENGTH BETWEEN I & J
C PCENT(1/2/3) = CENTER OF INTERSECTIONS (FROM POS)
C NUM(I) = NUMBER OF MOLECULES AT I
C PTRN(I,J) = TRN*NUM/RATE; ROW MULTIPLICATION ONLY
C RATE = SUMMATION OF PTRN; TOTAL RATE TO FIND ARR.
C        TIME
C IVENT(K) = TRANSN. FROM INTSXN #IVENT(1) TO #IVENT(2)
C ************************************************************************
      PROGRAM NETWORK
      REAL PCENT(3)
      INTEGER IVENT(2)
      CLICHE COMMONS
      PARAMETER (III=600)
      COMMON/IST/STOR(III),ISTR(III),NP(III)
      COMMON/NTWRK/POS(III,3)
      COMMON/TRANS/TRN(III,III)
      COMMON/PTRAN/PTRN(III,III)
      COMMON/REST/ICN(III,III),PLN(III,III),NUM(III)
      ENDCLICHE
      COMMONS
```

Listing Program I: NETWORK (continued)

```
C ** SPECIFY PARAMETERS
      IRUN=2
      ITYPE=2
      ICORD=199
      NPORE=5
      IMAX=III
C ** FTIM=MAX CPU TIME TO BE USED IN PROGRAM
C ** NRN=STARTING RANDOM NUMBER
C ** RANSET IS LIBRARY SUBROUTINE THAT RESETS THE
C    RANDOM
C ** NUMBER GENERATOR RANF
      FTIM=30.0
      NRN=647
      CALL RANSET(NRN)
C ** STORE EVENT DATA ON UNIT 11, NETWORK COORDS. ON 12
      OPEN (UNIT=11,FILE='EVEN26)
      OPEN (UNIT=12,FILE='POSI26)
C ** CREATE INTSXNS., STORE TYPE ON 11
      CALL INTERX(ITYPE,ICORD,IMAX,PCENT)
C ** INITIALIZE
      NMOLES=0
      NUM(1)=3000
      DO 400 I=2,IMAX
      DIST=SQRT(POS(I,1)**2+POS(I,2)**2+POS(I,3)**2)
      IF (DIST.LT.0.1)                     NUM(I)=1000
      IF ((DIST.LT.0.2).AND.(DIST.GT.0.1)) NUM(I)=500
      IF ((DIST.LT.0.3).AND.(DIST.GT.0.2)) NUM(I)=300
      IF ((DIST.LT.0.4).AND.(DIST.GT.0.3)) NUM(I)+100
      IF (DIST.GT.0.4)                     NUM(I)=0
      NMOLES=NMOLES+NUM(I)400  CONTINUE
      NMOLES=NMOLES+NUM(1)C
C ** CREATE INTERCONNECTIONS AND STORE POSITIONS ON 12C
      CALL CONNEX(NPORE,IMAX,AVLTGH)
      WRITE(12,25) NPORE,IMAX,AVLTGH,IRUN,NRN,ICORD
25    FORMAT(1X,I2,2X,I5,2X,F8.5,2X,I2,2X,I4,2X,I3)
      WRITE(12,26) PCENT(1),PCENT(2),PCENT(3)
26    FORMAT(1X,3(2X,E14.6))
      DO 100 I=1,IMAX
      WRITE(12,27) I,POS(I,1),POS(I,2),POS(I,3),NUM(I)
27    FORMAT(1X,I6,2X,3(E14.6,2X),2X,I9)
100   CONTINUE
      CLOSE (UNIT=12)
C
C
```

```
      TIME=0.0
      IVENT(1)=1
      IVENT(2)=1
      IFR=IVENT(1)
      ITO=IVENT(2)
      WRITE(11,420) ITYPE,ICORD,NMOLES,IRUN,NRN
420   FORMAT(1X,I3,2X,I3,2X,I7,2X,I2,2X,I4)
      WRITE(11,440) TIME,IVENT(1),IVENT(2)
440   FORMAT(1X, E12.6, 5X, I3, 2X, I3)
C
C ** DETERMINE EVENT
C
999   CONTINUE
      CALL TRANSN(IMAX,NMOLES,RATE,IFR,ITO)
C ** RANF HAS AN INTERNALLY GENERATED ARGUMENT
      RV=RATE*RANF( )
      CALL EVENT(IMAX,RV,IVENT)
C ** UPDATE NUMBER VECTOR
      IFR=IVENT(1)
      ITO=IVENT(2)
      NUM(IFR)=NUM(IFR)-1
      NUM(ITO)=NUM(ITO)+1
C ** DETERMINE TIME OF TRANSITION
      RV=RANF( )
      TINV=-LOG(RV)/RATE
      TIME=TIME+TINV
      WRITE(11,450) TIME,IVENT(1),IVENT(2)
450   FORMAT(1X,E12.6,5X,I3,2X,I3)
C ** STOPPING CRITERIA
C ** CALCULATE CPU USED BY INTERNAL ROUTINE QTIME
      CALL QTIME(ICPU,IO,MEM,FCPU,FIO,FMEM)
      CUMCPU=FLOAT(ICPU)/(60*10**6)
C ** ALSO CHECK FOR STEADY STATE
      IF(NUM(1).GT.NMOLES/IMAX) THEN
      IF(CUMCPU.LT.FTIM) GO TO 999
      ENDIF
      A=999999.0
      JJ=1
      KK=1
      WRITE(11,*) A,JJ,KK
      CLOSE (UNIT=11)
      STOP
      END
C
C ** GENERATE INTSECTIONS
C
```

Listing Program I: NETWORK (continued)

```
      SUBROUTINE INTERX(ITYPE,ICORD,IMAX,PCENT)
      REAL PCENT(3)
      COMMONS
      PCENT(1)=0.0
      PCENT(2)=0.0
      PCENT(3)=0.0
      DO 100 I=2,IMAX
      DO 200 J=1,3
      U=RANF( )
      IF (ITYPE.EQ.1) THEN
                    POS(I,J)=U-0.5
                    GO TO 150
      ENDIF
      IF (ICORD.EQ.199) THEN
          IF (J.EQ.1) V=2*U-U*U
          IF (J.EQ.2) V=U
          IF (J.EQ.3) V=U
          POS(I,J)=V-0.5
      ELSE
      V=2*U-U*U
      POS(I,J)=V-0.5
      ENDIF
150   CONTINUE
      PCENT(J)=PCENT(J)+POS(I,J)
200   CONTINUE
100   CONTINUE
      POS(1,1)=0.0
      POS(1,2)=0.0
      POS(1,3)=0.0
      PCENT(1)=PCENT(1)/IMAX
      PCENT(2)=PCENT(2)/IMAX
      PCENT(3)=PCENT(3)/IMAX
      RETURN
      END
C
C ** GENERATE CONNECTIONS AND DETERMINE TRANSITION
C    MATRIX
C
      SUBROUTINE CONNEX(NPORE,IMAX,AVLTGH)
      COMMONS
      DO 50 I=1,IMAX
      DO 60 J=1,IMAX
      ICN(I,J)=0
60    CONTINUE
      NP(I)=0
```

```
      ISTR(I)=0
      STOR(I)=0.0
50    CONTINUE
      TLGT=0.0
      DO 100 I=1,IMAX
      K=0
      DO 200 J=1,IMAX
      IF (J.EQ.I) GO TO 200
      IF (NPORE.EQ.NP(I)) GO TO 100
      IF (NPORE.EQ.NP(J)) GO TO 200
      IF ((ICN(I,J).NE.0).OR.(ICN(J,I).NE.0)) GO TO 200
      K=K+1
      PLN(I,J)=SQRT((POS(I,1)-POS(J,1))**2
1                 (POS(I,2)-POS(J,2))**2
2                 (POS(I,3)-POS(J,3))**2)
      PLN(J,I)=PLN(I,J)
      ISTR(K)-J
      STOR(K)=PLN(I,J)
200   CONTINUE
      NEED=NPORE-NP(I)
      IF (K.EQ.0) GO TO 100
      IF (K.EQ.1) THEN
                 NEED=K
                 GO TO 350
      ENDIF
      IF (NEED.GT.K) THEN
                     NEED=K
      ENDIF
      KMX=K-1
400   CONTINUE
C ** SORT
      IFLAG=0
      DO 300 KK=1,KMX
      IF (STOR(KK).GT.STOR(KK+1)) THEN
         IEMP1=ISTR(KK)
         TEMP2=STOR(KK)
         ISTR(KK)=ISTR(KK+1)
         STOR(KK)=STOR(KK+1)
         ISTR(KK+1)=IEMP1
         STOR(KK+1)=TEMP2
         IFLAG=1
      ENDIF
300   CONTINUE
      IF (IFLAG.EQ.1) GO TO 400
C **
350   CONTINUE
      DO 500 NN=1,NEED
```

Listing Program I: NETWORK (continued)

```
      ICN(I,ISTR(NN))=1
      NP(I)=NP(I)+1
      ICN(ISTR(NN),I)=1
      NP(ISTR(NN))=NP(ISTR(NN))+1
500   CONTINUE
      DO 600 NN=1,NEED
600   CONTINUE
      DO 700 KZ=1,IMAX
      ISTR(KZ)=0
      STOR(KZ)=0.0
700   CONTINUE
100   CONTINUE
C **
      NPTOT=0
      TLTGH=0.0
      DO 800 I=1,IMAX
      DSUM=0.0
      DO 900 J=1,IMAX
      IF (ICN(I,J).EQ.1) THEN
         NPTOT=NPTOT+1
         TLTGH=TLTGH+PLN(I,J)
         DSUM=DSUM+PLN(I,J)
      ENDIF
900   CONTINUE
      DO 950 J=1,IMAX
      IF (ICN(I,J).EQ.1) THEN
                        TRN(I,J)=2.0/(PLN(I,J)*DSUM)
         ELSE
         TRN(I,J)=0.0
      ENDIF
950   CONTINUE
800   CONTINUE
      AVLTGH=TLTGH/NPTOT
      NPOT=NPTOT/2
      RETURN
      END
C
C ** CALCULATE TRANSITION RATE
C
      SUBROUTINE TRANSN(IMAX,NMOLES,RATE,IFR,ITO)
      COMMONS
      IF (IFR+ITO.GT.2) GO TO 150
      RATE=0.0
      DO 100 I=1,IMAX
      DO 200 J=1,IMAX
```

```
      PTRN(I,J)=TRN(I,J)*NUM(I)
      RATE=RATE+PTRN(I,J)
200   CONTINUE
100   CONTINUE
      GO TO 600
150   CONTINUE
      DO 300 J=1,IMAX
      RATE=RATE-PTRN(IFR,J)-PTRN(ITO,J)
      PTRN(IFR,J)=TRN(IFR,J)*NUM(IFR)
      PTRN(ITO,J)=TRN(ITO,J)*NUM(ITO)
      RATE=RATE+PTRN(IFR,J)+PTRN(ITO,J)
300   CONTINUE
600   CONTINUE
      RETURN
      END
C
C ** DETERMINE EVENT
C
      SUBROUTINE EVENT(IMAX,RV,IVENT)
      INTEGER IVENT(2)
      COMMONS
      SUM=0.0
      DO 100 I=1,IMAX
      DO 200 J=1,IMAX
      SUM=SUM+PTRN(I,J)
      IF (SUM.GT.RV) THEN
          IVENT(1)=I
          IVENT(2)=J
          GO TO 300
      ENDIF
200   CONTINUE
100   CONTINUE
300   CONTINUE
      RETURN
      END
```

Listing Programm II: DIFFUSIVITY AND DRIFT

```
C ********************************************************************
C OBJECTIVE: READ EVENT AND POSITION DATA FROM
C PROGRAM "NETWORK", AND COMPUTE DIFFUSIVITIES
C AND DRIFT VELOCITIES.
C CENT(1/2/3) = CENTER OF MASS AT ANY TIME
C MSQDSi = MEAN SQUARE DISTANCE FROM CENTER OF MASS
C ********************************************************************
```

Listing Programm II: DIFFUSIVITY AND DRIFT (continued)

```
      DIMENSION POS(1000,3),NUM(1000),IVENT(2)
      DIMENSION PCENT(3),CENT(3)
      REAL MSQDSX,MSQDSY,MSQDSZ,MSOLDX,MSOLDY,MSOLDZ,
      MSQDIS
      COMMON/LENGTH/POS
      COMMON/CENTER/OCENTX,OCENTY,OCENTZ
C ** IMEAN: FIXES NUMBER OF EVENTS OR TIME STEPS
C    BETWEEN WHICH DIFFUSIVITIES AND VELOCITIES
C    WILL BE CALCULATED
     IMEAN=100
     OCENTX=0.0
     OCENTY=0.0
     OCENTZ=0.0
C ** READ NETWORK
      OPEN (UNIT=11,FILE='EVEN26)
      OPEN (UNIT=12,FILE='POSI26)
      OPEN (UNIT=13,FILE='DIFF26)
      OPEN (UNIT=14,FILE='VELO26)
      READ(12,*) NPORE,IMAX,AVLTGH,IRUN,NRN
      READ(12,*) PCENT(1),PCENT(2),PCENT(3)
      DO 100 I=1,IMAX
      READ(12,*) II,POS(I,1),POS(I,2),POS(I,3),NUM(I)
100   CONTINUE
C
C ** INITIALIZE
C
     READ(11,*) ITYPE,ICORD,NMOLES,IRUN
     WRITE(13,125)ITYPE,ICORD,NPORE,IMAX,NMOLES,AVLTGH,
    1             IRUN,
    1             IMEAN,NRN
125   FORMAT(1X,I2,2X,I4,2X,I3,2X,I5,2X,I7,2X,F9.7,2X,I2,2X,
    1        I4,2X,I4)
      WRITE(13,126) PCENT(1),PCENT(2),PCENT(3)
126   FORMAT(1X,3(F12.9,2X))
C
      MSOLDX=0.0
      MSOLDY=0.0
      MSOLDZ=0.0
      MSOLD=0.0
      TIOLD=0.0
      READ(11,*) TIME,IVENT(1),IVENT(2)
      ICNT=IMEAN
      CALL PARAM(IMAX,NUM,NMOLES,IVENT,CENT,ICNT,IMEAN,
    1             MSQDSX,MSQDSY,MSQDSZ,MSQDIS)
```

```
      OCENTX=CENT(1)
      OCENTY=CENT(2)
      OCENTZ=CENT(3)
      MSOLDX=MSQDSX
      MSOLDY=MSQDSY
      MSOLDZ=MSQDSZ
      MSOLD=MSQDIS
      TIOLD=TIME
      WRITE(13,121) TIME,OCENTX,OCENTY,OCENTZ
121   FORMAT(1X,F13.8,3(2X,F12.9))
C **
      ICNT=0
990   CONTINUE
      READ(11,*) TIME,IVENT(1),IVENT(2)
      ICNT=ICNT+1
      IF (TIME.GT.999.0) THEN
                          GO TO 999
        ELSE
      CALL PARAM(IMAX,NUM,NMOLES,IVENT,CENT,ICNT,IMEAN,
     1                  MSQDSX,MSQDSY,MSQDSZ,MSQDIS)
      IF (ICNT.NE.0) THEN
                      GO TO 260
        ELSE
C
      DIFFX=(MSQDSX-MSOLDX)/(2.0*(TIME-TIOLD))
      DIFFY=(MSQDSY-MSOLDY)/(2.0*(TIME-TIOLD))
      DIFFZ=(MSQDSZ-MSOLDZ)/(2.0*(TIME-TIOLD))
C
      VELOX=(CENT(1)-OCENTX)/(TIME-TIOLD)
      VELOY=(CENT(1)-OCENTY)/(TIME-TIOLD)
      VELOZ-(CENT(1)-OCENTZ)/(TIME-TIOLD)
C
      TIOLD=TIME
      MSOLD=MSQDIS
      MSOLDX-MSQDSX
      MSOLDY-MSQDSY
      MSOLDZ=MSQDSZ
      OCENTX=CENT(1)
      OCENTY=CENT(2)
      OCENTZ=CENT(3)
      WRITE(13,250) TIME,DIFFX,DIFFY,DIFFZ
250   FORMAT(1X,F13.8,3(1X,F13.8))
      WRITE(14,251) TIME,VELOX,VELOY,VELOZ
251   FORMAT(1X,F13.8,3(1X,F13.8))
      ENDIF
260   CONTINUE
      GO TO 990
```

Listing Program II: DIFFUSIVITY AND DRIFT (continued)

```
      ENDIF
999   CONTINUE
      CLOSE (UNIT=11)
      CLOSE (UNIT=12)
      CLOSE (UNIT=13)
      CLOSE (UNIT=14)
      STOP
      END
C
C *   CALCULATES CENTER OF MASS AND MEAN SQUARE
C     DISTANCES
C *   AROUND CENTER OF MASS
C
      SUBROUTINE PARAM(IMAX,NUM,NMOLES,IVENT,CENT,ICNT,
     1                 IMEAN,MSQDSX,MSQDSY,MSQDSZ,
     1                 MSQDIS)
      DIMENSION NUM(1000),IVENT(2),POS(1000,3),CENT(3)
      REAL MSQDSX,MSQDSY,MSQDSZ,MSQDIS
      COMMON/LENGTH/POS
      COMMON/CENTER/OCENTX,OCENTY,OCENTZ
      XDIS=0.0
      YDIS=0.0
      ZDIS=0.0
      MSQDIS=0.0
      NUM(IVENT(1))=NUM(IVENT(1))-1
      NUM(IVENT(2))=NUM(IVENT(2))+1
      IF    (ICNT.NE.IMEAN) THEN
                               GO TO 200
           ELSE
           DO 100 I=1,IMAX
           XDIS=XDIS+NUM(I)*POS(I,1)
           YDIS=YDIS+NUM(I)*POS(I,2)
           ZDIS=ZDIS+NUM(I)*POS(I,3)
100   CONTINUE
      CENT(1)=XDIS/NMOLES
      CENT(2)=YDIS/NMOLES
      CENT(3)=ZDIS/NMOLES
      DO 300 I=1,IMAX
      SQDSX=NUM(I)*((POS(I,1)-CENT(1))**2)
      SQDSY=NUM(I)*((POS(I,2)-CENT(2))**2)
      SQDSZ=NUM(I)*((POS(I,3)-CENT(3))**2)
      SQDIS=SQDSX+SQDSY+SQDSZ
      MSQDSX=MSQDSX+SQDSX
      MSQDSY=MSQDSY+SQDSY
      MSQDSZ=MSQDSZ+SQDSZ
```

```
      MSQDIS=MSQDIS+SQDIS
300   CONTINUE
      MSQDSX=MSQDSX/NMOLES
      MSQDSY=MSQDSY/NMOLES
      MSQDSZ=MSQDSZ/NMOLES
      MSQDIS=MSQDIS/NMOLES
      ICNT=0
      ENDIF
200   CONTINUE
      RETURN
      END
```

Listing Program III: EQUATIONS (6.44) AND (6.54)

```
C **********************************************************************
C OBJECTIVE: CREATE NETWORK OF PORES AND APPLY EQNS.
C (6.44), (6.54) TO CALCULATE DIFFUSIVITIES AND DRIFT
C VELOCITIES. NO RANDOM WALK GENERATED.
C THE FIRST PART OF THIS PROGRAM IS SIMILAR TO
  PROGRAM I.
C ITYPE: 1 = UNIFORM NETWORK; 2 = SKEWED NETWORK.
C ICORD: 199 = X-SKEWED, X COORD. 2U—U*U—0.5
C        999 & ITYPE=1 UNIFORM
C        999 & ITYPE=2 X,Y,Z-SKEWED, ALL COORDS.
C              2U—U*U—0.5
C NPORE = NUMBER OF PORES PER INTERSECTION
C IMAX = NUMBER OF INTERSECTIONS IN CUBE
C NMOLES = NUMBER OF MOLECULES, TOTAL
C POS(I,1/2/3) = COORDS. OF INTSXN. #I
C ICN(I,J) = INTERCONNXN. MATRIX BETWEEN INTSXN. I & J
C            0 = NO CONNXN, 1 = CONNXN.
C TRN(I,J) = TRANSITION INTENSITY BETWEEN INTSXN. I & J
C PLN(I,J) = PORE LENGTH BETWEEN I & J
C PCENT(1/2/3) = CENTER OF INTERSECTIONS (FROM POS)
C NUM(I) = NUMBER OF MOLECULES AT I
C PTRN(I,J) = TRN*NUM/RATE; ROW MULTIPLICATION ONLY
C RATE = SUMMATION OF PTRN; TOTAL RATE TO FIND ARR.
C TIME
C IVENT(K) = TRANSN. FROM INTSXN #IVENT(1) TO #IVENT(2)
C **********************************************************************
      PROGRAM NETWORK
      REAL PCENT(3)
      INTEGER IVENT(2)
      CLICHE COMMONS
      PARAMETER (III=600)
```

Listing Program III: EQUATIONS (43) AND (52) (continued)

```
      COMMON/IST/STOR(III),ISTR(III),NP(III)
      COMMON/NTWRK/POS(III,3)
      COMMON/TRANS/TRN(III,III)
      COMMON/PTRAN/PTRN(III,III)
      COMMON/REST/ICN(III,III),PLN(III,III),NUM(III)
      ENDCLICHE
      COMMONS
C ** SPECIFY PARAMETERS
      IRUN=2
      ITYPE=2
      ICORD=199
      NPORE=5
      IMAX=III
C ** NRN=STARTING RANDOM NUMBER
C ** RANSET IS LIBRARY SUBROUTINE THAT RESETS THE
C    RANDOM
C    NUMBER GENERATOR RANF
      NRN=647
      CALL RANSET(NRN)
C ** STORE COEFFICIENTS CALCULATED ON UNIT 11
      OPEN (UNIT=11,FILE='COEF26')
C ** CREATE INTSXNS., STORE TYPE ON 11
      CALL INTERX(ITYPE,ICORD,IMAX,PCENT)
C
C ** CREATE INTERCONNECTIONS AND STORE POSITIONS
C
      CALL CONNEX(NPORE,IMAX,AVLTGH)
      WRITE(11,25) NPORE,IMAX,AVLTGH,IRUN,NRN,ICORD
25    FORMAT(1X,I2,2X,I5,2X,F8.5,2X,I2,2X,I4,2X,I3)
      WRITE(11,26) PCENT(1),PCENT(2),PCENT(3)
26    FORMAT(1X,3(2X,E14.6))
C
      CALL PARAMS(NPORE,IMAX,AVLTGH,DIFFX,DIFFY,DIFFZ,
     1                  VELOX,VELOY,VELOZ)
      WRITE(11,27) DIFFX,DIFFY,DIFFZ
      WRITE(11,27) VELOX,VELOY,VELOZ
27    FORMAT(1X,3(2X,E14.6))
      CLOSE (UNIT=11)
      STOP
      END
C
C ** GENERATE INTERSECTIONS
C
      SUBROUTINE INTERX(ITYPE,ICORD,IMAX,PCENT)
      REAL PCENT(3)
```

```
      COMMONS
      PCENT(1)=0.0
      PCENT(2)=0.0
      PCENT(3)=0.0
      DO 100 I=2,IMAX
      DO 200 J=1,3
      U=RANF( )
      IF (ITYPE.EQ.1) THEN
                    POS(I,J)=U-0.5
                    GO TO 150
      ENDIF
      IF (ICORD.EQ.199) THEN
          IF  (J.EQ.1) V=2*U-U*U
          IF  (J.EQ.2) V=U
          IF  (J.EQ.3) V=U
          POS(I,J)=V-0.5
      ELSE
      V=2*U-U*U
      POS(I,J)=V-0.5
      ENDIF
150   CONTINUE
      PCENT(J)=PCENT(J)+POS(I,J)
200   CONTINUE
100   CONTINUE
      POS(1,1)=0.0
      POS(1,2)=0.0
      POS(1,3)=0.0
      PCENT(1)=PCENT(1)/IMAX
      PCENT(2)=PCENT(2)/IMAX
      PCENT(3)=PCENT(3)/IMAX
      RETURN
      END
C
C ** GENERATE CONNECTIONS AND DETERMINE TRANSITION
C     MATRIX
C
      SUBROUTINE CONNEX(NPORE,IMAX,AVLTGH)
      COMMONS
      DO 50 I=1,IMAX
      DO 60 J=1,IMAX
      ICN(I,J)=0
60    CONTINUE
      NP(I)=0
      ISTR(I)=0
      STOR(I)=0.0
50    CONTINUE
      TLGT=0.0
```

Listing Program III: EQUATIONS (43) AND (52) (continued)

```
      DO 100 I=1,IMAX
      K=0
      DO 200 J=1,IMAX
      IF (J.EQ.1) GO TO 200
      IF (NPORE.EQ.NP(I)) GO TO 100
      IF (NPORE.EQ.NP(J)) GO TO 200
      IF ((ICN(I,J).NE.0).OR.(ICN(J,I).NE.0)) GO TO 200
      K=K+1
      PLN(I,J)=SQRT((POS(I,1)-POS(J,1))**2
     1            (POS(I,2)-POS(J,2))**2
     2            (POS(I,3)-POS(J,3))**2)
      PLN(J,I)=PLN(I,J)
      ISTR(K)=J
      STOR(K)=PLN(I,J)
200   CONTINUE
      NEED=NPORE-NP(I)
      IF (K.EQ.0) GO TO 100
      IF (K.EQ.1) THEN
                  NEED=K
                  GO TO 350
      ENDIF
      IF (NEED.GT.K) THEN
                      NEED=K
      ENDIF
      KMX=K-1
400   CONTINUE
C ** SORT
      IFLAG=0
      DO 300 KK=1,KMX
      IF (STOR(KK).GT.STOR(KK+1)) THEN
         IEMP1=ISTR(KK)
         TEMP2=STOR(KK)
         ISTR(KK)=ISTR(KK+1)
         STOR(KK)=STOR(KK+1)
         ISTR(KK+1)=IEMP1
         STOR(KK+1)=TEMP2
         IFLAG=1
      ENDIF
300   CONTINUE
      IF (IFLAG.EQ.1) GO TO 400
C **
350   CONTINUE
      DO 500 NN=1,NEED
      ICN(I,ISTR(NN))=1
      NP(I)=NP(I)+1
```

```
      ICN(ISTR(NN),I)=1
      NP(ISTR(NN))=NP(ISTR(NN))+1
500   CONTINUE
      DO 600 NN=1,NEED
600   CONTINUE
      DO 700 KZ=1,IMAX
      ISTR(KZ)=0
      STOR(KZ)=0.0
700   CONTINUE
100   CONTINUE
C**
      NPTOT=0
      TLTGH=0.0
      DO 800 I=1,IMAX
      DSUM=0.0
      DO 900 J=1,IMAX
      IF (ICN(I,J).EQ.1) THEN
         NPTOT=NPTOT+1
         TLTGH=TLTGH+PLN(I,J)
         DSUM=DSUM+PLN(I,J)
      ENDIF
900   CONTINUE
800   CONTINUE
      AVLTGH=TLTGH/NPTOT
      NPOT=NPTOT/2
      RETURN
      END
C
C ** CALCULATE COEFFICIENTS
C
      SUBROUTINE PARAMS(NPORE,IMAX,AVLTGH,DIFFX,DIFFY,
     1                         DIFFZ,
     1                         VELOX,VELOY,VELOZ)
      COMMONS
      AVLTG2=AVLTGH**2
      DIFFX=0.0
      DIFFY=0.0
      DIFFZ=0.0
      VELOX=0.0
      VELOY=0.0
      VELOZ=0.0
      DO 100 I=1,IMAX
      DENO=0.0
      DENX=0.0
      DENY=0.0
      DENZ=0.0
      DO 200 J=1,IMAX
```

Listing Program III: EQUATIONS (43) AND (52) (continued)

```
      IF (ICN(I,J).EQ.0) GO TO 200
      RLEN=SQRT((POS(J,1)-POS(I,1))**2
     1          +((POS(J,2)-POS(I,2))**2
     2          +((POS(J,3)-POS(I,3))**2)
      DENO=DENO+RLEN
      DENX=DENO
      DENY=DENO
      DENZ=DENO
200   CONTINUE
      DO 300 J=1,IMAX
      IF (ICN(I,J).EQ.0) GO TO 300
      XL=POS(J,1)-POS(I,1)
      YL=POS(J,2)-POS(I,2)
      ZL=POS(J,3)-POS(I,3)
      RLEN=SQRT((POS(J,1)-POS(I,1))**2
     1          +((POS(J,2)-POS(I,2))**2
     2          +((POS(J,3)-POS(I,3))**2)
      DDX=XL/RLEN
      DDY=YL/RLEN
      DDZ=ZL/RLEN
      DIFFX=DIFFX+RLEN*(DDX**2)/DENX
      DIFFY=DIFFY+RLEN*(DDY**2)/DENY
      DIFFZ=DIFFZ+RLEN*(DDZ**2)/DENZ
      VELOX=VELOX+DDX/DENX
      VELOY=VELOY+DDY/DENY
      VELOZ=VELOZ+DDZ/DENZ
300   CONTINUE
100   CONTINUE
      DIFFX=DIFFX/IMAX
      DIFFY=DIFFY/IMAX
      DIFFZ=DIFFZ/IMAX
      VELOX=-2.0*NPORE*VELOX/IMAX
      VELOY=-2.0*NPORE*VELOY/IMAX
      VELOZ=-2.0*NPORE*VELOZ/IMAX
      RETURN
      END
```

7. Epilogue

As stated in the preface of this monograph, a quantitative treatment can facilitate computer-aided design of controlled-release devices or systems. In theory, the mathematical models presented in this monograph could be employed to design these devices. In reality, however, the resultant device cannot be expected to perform optimally over the whole range of possible "environmental" conditions.

The preceding chapters have indicated that a controlled-release device often interacts with its surrounding environment in a highly complex manner. Thus it may be nearly impossible to develop a mathematical model that can precisely or always predict the device's behavior or performance. Even if such a model can be developed, its solution with conventional techniques may be unwieldy; consequently, approximations may be required. Although the resultant approximate model may suffice in practice, certain situations may exist outside its bounds, as a rule rather than as an exception. Nevertheless, the expert designer or user often possesses heuristic models of controlled-release devices, which cannot be expressed mathematically, but which may be employed for a variety of purposes under various conditions. For example, the designer may resort to heuristic models to examine biocompatibility of polymer materials, to select the types of adhesive and backing polymer layer for transdermal patches, the linkage between active agent and polymer backbone for chemical-reaction controlled release, the size and distribution of microparticulates in injectibles, and the arrangement of the semi-permeable membrane in osmotic devices. The actions of the expert designer appear to be shaped by perception and judgement, and honed by the experience gained in dealing with the devices. Such perception and judgement are valuable assets which, if extracted and applied properly, can form the basis for developing effective and practical models for controlled-release systems.

Research in the area of Knowledge Engineering (see, e.g., Barr and Feigenbaum, 1981), a sub-field of Artificial Intelligence, explicitly recognizes the power of expert knowledge. An expert designer or operator of any system uses pragmatic know-how comprising heuristics, rules-of-thumb, and other empirical associations based on the expert's perception and understanding of the process system and its subsystems. This expertise is extremely valuable, although it can frequently be expressed only in qualitative and imprecise terms. The aim of Knowledge Engineering is to extract this knowledge from the expert and to harness it in solving complex problems in a real-world environment. Such a system, termed an expert system, derives its power solely from the knowledge embedded within it. Since several systems have been rated as performing at or above the level of human experts, numerous industrial organizations and governmental agencies have increasingly expressed interest in the application of expert systems.

Clearly, expertise pertaining to a complex system, e.g., a controlled-release device, involves several different kinds of knowledge. Since an expert system derives its power from the knowledge emodied in it, a faithful representation of all concepts and intentions of the expert is essential. Recently, the designers of Artificial Intelligence language have recognized this need for multiple forms of knowledge representation by creating programming environments, such as KEE (Fikes and Kehler, 1985) and LOOPS (Bobrow and Stefik, 1982), that integrate several knowledge programming formalisms. It would be highly desirable, therefore, that in the foreseeable future both mathematical and heuristic models be incorporated into these formalisms for the optimal design of controlled-release devices. For additional information on the concepts, theories and methodologies underlying the field of Artificial Intelligence or expert systems, readers are referred to available books and monographs on the subject (see, e.g., Nilsson, 1980; Rich, 1983; Winston, 1984).

References

Barr, A., Feigenbaum, E. A.: The Handbook of Artificial Intelligence, Vol. I, pp. 198–199, William Kaufman, Los Altos, California (1981)

Bobrow, D. G., Stefik, M.: The LOOPS Manual, Technical Report KB-VLSI-81-13, Knowledge Systems Area, Xerox Palo Alto Research Center (1981)

Fikes, R., Kehler, T.: The Role of Frame-Based Representation in Reasoning, Communications of the ACM *28*, 904–920 (1985)

Nilsson, N. J.: Principles of Artificial Intelligence, Tioga Publishing, Palo Alto, CA (1980)

Rich, E.: Artificial Intelligence, McGraw-Hill, New York, NY (1983)

Winston, P. H.: Artificial Intelligence, 2nd Edition, Addison-Wesley, Reading, MA (1984)

Author Index

Abramson, I. S. 168, 200
Ackerman, J. L. 24, 85
Ahn, B. J. 123, 152
Akanni, K. A. 168, 200
Alfrey, T. 112, 113, 116, 123, 144, 152
Allan, G. G. 95, 96, 100, 102, 107, 108
Amore, P. D. 5, 8
Andersen, C. G. 97, 108
Anderson, J. M. 97, 108
Apicella, A. 126, 154
Aris, R. 19, 83, 167, 200
Ashare, E. 79, 83
Astarita, G. 112, 116, 120, 123, 152, 154
Attwood, D. 41, 88
Aulabaugh, A. E. 6, 8, 95, 97, 107
Ayres, J. W. 78, 83, 86

Bagley, E. 115, 152
Baird, B. R. 112, 136, 152
Baker, R. W. 3, 7, 27, 37, 38, 40, 41, 43, 51–53, 62, 71, 82, 83, 89, 91–94, 97, 100, 105, 107, 127, 152, 162, 163, 166
Balazs, A. C. 20, 83
Bamba, M. 125, 152
Barr, A. 225, 226
Barrer, R. M. 18, 23–25, 83
Barrie, J. A. 2, 59, 83
Batz, H. G. 102, 107
Baudys, M. 101, 108
Bearman, R. J. 141, 152
Bechgaard, H. 60, 83
Beer, J. W. 95, 96, 100, 102, 107
Berens, A. R. 123, 153
Berg, J. 44, 83
Bernier, G. A. 115, 153
Berner, B. 59, 83
Bharucha-Reid, A. T. 141, 153, 183, 200
Bhatia, S. K. 169, 173, 177, 189, 197, 198, 200
Bird, R. B. 13, 83
Bixler, J. 19, 86, 115, 116, 154
Blake, D. R. 93, 109
Bobeck, H. 164–166
Bobrow, D. G. 226
Bodde, H. E. 6, 8, 128, 155
Boyer, R. F. 23, 83
Brooke, D. 39, 42, 76, 83
Brooks, T. W. 83, 179
Brott, B. C. 5, 8
Bruck, S. D. 157, 166
Bueche, F. 24, 25, 85
Bunge, R. G. 38, 87
Buri, P. 82, 85, 139, 151, 155
Burrell, H. 29, 83
Butt, J. B. 19, 84, 170, 200
Byron, P. R. 26, 83

Cable, H. C. 101, 107
Calef, D. F. 20, 83
Caplan, S. R. 14, 86
Carbonell, R. G. 169, 201
Cardarelli, N. F. 31, 87
Cardinal, J. R. 38, 77, 85
Carslaw, H. S. 57, 59, 63, 83
Carstensen, J. T. 81, 84, 125, 152
Case, R. D. 6, 8, 95, 97, 107
Chandrasekaran, S. K. 78, 83, 163, 166
Chazov, E. I. 92, 99, 108
Chien, Y. W. 29, 32, 44, 58, 69, 79, 83, 87
Christensen, F. N. 60, 83
Christopherson, J. 31, 36, 85
Chou, S. T. 168, 200, 201
Clarke, M. 62, 85
Cobby, J. 76, 84
Cohen, D. S. 16, 84, 122, 153
Collett, J. H. 41, 88
Colton, C. K. 15, 20, 87, 129, 130, 156
Cooney, D. O. 102, 103, 107
Cousin, M. J. 95, 96, 100, 102, 107
Cowsar, D. R. 2, 7, 28, 85
Crank, J. 13, 14, 51, 53, 62–65, 72, 74, 84, 112, 120–123, 147, 153
Curran, P. F. 158, 166
Currie, J. A. 19, 84

Dalby, R. N. 26, 83
Datyner, A. 16, 86, 177, 201
Davidson, G. W. R. 119, 127, 153
DeBoer, A. G. 6, 8, 127, 128, 155
DeGroot, S. R. 12, 84
DeJong, G. 168, 172, 189, 200
DeLeede, L. G. T. 128, 155
Deutch, J. M. 20, 83
DeVries, M. A. 128, 155
DiBenedetto, A. T. 16, 84
Doelker, E. 82, 85, 139, 151, 152, 155, 198, 200
Duda, J. L. 16, 17, 22, 84, 88, 117–119, 135, 136, 147, 152, 156, 177, 201
Duggirala, S. K. 168, 200
Duncan, R. 6, 7, 95, 97, 101, 102, 107
Dykes, M. K. 6, 7, 95, 97, 107

Eckenhoff, B. 160, 166
Edelman, E. 164–166
Eldridge, J. 93, 109
Elking, D. 32, 33, 86
Elworthy, P. H. 19, 87
Evans, J. W. 168, 170, 200, 201
Evans, R. B. 19, 84

Fan, L. T. 53, 60, 61, 84, 87, 106, 108, 141, 142, 144, 146–150, 152, 153, 156, 168, 182, 194, 200, 201
Fan, L. S. 60, 61, 84
Fang, S. M. 97, 108
Fatt, I. 170, 200
Feigenbaum, E. A. 225, 226
Feijen, J. 97, 108
Feld, W. A. 6, 8, 95, 97, 107
Felder, R. M. 14, 84
Feng, C. 168, 190, 200
Fessi, H. 81, 84
Feynman, R. P. 141, 153
Fikes, R. 229
Flynn, G. L. 24, 29, 53, 54, 58, 59, 84
Folk, G. M. 112, 153
Folkman, J. 164, 166, 199, 200
Fooladi, M. 97, 98, 108
Foster, R. N. 19, 83, 170, 200
Fox, R. O. 193, 200
Franson, N. M. 112, 127, 136, 153
Franzmann, G. 96, 107
Frisch, H. L. 15, 17, 84, 88, 112–116, 119, 121, 123, 124, 129, 134, 135, 140, 141, 144, 147, 152, 153, 156
Fu, J. C. 63, 84
Fuhrmann, J. 20, 84
Fujita, M. 16, 84
Furuhata, K. 27, 85
Furusawa, T. 15, 20, 84

Gale, R. M. 30, 50, 86, 87, 92, 93, 107, 163, 166
Ganderton, D. 87
Garcin, M. 130, 155
Garmon, W. E. 130, 155
Gavalas, G. R. 19, 84, 170, 200
Geddes, A. L. 14, 84
Ghannam, M. M. 58, 87
Gienger, G. 68, 69, 77, 84
Gillett, J. W. 35, 36, 85
Good, W. R. 125, 127, 153
Gostoli, C. 112, 124, 153
Greenkorn, R. A. 19, 85, 170, 200
Gregonis, D. E. 97, 108
Graham, N. B. 127, 153
Gülcat, U. 113, 123, 156
Gupta, D. V. S. 75, 85
Gürkan, T. 113, 123, 156
Gurnee, E. F. 112, 113, 123, 144, 152
Gurny, R. 82, 85, 139, 152, 155, 198, 200

Haas, Ch. 20, 86
Hagemeir, C. 63, 84
Haleblian, J. 31, 35, 36, 85
Halpern, B. D. 24, 85
Hansch, C. 33, 86
Hansen, F. Y. 60, 83
Haring, R. E. 19, 85, 170, 200
Harland, R. S. 167, 200
Harr, J. R. 36, 85
Harrigan, S. E. 93, 107
Harris, F. W. 6, 8, 95, 97–100, 102, 107
Heller, J. 91–93, 107
Hellwege, K. H. von 24, 85
Higuchi, T. 69, 70, 73, 74, 76, 77, 80, 85, 159, 160, 166
Higuchi, W. I. 70, 73, 74, 76, 77, 80, 86, 87
Hildebrand, J. H. 29, 85
Himmelstein, K. J. 104, 108
Hoffman, V. 98, 107
Holley, R. H. 115, 153
Hopfenberg, H. B. 103, 108, 112, 113, 116, 117, 119, 123, 125–127, 136, 152–154
Howell, J. M. 119, 123, 156
Howes, J. F. 93, 108
Hsieh, D. S. T. 76, 85, 86, 119, 123, 156, 164, 166
Hsu, K. C. 112, 125–127, 136, 154
Huber, S. 44, 83
Huvard, G. S. 14, 84
Hwang, S. T. 38, 87
Imanidis, G. 2, 86

Jacobs, H. 38, 77, 85
Jacques, C. H. M. 114, 154
Jaeger, J. C. 57, 59, 62, 83, 85
Jarzebski, C. M. 117–119, 147, 156
Jeffcoat, R. 93, 108
Johnson, M. F. L. 19, 85, 168, 189, 190, 197, 200
Joshi, S. 116, 123, 152, 154
Junginger, H. E. 6, 8, 128, 155

Kambour, R. P. 115, 153
Katchalsky, A. 158, 166
Kehler, T. 229
Kelley, F. N. 24, 25, 85
Kim, S. 19, 84, 170, 200
Kim, S. H. 38, 77, 85
Kim, S. W. 38, 77, 85, 97, 108
Kimmel, G. L. 93, 108
King, K. F. 44, 83
Kinneberg, D. J. 193, 201
Kirkwood, J. G. 141, 152
Knappe, W. 24, 85
Knoch, A. 68, 69, 77, 84
Kojima, Y. 27, 85
Kopec, L. 24, 85
Kopecek, J. 6, 7, 91, 95, 97, 101, 102, 107, 108
Korostoff, E. 24, 85
Kosmeyer, R. W. 82, 85, 110, 125, 127, 128, 135, 136, 149, 152, 154
Kosho, C. 128, 155
Kost, J. 164–166
Kostka, V. 101, 108
Kubin, M. 54, 56, 87
Kunica, E. 164, 166
Kuu, W. Y. 76, 85
Kwei, T. K. 15, 88, 112, 121, 136, 140, 144, 153, 154, 156
Kydonieus, A. F. 2, 3, 8

Lacey, R. E. 28, 85
Lamaze, C. E. 129, 130, 156
Lambert, H. J. 32, 44, 83
Langer, R. 3, 5, 6, 8, 15, 20, 38, 76, 82, 83, 85–87, 89, 91, 93, 108, 158, 164–166, 199, 200

Lapidus, H. 125, 127, 141, 154, 155, 183, 199, 200, 201
Lawyer, K. R. 98, 108
Leafe, T. 91, 109
Lee, E. S. 38, 77, 85
Lee, P. I. 66, 67, 72, 85, 103, 108, 125, 127, 129, 131, 133, 152, 154
Leeper, H. M. 159, 160, 166
Leighton, R. B. 141, 153
Leo, A. 33, 86
Leong, K. W. 5, 8, 165, 166
Leuenberger, H. 20, 86
Levine, J. D. 59, 83
Lightfoot, E. N. 12, 14, 83, 86
Lindstrom, F. T. 36, 70, 83, 85, 86
Ling, H. C. 135, 136, 152, 156
Linhardt, R. J. 164–166
Lipper, R. A. 76, 86
Litovitz, T. A. 16, 85
Lloyd, J. B. 101, 107
Lloyd, W. G. 112, 113, 123, 144, 152
Lohe, P. 24, 85
Long, F. A. 115, 123, 152, 154
Lonsdale, H. K. 27, 37, 40, 41, 52, 53, 62, 70, 83, 105, 107
Lordi, N. G. 125, 127, 154, 199, 200
Lotan, N. 91, 109
Lustig, S. R. 150, 152, 154

Macedo, P. B. 16, 86
Makarewicz, P. J. 119, 155
Malinauskas, A. P. 170, 201
Manning, J. R. 169, 183, 201
Mares, S. F. 44, 83
Marletta, M. 6, 8
Marty, J. P. 81, 84, 125, 152
Mason, E. A. 19, 84, 201
Mayersohn, M. 76, 84
McCarthy, D. A. 93, 108
McCormick, C. L. 97, 98, 108
McGregor, R. 15, 18, 86, 112, 154, 177, 201
McNeill, M. E. 127, 153
McSpadden, S. K. 71, 72, 74, 86
Meadows, D. L. 130, 155
Meares, P. 21, 22, 86
Merkle, H. P. 68, 69, 77, 84
Merril, E. W. 129, 130, 156
Meyer, F. 93, 109
Michaels, A. S. 19, 30, 59, 83, 86, 115, 154
Mikels, R. A. 95, 96, 100, 102, 107
Mikulecky, D. C. 14, 86
Miller, E. S. 81, 86
Mitra, S. 97, 108
Miyanami, K. 53, 60, 61, 84, 87, 106, 108
Miyasaka, K. 27, 85
Mo, W. T. 170, 201
Mount, D. A. 36, 84
Moyer, D. L. 63, 84
Moynihan, H. J. 130, 155
Mueller, N. 31, 36, 85

Nagy, M. 120, 154
Nardse, Y. 19, 88, 170, 201
Nassar, R. 168, 200, 201
Neogi, A. N. 95, 96, 108
Neogi, P. 95, 108, 123, 154
Ng, K. 31, 35, 85
Nicolais, L. 120, 152
Nilsson, N. J. 226
Noecker, R. 165, 166

Omar, M. A. 141, 155
Ouano, A. C. 122, 156

Pace, R. J. 16, 86, 177, 201
Park, G. S. 14, 16, 84, 86, 112, 155
Parrish, W. R. 123, 152
Pate, W. T. 193, 201
Paul, D. R. 71, 72, 74, 77, 83, 86, 130, 155
Peppas, N. A. 15, 80–82, 85, 86, 90, 93, 108, 110, 112, 119, 123, 127, 128, 130, 136, 139, 142, 147, 149, 150, 151, 153, 155, 156, 158, 166, 167, 200
Peterlin, A. 19, 86, 114, 121, 122, 155, 167, 201
Petersen, E. E. 19, 86
Petersen, R. V. 97, 108
Petropoulos, J. H. 123, 124, 155
Pines, E. 199, 201
Pismen, I. M. 19, 86, 168, 177, 201
Pitcher, W. H. 15, 20, 87
Pitt, C. G. 93, 94, 108
Pittman, C. U. 98, 107
Pohl, J. 101, 108
Pontius, R. B. 14, 84
Prather, R. 29, 86
Prins, W. 199, 201
Puisieux, F. 81, 84, 125, 152

Rajamani, K. 193, 200
Ramachandran, K. S. 98, 108
Rebenfeld, L. 119, 155
Reid, R. C. 65, 86
Reinhart, C. T. 130, 155
Rejmanova, P. 101, 106, 107
Rhine, W. D. 38, 39, 42, 76, 85, 86
Rich, E. 226
Richman, D. 123, 154
Ringsdorf, H. 96, 98, 102, 107, 108
Rodin, J. O. 92, 107
Rogers, C. E. 12, 25, 27, 33, 34, 87, 116, 155
Rogers, M. G. 24, 83
Roorda, W. E. 6, 8, 127, 128, 155
Roseman, T. J. 31, 41, 43, 53, 54, 58, 59, 70, 73, 74, 77, 80, 84, 87
Rosen, H. B. 164–166
Roussis, P. P. 123, 124, 155
Rowe, R. C. 19, 87
Rudolph, F. B. 137–139, 149, 152, 155
Runkel, N. 31, 36, 85
Ryan, D. 169, 201

Saffman, P. G. 168, 201
Saleeby, D. E. 127, 153
Saltzman, W. M. 20, 38, 82, 87
Sands, M. 141, 153
Sarti, G. C. 112, 123, 124, 152, 153
Satterfield, C. N. 15, 20, 87, 168, 190, 191, 201
Scheidegger, A. E. 168, 170, 201

Schindler, A. 93, 94, 108
Schwope, A. D. 65, 86, 93, 108
Scott, R. L. 29, 85
Seganova, A. 98, 107
Segot-Chicq, S. 127, 155
Seinfeld, J. H. 141, 155, 183, 201
Sfirakis, A. 116, 155
Shin, S. H. 168, 183, 194, 200
Shippy, R. L. 38, 87
Sideman, S. 91, 109
Sidman, K. R. 65, 86
Siegel, R. A. 20, 83, 87
Silberberg, A. 199, 201
Sinclair, J. L. 123, 147, 149, 155
Singh, S. K. 141, 142, 144-152, 155, 156, 193, 194, 201
Skirrow, G. 24, 25, 83
Sladek, K. J. 123, 156
Smirnov, V. N. 92, 99, 108
Smith, J. M. 15, 19, 20, 84, 88, 123, 152, 168, 170, 189, 190, 201
Smith, K. A. 129, 156
Solomon, O. 84
Somasekharan, K. N. 68, 87
Spacek, P. 54, 56, 87
Sparks, R. E. 75, 85
Speiser, P. 4, 8
Spencer, R. S. 23, 83
St. Clair, A. D. 36, 85
Stanley, E. A. 122, 153
Stannett, V. 21, 22, 33, 87, 112–114, 116, 117, 127, 136, 153, 154
Stefik, M. 226
Stewart, W. E. 13, 19, 83, 85, 168, 189, 190, 197, 200
Subramanian, R. V. 68, 87
Sukhatme, V. 76, 86
Sun, Y. 58, 87
Sundelöf, L. O. 33, 34, 88
Swanson, D. 163, 166
Swenson, D. W. 79, 83
Szwarc, M. 32, 87

Taverdet, J. L. 24, 87
Theeuwes, F. 50, 87, 158-161, 166
Theodorou, D. 170, 201
Thomas, N. 112, 113, 124, 136, 144, 147, 156
Thombre, A. G. 104, 106, 108, 162, 166
Till, D. E. 65, 86
Tischenko, E. G. 92, 99, 108
Tojo, K. 32, 33, 46, 53, 58, 60, 61, 73, 84, 87, 105, 108
Too, J. R. 168, 201
Torchilin, V. P. 92, 99, 108
Tosun, I. 123, 156
Tranter, C. J. 63, 87
Trescony, P. V. 91, 107
Tu, Y. O. 122, 156
Turnbull, D. 16, 83
Tuttle, M. E. 40, 41, 83

Urdahl, K. G. 127, 156
Usteri, M. 20, 86

Valia, K. H. 32, 87
van Amerongen, G. J. 24, 26, 28, 30, 88
Van Brunt, J. 3, 5, 8
van Kampen, N. G. 141, 156, 183, 201
Vergnaud, J. M. 24, 87
Vieth, W. R. 19, 85, 119, 123, 156
Vrentas, J. S. 16, 17, 88, 117–119, 135, 136, 147, 152, 156, 177, 201

Wakad, N. 19, 88, 170, 201
Wakao, N. 19, 88, 170, 201
Walker, G. C. 76, 84
Wall, M. E. 93, 108
Wang, C. T. 168, 189, 201
Wang, F. H. L. 134, 152, 156
Wang, T. T. 15, 88, 120, 121, 140, 144, 153, 156
Washkuhn, R. J. 39, 42, 76, 83
Watson, G. M. 19, 84
Weber, L. J. 36, 85
Wei, J. 170, 201
Weignann, H. D. 119, 155
Whitaker, S. 160, 169, 201
Wilkes, G. L. 119, 155
Williams, J. L. 21, 22, 87
Windle, A. H. 112, 113, 116, 120, 124, 136, 144, 147, 156
Winston, P. H. 226
Winzap, S. 20, 86
Wirbrant, A. 33, 34, 88
Wise, D. L. 93, 108
Wong, P. 59, 83
Wong, P. S. L. 30, 86
Wood, J. M. 41, 88
Woodland, J. H. R. 93, 109
Wright, J. 163, 166

Yaacobi, Y. 91, 109
Yalkowsky, S. H. 41, 43, 53, 54, 58, 59, 76, 84, 85
Yariv, A. 141, 156
Yasuda, H. 129, 130, 156
Yilmaz, L. 113, 156
Yolles, S. 93, 109
Yum, S. I. 160, 161, 166

Zimmermann, I. 38, 88
Zupko, H. M. 112, 121, 136, 140, 154
Zweidinger, R. 93, 108
Zygourakis, K. 19, 88

Subject Index

Active agent, decay of 60
  dispersed 2, 4, 36, 39, 69, 107
  dissolved 2, 4, 11, 36, 39, 61, 69, 70, 71
  loading of 34, 61, 69, 81
  solubility of 28
Amorphous regions, see Polymer
Anisotropic system 169, 183, 186, 197
Anomalous transport 112, 114, 116, 127, 140, 149
Approximate solution, heat balance integral method 66, 131
  long-time 62
  short-time 62
Artificial intelligence 225, 226

Blocking factor 19
Boundary layer 40, 56, 58, 60, 68, 73
Bulk erosion 5, 90, 92
Burst effect 53, 61

Case-I diffusion 112, 115, 140
Case-II transport 112, 119, 122, 126, 140, 144, 147
Characteristic diffusion time 117, 118
Characteristic relaxation time 117, 118, 150
Chemical immobilization 89, 94
Chemical potential 12, 121, 157
Chemical reaction controlled release, see Release
Computer-aided design 7, 225
Convective velocity 121, 140, 144, 169, 183, 187, 191, 194, 196
Copolymerization 95, 97, 99
  graft 24
Crosslinking 23, 92, 110, 130
Crystalline regions, see Polymer

Deborah number 117, 124, 127, 151
Detour ratio 19
Devices, controlled-porosity osmotic 162
  erodible 5, 89, 102
  externally-modulated 164
  hemispherical monolithic 38
  hollow-fiber 79
  laminated monolithic 38, 77
  laminated reservoir 58, 68
  magnetically-modulated 164
  matrix, see Matrix devices
  monolithic, see Monolithic devices
  osmotic pumping 6, 157
  osmotically-activated monolithic 162
  reservoir, see Reservoir devices
  sectioned-cylinder monolithic 39, 76
  swellable 110
  ultrasound-modulated 165
Diffusion, Fickian 14, 17, 112, 115, 119, 127, 130, 132, 134, 147
  non-Fickian 112, 123
Diffusion, Theory of
  Approximate molecular 16
  Free-volume 16, 177
  Statistical mechanical 177
Diffusion coefficient, see Diffusivity
  Activity-based 12
  Self- 12, 13, 16, 117
  Effective, see Effective diffusivity
  History-dependent 120
  Intrinsic 12, 14
  Mutual (inter)- 13, 17, 117
  Stress-dependent 123
Diffusion - controlled release, see Release
Diffusion - dissolution controlled release, see Release
Diffusion equation, generalized form of 14, 121, 140, 184, 197
Diffusivity 11, 20, 120, 122, 129, 135, 151, 177, 190
  Knudsen 169, 177, 190
  tensor 186
Diffusivity, effect of crosslinking 23, 130
  crystallites 24
  fillers 24
  graft copolymerization 24
  plasticizers 23, 24
  solute molecular orientation 27

solute molecular shape 27
solute molecular size 22, 27
solute molecular structure 28
thermal history 26
Dissolution 9, 13, 28, 37, 48, 60, 78, 81, 92, 102, 122, 198
Drift velocity 169, 183, 187, 191, 194, 196
stress-induced 121, 140, 144
Drug targeting 90, 101

Effective diffusivity 18, 82, 121, 168, 177, 184, 188, 191, 194, 197, 199
Enzymatic cleavage 6, 90, 101
Equilibrium swelling factor 112, 145
Erodible devices, see Devices
Erosion controlled release, see Release
Ethyl cellulose 5, 199
Ethylene-vinyl acetate copolymers 3, 163, 199
Expert systems 226
Externally-modulated devices, see Devices
Extraction front 69, 71, 74, 77

Fick's law 11, 14
Fillers 24, 30
Finite sink 53, 54, 62, 64, 65, 75
Free volume, hole 16, 17, 23
interstitial 16
Free-volume theories, see Diffusion, Theory of

Gel 15, 110, 136, 199
Diffusion in 17, 129
Gel-glassy polymer interface 112, 115, 121, 126, 136, 151
velocity 114, 124, 138, 143
Geometry, of device 38, 39, 76
of obstructing particles 18
Glass-to-gel transition 6, 110, 128, 141, 147
Glass transition temperature 6, 22, 110, 112, 116
Glycolic acid 5, 93
Graft copolymerization 24

Heat balance integral method, see Approximate solution
Hemispherical monolithic devices, see Devices
Heterogeneous erosion 5, 90, 93, 105
Heterogeneous polymer 15, 17, 167, 177, 183, 184, 188, 197
Hollow-fiber devices, see Devices
Homogeneous erosion 5, 90, 92
Homogeneous polymer, see Polymer

Homopolymerization 95, 98, 100
Hydrodynamic diffusion layer 40, see also Boundary layer
Hydrogels 6, 79, 99, 125, 127, 199
Hydrolytic cleavage 6, 90, 101
Hydroxypropyl cellulose 5

Intrinsic mobility 12

Knowledge engineering 225

Labile bonds 91, 93, 101
Lactic acid 5, 93
Laminated devices, see Devices
Low-temperature casting 5, 164

Magnetically-modulated devices, see Devices
Maleic anhydride 5, 92
Master equation 169, 172, 175, 184, 193, 197
Matrix devices 2, 9, 11, 28, 31, 36, 47, 49, 51, 61, 64, 68, 78, 81, 90, 105, 162, 198
Matrix-medium interface mass transfer 68
flux continuity 139
Methyl vinyl ether 5, 92
Microencapsulation 4
Mobility coefficient 121
Monolithic devices 2, 9, 11, 28, 31, 36, 47, 49, 51, 61, 64, 68, 78, 81, 90, 105, 162, 198
Monolithic dispersions, complex 37
simple 37, 39
Monolithic matrix systems 38
Monte Carlo simulation 20, 167, 169, 192, 198
Moving boundary 80, 123, 136, 163

Osmosis 3, 162
Osmotically-activated monolithic device, see Devices
Osmotic device, controlled-porosity, see Devices
Osmotic pressure 6, 157
Osmotic pumping devices, see Devices
Osmotic pump, elementary 7, 157, 159
Higuchi-Leeper 159
Theeuwes miniature 160

Particle size, effect on solubility 31
Partition 9, 111, see also Partition coefficient
Partition coefficient 20, 26, 28, 31, 54, 58, 68, 77, 134, 138
Pendant chain 6, 90
Penetrant 17, 111, 116, 122, 129, 130, 132
Permeability 26, 30, 50, 59, 158, 165

Percolation theory 20
Physical immobilization 89
Plasticizers 23
Poly (vinyl pyrrolidone) 5, 92
Polyagents 90, 95
Polyanhydrides 5, 93
Polyethylene 3, 33, 34
Polymer, amorphous regions of 17, 20, 91, 167, 183
  continuum 9, 37, 69, 79
  crystalline regions of 17, 20, 24, 27, 91, 167
  heterogeneous, see Heterogeneous polymer
  homogeneous 15
  porous 15, 80, 167
  relaxation of 110, 114, 120, 140, 144, 147
  semi-crystalline 17, 20, 110, 167
  stress cracking of 115, 128
Polymer-drug combinations 98, 101
Polymer-pesticide combinations 98
Polyurethane 3, 43
Pore-diffusion controlled release, see Release
Pore parameter distribution 171
Pore size distribution 82
Porosity 18, 24, 80, 162, 168, 188, 191
Probability density function 171, 173, 180, 184, 185, 202
Probability distribution function 172
Probabilistic theory 168
Psuedo-case-II 127
Psuedo-convective contribution 121, 140
Psuedo-Fickian 114, 120
Psuedo-super-case-II 127

Random walk 167, 189, 197
Relaxation of polymer, see Polymer
Release, chemical reaction controlled 3, 6, 89
  diffusion controlled 3, 9, 44, 69
  diffusion-dissolution controlled 77, 82, 198
  erosion controlled 3, 5, 89, 102
  osmotically controlled 3, 157, 160, 162
  sustained 2
  swelling controlled 3, 6, 110, 125, 130, 199
Reservoir devices 3, 5, 9, 28, 36, 44, 49, 51, 58, 60, 68, 90, 105
Reservoir fluid 39

Sectioned-cylinder monolithic devices, see Devices
Self-diffusion coefficient, see Diffusion coefficient
Semi-crystalline polymer, see Polymer
Sigmoid behavior 114, 120
Silicone rubber 3, 5, 27, 38
Solubility parameter 29
Solution-diffusion mechanism 9, 69, 79
Solvent-casting 5, 164
Statistical mechanical theories, see Diffusion, Theory of
Stefan problem 122
Stochastic modeling 141, 167, 169, 183
Stress cracking, see Polymer
Super case-II transport 113, 123, 127
Surface erosion 5, 90, 93, 105
Sustained release, see Release
Swellable polymer matrix 2, 110, 125
Swellable release device, see Devices
Swelling controlled, release, see Release
Swelling interface number 127, 157
Swelling stress 116, 121, 123, 141

Targeting residues 102
Temperature, glass transition, see Glass transition temperature
  -penetrant activity diagrams 118
  surrounding 62
Thermal history 26
Time-lag 45, 51, 105
Tortuosity 18, 24, 80, 167, 190, 199
Transference 50
Transport coefficients 184, 188, 192, 195, 199
Two-stage sorption 114

Ultrasound - modulated devices, see Devices

GPSR Compliance

The European Union's (EU) General Product Safety Regulation (GPSR) is a set of rules that requires consumer products to be safe and our obligations to ensure this.

If you have any concerns about our products, you can contact us on ProductSafety@springernature.com

In case Publisher is established outside the EU, the EU authorized representative is:

Springer Nature Customer Service Center GmbH
Europaplatz 3
69115 Heidelberg, Germany

Batch number: 10620878

Printed by Printforce, the Netherlands